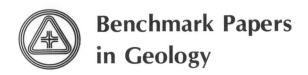

Benchmark Papers
in Geology

Series Editor: Rhodes W. Fairbridge
Columbia University

Additional volumes in preparation

**Benchmark Papers
in Geology / 27**

A BENCHMARK® Books Series

PERIGLACIAL PROCESSES

Edited by

CUCHLAINE A. M. KING
University of Nottingham, England

Dowden, Hutchinson & Ross, Inc.

STROUDSBURG, PENNSYLVANIA

Distributed by

HALSTED
PRESS

A Division of
John Wiley & Sons, Inc.

Copyright © 1976 by **Dowden, Hutchinson & Ross, Inc.**
Benchmark Papers in Geology, Volume 27
Library of Congress Catalog Card Number: 75-33696
ISBN: 0-470-47846-2

78 77 76 1 2 3 4 5
Manufactured in the United States of America.

LIBRARY OF CONGRESS CATALOGING IN PUBLICATION DATA
Main entry under title:
Periglacial processes
   (Benchmark papers in geology/27)

   Bibliography: p. 441
   Includes indexes.
   1. Frozen ground—Addresses, essays, lectures. 2. Cold regions—
Addresses, essays, lectures. I. King, Cuchlaine A. M.
GB641.P38   551.3'8   75-33696
ISBN 0-470-47846-2

Exclusive Distributor: **Halsted Press**
A Division of John Wiley & Sons, Inc.

ACKNOWLEDGMENTS AND PERMISSIONS

ACKNOWLEDGMENTS

GEOLOGICAL SOCIETY OF AMERICA—*Bulletin of the Geological Society of America*
    Classification of Patterned Ground and Review of Suggested Origins
    Rock Glaciers in the Alaska Range

UNIVERSITY OF CHICAGO PRESS—*The Journal of Geology*
    Solifluction, A Component of Subaerial Denudation

PERMISSIONS

The following papers have been reprinted and/or translated with the permission of the authors and copyright holders.

AMERICAN ASSOCIATION FOR THE ADVANCEMENT OF SCIENCE—*Science*
    Particle Sorting by Repeated Freezing and Thawing
    Stone Migration by Freezing of Soil

AMERICAN JOURNAL OF SCIENCE (YALE UNIVERSITY)—*American Journal of Science*
    Characteristics of Beaches Formed in Polar Climates
    The Effects of Ice-Push on Arctic Beaches
    Protalus Ramparts on Navajo Mountain, Southern Utah

ARCTIC INSTITUTE OF NORTH AMERICA
    Arctic
    Thermal Contraction Cracks in an Arctic Tundra Environment
    The Periglacial Environment
    Earth and Ice Mounds: A Terminological Discussion

FRIEDRICH VIEWEG & SOHN GMBH—*Geoforum*
    Ice Wedge Casts and Past Permafrost Distribution in North America

GEOGRAFISKE SELSKAB—*Geografiska Tidsskrift*
    The Wind Erosion in Arctic Deserts

GEOLOGICAL SOCIETY OF AMERICA—*Geological Society of America Special Paper 70*
    Mechanics of Thermal Contraction Cracks and Ice-Wedge Polygons in Permafrost

INFORMATION CANADA—*Geographical Bulletin (Canada)*
    Pingos of the Pleistocene Mackenzie River Delta Area

MACMILLAN JOURNALS LTD.—*Nature*
    Fossil Pingos in the South of Ireland

Acknowledgments and Permissions

NORSK POLARINSTITUTT—*Norsk Polarinstitutt*
Talus Slopes and Mountain Walls at Tempelfjorden, Spitzbergen: A Geomorphological Study of the Denudation of Slopes in an Arctic Locality

REGENTS OF THE UNIVERSITY OF COLORADO—*Arctic and Alpine Research*
Downslope Soil Movement in a Colorado Alpine Region: Rates, Processes, and Climatic Significance

REVUE GÉOGRAPHIQUE DES PYRÉNÉES ET DU SUD-OUEST—*Revue Géographique des Pyrénées et du Sud-Ouest*
Les Grèzes litées de Charente

ROYAL GEOGRAPHICAL SOCIETY—*The Geographical Journal*
Snow-Patch Erosion in Iceland

SWEDISH GEOGRAPHICAL SOCIETY—*Geografiska Annaler*
Avalanche Boulder Tongues in Lappland: Descriptions of Little-Known Forms of Periglacial Debris Accumulations
Block Fields, Weathering Pits and Tor-Like Forms in the Narvik Mountains, Nordland, Norway
Location, Morphology and Orientation of Inland Dunes in Northern Sweden

UNIVERSITY OF CHICAGO PRESS—*The Journal of Geology*
Loess, Pebble Beds and Boulders from Glacial Outwash of the Greenland Continental Glacier
The Mechanics of Frost Heaving
A Model for Alpine Talus Slope Development by Slush Avalanching

YORKSHIRE GEOLOGICAL SOCIETY—*Proceedings of the Yorkshire Geological Society*
Some Periglacial Problems

SERIES EDITOR'S PREFACE

The philosophy behind the "Benchmark Papers in Geology" is one of collection, sifting, and rediffusion. Scientific literature today is so vast, so dispersed, and, in the case of old papers, so inaccessible for readers not in the immediate neighborhood of major libraries that much valuable information has been ignored by default. It has become just so difficult, or so time consuming, to search out the key papers in any basic area of research that one can hardly blame a busy man for skimping on some of his "homework."

This series of volumes has been devised, therefore, to make a practical contribution to this critical problem. The geologist, perhaps even more than any other scientist, often suffers from twin difficulties—isolation from central library resources and immensely diffused sources of material. New colleges and industrial libraries simply cannot afford to purchase complete runs of all the world's earth science literature. Specialists simply cannot locate reprints or copies of all their principal reference materials. So it is that we are now making a concerted effort to gather into single volumes the critical material needed to reconstruct the background of any and every major topic of our discipline.

We are interpreting "geology" in its broadest sense: the fundamental science of the planet Earth, its materials, its history, and its dynamics. Because of training and experience in "earthy" materials, we also take in astrogeology, the corresponding aspect of the planetary sciences. Besides the classical core disciplines such as mineralogy, petrology, structure, geomorphology, paleontology, and stratigraphy, we embrace the newer fields of geophysics and geochemistry, applied also to oceanography, geochronology, and paleoecology. We recognize the work of the mining geologists, the petroleum geologists, the hydrologists, the engineering and environmental geologists. Each specialist needs his working library. We are endeavoring to make his task a little easier.

Each volume in the series contains an Introduction prepared by a specialist (the volume editor)—a "state of the art" opening or a summary of the object and content of the volume. The articles, usually some twenty to fifty reproduced either in their entirety or in significant extracts, are selected in an attempt to cover the field, from the key papers of the last century to fairly recent work. Where the original works are in foreign languages, we have endeavored to locate or commission translations. Geologists, because of their global subject, are often acutely aware of the oneness of our world. The selections cannot, therefore, be restricted to any one country, and whenever possible an attempt is made to scan the world literature.

To each article, or group of kindred articles, some sort of "highlight commentary" is usually supplied by the volume editor. This commentary should serve to bring that article into historical perspective and to emphasize its particular role in the growth of the field. References, or citations, wherever possible, will be reproduced in their entirety—for by this means the observant reader can assess the background material available to that particular author, or, if he wishes, he, too, can double check the earlier sources.

A "benchmark," in surveyor's terminology, is an established point on the ground, recorded on our maps. It is usually anything that is a vantage point, from a modest hill to a mountain peak. From the historical viewpoint, these benchmarks are the bricks of our scientific edifice.

RHODES W. FAIRBRIDGE

CONTENTS

Contents

PART III: SNOW ACTION

PART IV: WIND ACTION

PART V: FLUVIAL ACTION

Contents

PART VI: MARINE ACTION

CONTENTS BY AUTHOR

PERIGLACIAL PROCESSES

INTRODUCTION

As the pressure of development and exploitation cause increasing spread into the more remote areas in which climatic conditions become increasingly restricting, it becomes more and more important to understand the natural processes that take place in these harsher environments. The periglacial zone is one such environment into which modern development is spreading remorselessly. The mineral wealth and, in particular oil reserves are being sought and exploited as reserves in more accessible areas become depleted. To reduce to a minimum the disturbance of the delicate environment of the periglacial areas it is essential that the natural processes that operate in these areas to produce the characteristic landforms be fully understood. This volume brings together some of the most important work that has been done in periglacial areas and illustrates the wide range of phenomena that are associated with the periglacial regions.

The book is divided into six sections, each of which deals with a major process. One of the most distinctive processes of the periglacial area is that associated with frost action and the ground ice to which it gives rise. Part I is devoted to the processes associated with ground freezing and thawing and the special forms which these frost processes produce. These forms include the patterned ground features that have for long intrigued geomorphologists. Part II is concerned with the processes of mass wasting in a periglacial area, which produce a considerable range of recognizable land features. The processes of snow action are considered in Part III, including both the passive and active,

1

which are, respectively, nivation and avalanche activity. Part IV is devoted to wind action in a periglacial environment and considers both finer and coarser material and erosional effects; thus loess, dunes, and wind erosion are covered. The controls of the periglacial environment on river action are considered in Part V. Seasonal variation is one of the major elements that controls fluvial action in periglacial areas. Part VI is concerned with marine action in a periglacial area.

Many special features characteristic of the periglacial environment owe their qualities to the effects of ground ice in one or other of its many forms. There is, however, a distinction between permafrost areas and periglacial ones. Not all periglacial areas are necessarily underlain by permafrost, although many periglacial features are formed best in the permafrost zone. Thus the periglacial area is more extensive than the permafrost zone.

There are two main environments in which periglacial forms are strongly developed. The major area of periglacial activity is found in high latitudes. Originally, the term was applied mainly to those areas adjacent to present-day ice masses, the actual term "periglacial" being introduced by Łoziński in 1909. The idea of periglacial processes had, however, been considered in the nineteenth century, for example by Geikie in *The Great Ice Age and Its Relation to the Antiquity of Man*, published in 1874, in which he described the rubble drift and head deposits in Britain and suggested that they had formed under a cold climate. It is, however, difficult to define the periglacial zone in terms of climatic conditions, although several attempts have been made, of which Peltier's (1950) is one of the best-known. He suggested that the periglacial zone extends to those areas where the mean annual temperature is between −1 and −15°C and the precipitation varies between 120 and 1400 mm. These limits do not define the periglacial zone very satisfactorily. They do, however, indicate that there is a considerable range of conditions in which periglacial processes may operate and periglacial landforms be created. In high-latitude areas, periglacial processes overlap considerably with the existence of permafrost, a term implying that the ground remains frozen at depth throughout the year, the depth ranging up to a permafrost thickness exceeding 600 m.

The term "permafrost" was proposed by Muller in 1947. Three zonal types of permafrost have been suggested: continuous permafrost, which occurs in areas where only very small patches, for example under deep lakes, remain unfrozen at depth; discontinuous permafrost zones, which contain larger

2

patches of unfrozen ground, particularly where the ground insulation is good so that winter cold cannot penetrate; and sporadic permafrost, which occurs where the ground remains frozen at depth only in those areas that are most exposed to low temperatures and the ground is susceptible to freezing. These small patches are probably relict and in the process of melting under present conditions. One fifth of the land surface is underlain by permafrost; the largest areas are found in the USSR, where nearly half the land area is under permafrost, and in Canada, which is also about half underlain by permafrost. Other extensive areas of permafrost occur in Alaska, Greenland, and Antarctica.

It is difficult to suggest a precise relationship between the occurrence of permafrost and climatic conditions, although in general the southern limit of permafrost of the discontinuous type in Canada is associated with the −4°C isotherm; in the Yukon the value of −1°C provides a better fit. In many of the colder parts of the north, permafrost is actively growing under present conditions, as shown by the freezing of newly deposited silts. Elsewhere, near the margin of the permafrost zone, the ground ice is becoming degraded. The areas in which ground ice is now forming or thickening are called "zones of aggrading permafrost"; those in which permafrost is melting are known as "degrading zones." In between are the zones in equilibrium with present-day conditions. The permafrost areas of the present are not necessarily those that were left uncovered by Pleistocene ice, as some previously glaciated areas are now known to be underlain by permafrost. However, much remains to be learned in detail concerning the distribution and depth of permafrost and the climatic and other conditions that allow it to form or cause its deterioration. The relationships are complex and depend on many variables, including climate and environmental ones, such as soil type, vegetation, and relief.

The other major zone in which periglacial processes can operate effectively is in the high-altitude areas in all latitudes. These areas are often characterized by low temperatures and heavy precipitation, which often falls in the form of snow, especially in the winter months. There is an important distinction between the climate of high-latitude and high-altitude areas in their effect on periglacial processes. In the high-latitude areas the climatic contrasts are mainly of a seasonal nature, with a long, cold, dark winter being followed by a short, cool, but continuously light summer. In high-altitude areas, on the other hand, the extremes of climate are often mainly of a diurnal character, a contrast that

3

becomes increasingly marked as the latitude becomes lower. Thus in the high mountains of the Andes, the African equatorial mountains, and the Himalayas, for example, there are very great contrasts in temperature between day and night. This contrast exerts an important control on many of the processes that operate to produce the periglacial landforms that depend on changes of temperature across the freezing point. Permafrost is not widespread in the high-altitude zones, although it can occur sporadically under favorable conditions of soil and exposure, for example in the Rocky Mountains of Colorado, where high winds prevent snow accumulation on open slopes.

Owing to the major shifts of the climatic belts and associated geomorphological processes during the Pleistocene and Holocene periods, the zone of periglacial activity has also varied over time. Thus the extent of the periglacial zone during the colder periods of the Pleistocene was very different from the present distribution of periglacial processes. Many places that were at one time under the major Pleistocene ice sheets have at some period in the recent past been under the influence of periglacial processes, whereas those areas that were marginal to the Pleistocene ice sheets at their maxima are now at least in part in a temperate climate. These marginal areas were under periglacial conditions when the ice sheet was at its maximum. Thus some areas show a change from glacial to periglacial conditions, while others show a change from periglacial to temperate conditions.

Some areas may have had a periglacial climate for much of the Pleistocene and subsequent periods. These areas include those that have too low a precipitation to nourish an ice sheet, even when temperature conditions are sufficiently severe. This condition applies to parts of Siberia beyond the limit of the Pleistocene ice sheets, to small areas in northwest Canada and eastern Alaska, and to some of the Canadian Arctic islands. Many areas, however, have experienced a major change of geomorphological process with the waxing and waning of the major ice sheets in the northern hemisphere. These areas show varying degrees and types of polygenetic landforms, which include an element of periglacial activity.

The study of currently acting periglacial processes provides a valuable basis from which it is possible to recognize past periglacial activity in the landforms of the polygenetic zones, especially those marginal to the former larger ice sheets. The recognition of fossil periglacial features is increasing now that those currently forming are better described and understood. Often a useful palaeoclimatic implication can be derived from the study of fossil periglacial features. Some of these features are associated with permafrost conditions, and they will only form under fairly extreme conditions. Where they are found in fossil form, they, therefore, provide clear evidence of former harsher conditions. The widespread occurrence of fossil ice-wedge or tundra polygons, for example, in areas beyond the range of glacial ice in Europe and North America indicates that these areas were previously underlain by permafrost, and that they must have experienced temperatures at least $13\frac{1}{2}$°C colder than those occurring at present. This estimate is based on the view, expressed by Péwé (1966), that these features can only form where the mean annual temperature is lower than -6°C. Several other features that now only form effectively in a periglacial zone can be recognized in a fossil form in areas now having a temperate climate. These features include pingos, altiplanation terraces, sorted slope deposits, and various types of periglacial sorting, solifluction, and block fields.

The most significant geomorphological processes in periglacial areas are determined by the environmental characteristics. These include the climate, soil, and vegetation, which interact to a considerable extent. The climate is harsh, one of its major features being low temperatures, especially during the winter in the high latitudes and at night at high altitudes. Another important aspect of the climate is the degree of diurnal or seasonal warming, especially the number of times the freezing point is crossed. There tends to be one major freeze–thaw cycle with an annual time scale in the high latitudes. The numbers of times that the freezing point is crossed is at a maximum at ground level and diminishes rapidly with depth. Thus observations at Resolute in the Canadian Arctic indicated 18 freeze–thaw crossings at ground level, but only one at a depth of 2.5 cm.

The depth to which the summer thaw penetrates is an important aspect; this depth is the "active zone" and in it material is particularly subject to disturbance by periglacial processes. The base of the active zone is called the "permafrost table" in those areas where permafrost occurs; it forms an impermeable layer, rendering the active layer one of high moisture content in suitable soils. Thus the character of the soil or surface material plays a very important part in the operation of periglacial processes. Periglacial surface materials are subdivided into frost-susceptible and non-frost-susceptible according to their reaction to thaw–freeze processes. Non-frost-susceptible types are the very fine

5

clays and coarse sands and gravels, in which the low capillary force prevents the drawing up of water from below to freeze in the soil. Silts, on the other hand, are frost-susceptible materials because water can be readily drawn up into them to form ground ice as the winter cold penetrates downward. The periglacial areas are generally covered by tundra vegetation, with its characteristic low growth and poor holding capacity, thus rendering slopes liable to the many processes of mass movement that occur in the periglacial zone. Snow is another feature of the periglacial environment that gives rise to specific features.

Part I
FROST ACTION AND GROUND-ICE FORMATION

Editor's Comments
on Papers 1 Through 14

PROCESSES

Part I is divided into two parts; the first is devoted to the processes that take place as a result of frost action and ground-ice formation, and the second with the resulting morphological features created by these processes.

As early as 1914, Högbom wrote about frost processes and recognized the importance of frost in causing heaving in suitable soils and the mechanical breakup of rocks by expansion due to the freezing of water in cracks. The process of frost shattering of rocks is not, however, a simple one and the details of the process have still to be elucidated. Problems associated with the process include the creation of the crack as an initial step.

One process that is significant in this respect is the effect of pressure release, which operates in rocks when they are relieved of load, for example by deglaciation or glacial erosion, whereby rock is replaced by ice as the glacier eats into its bed. A good example of the effect of pressure release in creating joints that have been subsequently exploited by the freezing process can be seen in the Rapakivi granites of the Åland Islands off southwest Finland. These massive granites have been broken up into thin sheets parallel to the surface as a result of the rapid melting of the Scandinavian ice sheet, which was 3000 m thick in this vicinity.

Even if cracks are available in the rocks, the theoretical maximum pressures as water freezes in them will rarely be produced, owing to gaps. Impurities and air in the water also mean that temperatures must be considerably lower than freezing point to allow ice formation. A number of experiments have been carried out in laboratory conditions to establish the processes that lead to rock disintegration under thaw–freeze conditions, but much remains to be learned of the details of the process. It is, nevertheless, probably one of the most widespread and potent processes in areas of exposed rock outcrops where sufficient moisture is available, as it has been shown that moisture facilitates the process and is essential to it. The rock type also plays a part in the operation of the process, largely through the medium of the joints and bedding planes that provide the cracks into which moisture can penetrate before freezing takes place.

It is generally thought that mechanical weathering due to freezing is the dominant process leading to rock disintegration in a periglacial area, but it seems likely that chemical weathering can also be effective in suitable rocks. Mineral analysis of sandstones undergoing exfoliation in Spitzbergen suggests chemical changes. Limestones in western Baffin Island show considerable surface solution, giving a very roughened surface, which affects the thinly bedded platy rocks.

The results of the activity of frost in periglacial areas are considered in Part II, in which some of the features that are formed essentially of frost-rived material are considered. These features include screes, rock glaciers, block and boulder fields, and protalus ramparts. The process is also important in glacierized regions in providing much of the supraglacial load in areas where nunataks and hill slopes project above the ice surface.

Another important way in which frost operates in the periglacial environment to produce special forms is by the process of frost heaving. The mechanics of the process of frost heave were investigated in 1930 by S. Taber in a classic paper (Paper 1). He shows that more is required to account for the extent of frost heaving than the simple freezing of water within the pore spaces in sediments. He demonstrated that water is drawn up into the freezing sediment from below as the freezing plane slowly penetrates downward with the lowering temperatures at the onset of winter. Thus his work is associated with the formation of ground ice. Under favorable conditions the amount of ground ice in a frost-susceptible soil can amount to over 100 percent of the material volume. This occurs particularly where silts are underlain by sands and gravels that have access to a copious supply of water, which is drawn up into the silts to form ground-ice lenses as freezing penetrates downward.

Ground ice can occur in a number of different forms. Table 1, a classification based on work by Shumskii (1964), illustrates the variety of types that have been recognized.

Table 1   Classification of ground-ice types

1. Soil ice	a. Needle ice or pipkrake
	b. Segregated ice (sometimes called Taber ice)
	c. Pore ice filling pore spaces (interstitial ice)
2. Vein ice	a. Single veins
	b. Ice wedges
3. Intrusive ice	a. Pingo ice
	b. Sheet ice
4. Extrusive ice	Icings or aufeisen—formed subaerially
5. Sublimation ice	Formed in cavities by crystallization from water vapor
6. Buried ice	Buried icebergs, buried glacial ice, etc.

Needle ice, sometimes called pipkrakes, forms at the soil surface when the temperature falls below the freezing point and there is much surface moisture. The individual needles can be up to about 10 cm in length after one night of freezing. Small pebbles or sand grains are often lifted up on the thin vertical ice needles, which then collapse downslope as they melt during the day, causing considerable movement of the particles. The process operates particularly effectively in those high-altitude areas that are subject to considerable extremes of temperature diurnally and where precipitation can be heavy, providing adequate moisture for ice-needle formation.

Perhaps one of the most important types of soil ice is segregated ice, which is sometimes called Taber ice, to commemorate his great contribution to the study of this type of soil ice. It is a type of ice characteristic of supersaturated conditions, and occurs as layers or lenses of pure ice within the sediments. Segregated ice will only form in frost-susceptible soils into which water can be drawn by tension as freezing takes place. Capillarity is not involved because there is no free water surface; the water moves up from below toward the downward-penetrating freezing front. The other form of soil ice in this category is interstitial ice. This is ice frozen within the pore spaces between the grains. It renders the material impermeable and sets a limit to the growth of the ice layers or lenses. Thus there is a tendency for the layers of soil ice to become thicker and farther apart downward as interstitial ice grows more slowly at depth, as the rate of frost penetration slows down. Ice lenses may grow to a thickness of 4 m under favorable conditions, and can amount to 10 times the dry weight of the material in which they form. These figures come from observations in Alaska and the Mackenzie delta area.

One of the most important effects of segregated ice formation is the ground heaving to which it gives rise. The heaving is also associated with the differential movement of particles of different size and shape; it is, therefore, also responsible for the sorting of material, which is a very conspicuous aspect of periglacial processes. Before considering the heaving associated with segregated ice formation, some other forms of ground ice will be briefly described.

Vein ice will be considered in connection with the formation of ice wedges and the process of contraction that is thought to be responsible for their formation. Another type of ground ice, intrusive ice, gives rise to the ice-cored mounds called pingos that occur widely in the Mackenzie delta area and in parts of Greenland; fossil forms have been identified, for example, in Wales and the Isle of Man in Britain. The formation of these intriguing blisters will be further examined in the next section. Extrusive ice gives rise to features such as icings or aufeisen, to use the term given to these ice masses by Middendorff (1859) and exemplified by Leffingwell in his classic description of the Canning River district of north Alaska in 1919 (Paper 4). The aufeisen are found mainly in rivers and result from the rapid freezing of water held up by the growth of anchor ice at the base of the river and frazil ice at its surface. When the water pressure builds up enough to allow it to flood out and freeze, an icing is formed. It is characteristically formed of candle ice, which consists of vertical crystals 1 or 2 cm in diameter and up to several tens of centimeters in length. Alternate flooding and freezing can lead to layers of candle ice up to 4 m thick and covering up to 8 km².

Another environment in which aufeisen can form is in association with perennial springs draining from deep kettle holes. Such activity occurs in Baffin Island, where deep kettle holes remain unfrozen at depth; water escapes periodically through the gravelly sediments during the winter period and freezes to form an extensive mass of candle ice around the spring source. The kettle holes associated with the formation of aufeisen have no visible outlet, but clearly drain periodically, as changes of water level are evident in the distribution of lichen on the rocks around the kettle hole. The ice reached a total thickness of at least 3 m, occupying a gully about 1 km long. Individual layers of candle ice were up to 42 cm thick, the length of the candle-ice crystals. Thus a source of water throughout the winter period is an essential feature of aufeisen.

Sublimation ice and buried ice are not of significance from the point of view of periglacial processes.

The processes of frost heave are particularly important in periglacial phenomena, as they are instrumental in producing the sorting of material and in aiding many of the periglacial processes that operate on slopes, which will be considered in Part II. The early work by Taber has already been mentioned in this connection, but subsequent work has thrown further light on the processes associated with frost heaving and the resultant periglacial features. Papers 2 and 3, by A. E. Corte and C. W. Kaplar, which deal with the effects of frost heaving on the sorting of material, exemplify some of this work. Coarse material freezes more rapidly than fine material, as shown, for example, by the experiments carried out in a laboratory by Philberth (1964). The differential freezing of mixed material sets up pressures within the sediment, resulting in the sorting process. The larger stones, especially if they are tabular or downward-pointing wedges, are lifted by ice forming beneath them, leaving voids when the ice melts. Finer material may migrate into the voids, thus leaving the stone in a raised position. With repetition of the process the stones eventually emerge at the surface, a process which could account for the superstition in some cold areas that stones grow in the soil. Once on the surface the stones freeze more rapidly than surrounding finer material, pushing the fines away from the freezing centers and causing further segregation by particle size. At one time it was thought that the pressure came mainly from the fine material, but more recent work has established that the coarser material forms the freezing centers from which the fine material is ejected.

The actual process of frost heave has been monitored instrumentally in recent years by means of a variety of equipment. The "bedstead" provides one method. A frame is set up in a rigid position and loose rods resting on the surface are fixed to it in such a way as to be free to move up with the rising ground surface. The rate, time, and degree of heave can then be recorded by repeated observations against an index mark on the apparatus. A detailed study by James (1971) of diurnal frost heave in a temperate climate revealed a slow uplift over the period from 0200 to 1100 hours, followed by a rapid drop to 1500 hours on March 9, 1970. There was a close parallelism between the graphs of frost heave and the temperature curve, with a maximum heave of nearly 12 mm when the temperature fell to $-7°C$ over one diurnal cycle.

Frost heave is much greater when it operates over the annual cycle characteristic of the high-latitude permafrost areas. This type of situation was investigated by Corte (Paper 2), who made many of his observations near Thule in west Greenland. In this area he found two distinct types of periglacial processes operating. In one, sorting was an active effect of the process. In the areas undergoing this type of disturbance the material showed signs of sorting, and the bedding, revealed by removing the surface active layer by a bulldozer, was contorted. Amorphous masses of ground ice were found and the material was considerably finer in size than in the neighboring undisturbed area. A sedimentary analysis showed that the disturbed zone contained 5 percent fines (less than 0.074 mm in diameter) in the surface, 11 percent in the middle, and 14 percent at the bottom of the active layer, whereas in the undisturbed zone there was only 2 percent of fines throughout the active layer, in which the horizontal stratification was not disturbed. The difference in type of material was shown to be of fundamental importance in accounting for the different response to the periglacial processes. The disturbed material was frost-susceptible; the coarser undisturbed material was non-frost-susceptible. It is in the frost-susceptible material that ground ice can form, and this results in effective heaving and the differential sorting of material, with the formation of the features to be considered in later sections.

The work by Corte and Kaplar indicates how frost heaving results in sorting. To understand these processes, it is necessary to observe the degree of heaving in the field; laboratory experiments can also yield valuable evidence. Observations of environmental characteristics are also essential, and these include the nature of the sediment, the relief, the climate, the water supply, and the vegetation and soil character.

Corte found that in the non-frost-susceptible soils sorting did not occur; instead, he found that the undisturbed sediments were cut by ice veins and wedges. Papers 5 and 6 by A. H. Lachenbruch and D. E. Kerfoot support and confirm the early work of E. Leffingwell (Paper 4), who in 1915 first satisfactorily explained the occurrence and formation of ground ice wedges. He described the ice wedges in north Alaska, suggesting that they are formed as a result of thermal contraction under conditions of extremely low temperatures. At very low temperatures the ground contracts, and audible cracking as a result of the tensional stresses set up has been reported.

A more recent review of this process is given by Kerfoot in Paper 6; in Paper 5 Lachenbruch attempts successfully to provide a theoretical physical account of the operation of this important process, which gives rise to one of the most easily recognized types of patterned ground. The large tundra polygons or ice-wedge polygons formed by the process of contraction in extremely cold conditions are only found in the permafrost regions. Contraction cracking on a small scale can occur elsewhere, for example, in New Hampshire in an exceptionally cold spell in the winter of 1958 when the December temperature was −15°C mean value. The cracks were up to 5 mm wide and formed a polygonal pattern. However, it is only in permafrost areas that the process can continue for long enough to produce the wide ice wedges associated with tundra polygons. Such ice wedge tapers downward, but may reach a width of 10 m on the surface and penetrate more than 10 m downward, thus representing one of the most massive forms of ground ice. The widest wedges may take up to a 1000 or more years to form; each year the ice wedge cracks open, as it represents a weaker zone within the ground. Melt water freezes in the newly opened crack in the spring, thus gradually increasing the width of the ice wedge.

Paper 7, by T. L. Péwé, stresses the importance of ice wedges when they are found in fossil form as palaeoclimatic indicators. Péwé suggests that ice wedges can only grow actively where the mean annual temperature is lower than −6°C. At higher temperatures the ice slowly melts and is replaced by sediment, thus maintaining the wedge form as a recognizable feature and providing clear evidence that the area was formerly under a periglacial climate in which ice wedges could form. The location of fossil ice wedges in the English Midlands, for example, shows that this area had a temperature at least 13.5°C colder when the ice wedges were forming. Fossil ice wedges have been found widely in Europe and North America well to the south of the limit of the Pleistocene ice sheets, as well as in the area which they covered at their maximum extent.

Ice wedges often show up very well on aerial photographs as a result of minor differences in soil type over the ice wedges that outline the polygonal pattern. The ice wedge when melted often forms a slight depression in which moisture can accumulate; thus crop marks are conspicuous at certain stages of growth, and these can be easily recognized from the air. They usually are easier to appreciate from an aerial view than from the ground.

FORMS

Mention has already been made of some of the distinctive forms that the processes considered in the last section give rise to. This section is devoted to a further discussion of the forms. The section is divided into four subdivisions, which deal respectively with (1) patterned ground, (2) pingos, (3) palsas and related features, and (4) thermokarst. The first of these subdivisions includes a wide range of patterns that have given rise to much speculation, and many theories have been advanced to account for some of the patterns, such as the sorted polygons. Nearly all theories agree, however, that the patterns are related to the processes associated with freezing and thawing and with the formation of ground ice in its various forms. Thus the features can be explained in terms of the processes considered in the previous section.

Before the individual patterns are considered it will be useful to provide a classification of the patterns. The classification by A. L. Washburn in Paper 8 is a convenient and much-used one. In Paper 9 he has attempted to relate the pattern to the processes. Washburn has played a very active part in the increase in knowledge of periglacial processes and is particularly interesting when dealing with the generation of the patterns that characterize the periglacial zones, although, as he points out in Paper 9, the patterns are not entirely restricted to the periglacial zones. However, most of them form best under periglacial conditions. The table in which he relates the type of pattern to the formative processes contains a fair number of queries. This illustrates that there is still much to be learned concerning the genesis of the patterns. Washburn is helping to fill in these gaps by means of experimental methods, which are being carried out in the cold laboratories in the Quaternary Research Center of the University of Washington, where experiments can be carried out under varying temperature conditions to establish the way in which the processes operate. Fieldwork is also essential to enable the experimental results to be applied to the real situation under the more complex natural conditions. Washburn has himself had wide field experience of the periglacial zone, having done much work in Greenland, various parts of Arctic North America, and elsewhere.

The major point concerning the classification of patterned ground is the distinction between the sorted and nonsorted patterns. Each category of pattern is found in both sorted and nonsorted form. The range of patterns is subdivided on the basis of geometrical form. They include circles, nets, polygons, garlands, steps, and stripes. The distinction between the circles, nets, and polygons is related to the mutual interaction of the features. Where they develop in isolation, circles tend to form, but where they interact, then nets or polygons will form. The distinction between nets, polygons, and circles, on the one hand, and garlands, steps, and stripes, on the other hand, is based on the degree of land slope. The former features are confined largely to flatter areas; steps and garlands form on moderate slopes and the stripes on rather steeper slopes, up to a gradient of about 30°, beyond which patterns cannot form.

Patterned Ground

The basic difference between sorted and nonsorted patterns is related to the processes that form them. It is generally true that the sorted patterns require some form of disturbance to the original material in which they form, and this disturbance can best be brought about by processes associated with the formation of ground ice in the form of segregated ice. The features will form best in mixed material that includes a reasonable proportion of fine material, because, as has already been pointed out, it is this material that is frost-susceptible. A good water supply is also necessary or it will not be possible for the thick ice lenses and layers that produce the freezing pressures, which cause the sorting, to form. Another essential feature is the occurrence of a number of freeze–thaw cycles, because it is the alternation of freezing and thawing that causes the particles to move relative to each other. For this reason the sorted patterns are not found only in the permafrost areas of high latitudes where the freeze–thaw cycle is largely annual. In fact, the sorted features are often formed in the high-altitude cold areas where the diurnal cycle is important. Permafrost is not necessary for the sorting of material. Patterns do, however, tend to be rather larger and more extreme as the conditions get harsher.

The normal pattern of the sorted circles and polygons is for the border to consist of the larger stones, which are often arranged on edge, with their long axes parallel to the border of the polygon. The finer material is concentrated into the center of the feature, which is usually bulged up at the beginning of the thaw season as the fine material is forced into the polygon center from the freezing pressure center in the coarser border. This upbulge helps to move the stones down to the relatively depressed coarser borders. In section the stones of the borders are found to extend downward as far as the base of the active layer generally, where they occur on permafrost.

At the limit of periglacial conditions severe enough for sorting to occur, the features tend to be small and to extend only a short distance below the surface. There are examples of these fairly small features, often only a few tens of centimeters across, on the hills of the British Isles. They have been observed on the Tinto Hills of Scotland, the Lake District hills in England, and the Snowdonian area of north Wales.

An experiment on the small polygons on the Tinto Hills showed that the features are forming under present conditions. They were artificially destroyed by stirring up the mixed material in which they had formed. These features only penetrated 10 to 15 cm below the surface, but they reformed after 1 or 2 years. The formation of ground ice has also been observed to take place during cold spells on the higher hills of the Lake District. Thus even in temperate areas some periglacial features can form at the higher elevations, where winter conditions are reasonably severe and freeze–thaw activity is considerable.

The same type of sorting processes that are discussed by Washburn and in Paper 10, by Corte, operate on slopes as on flatter ground, but the resulting pattern differs. R. F. Black, another influential worker in periglacial problems, indicates in Paper 14 the importance of ground conditions in determining the type of periglacial pattern; he is also concerned with the use of aerial photographs in the interpretation of the patterns and the conditions that give rise to them.

Many of the most conspicuous polygonal features as seen on aerial photographs are large tundra or ice-wedge polygons. These features, as already suggested, are found only in an active state of the permafrost zone. They are the surface manifestation of the pattern of ice wedges that form by contraction under extreme cold. They can form either in a polygonal pattern, with sides meeting at approximately 120°, or in an orthogonal pattern of right angles. According to the analysis of Lachenbruch in Paper 5, the type of pattern is determined by the sort of material in which it occurs. Thus he considers that the orthogonal pattern is found normally in heterogeneous material or in a plastic medium. The size of the polygon depends on the magnitude of the thermal stress. Some of the orthogonal patterns are random, and others are preferentially arranged; for example, one side of the rectangles may be parallel to a retreating water surface, such as would occur in a region undergoing isostatic uplift following deglaciation.

An example of this type of orthogonal ice-wedge pattern occurs on Foley Island off the west coast of Baffin Island; the cracks are 35 m apart, forming double or triple ridges of limestone pebbles. Where the material is homogeneous and subject to uniform cooling as a result, the pattern tends to be polygonal, with sides meeting at angles of 120°. Under these conditions the cracks appear to propagate laterally until a limiting velocity causes bifurcation, and velocity again increases until the limit. These tundra polygons are not associated with sorting or disturbance of the active layer, as shown in the study by Corte of the relationship among the patterns, the structure of the active layer, and the distribution of ground ice in the permafrost area around Thule, Greenland. In the true tundra polygons the pattern is that of the ice wedges which outline the polygons, and no sorting of the material is indicated. They, therefore, tend to form on non-frost-susceptible soils, and even at times in solid bedrock. An ice wedge in solid limestone was seen on Foley Island.

Washburn considers a wide range of patterns in Papers 8 and 9. The effect of slope is the criterion that distinguishes between the two groups of patterns. The steps are arranged parallel to the contours on the slope, but as they are mainly related to processes of mass movement on slopes, a consideration of these features will be deferred to the next section. The steps can be sorted or not according to whether they end in a vegetated front or whether rocks accumulate at the front to form a rock-banked step. The same distinction can be applied to the stripes, which are arranged at right angles to the contours, with alternating stripes of finer and coarser material in the sorted type and alternating stripes of bare ground and vegetation in the nonsorted variety.

The sorted type is probably more common and is related to the sorted polygon or circle, in that the same processes probably operate to form both features, with the effect of gravity due to the slope causing the difference in pattern. In the sorted stripes there must be an element of sorting across the slope whereby the coarse particles accumulate away from the finer, which are again thrust out from the freezing centers in the coarse stripes. Once the separation of material has started, the slope helps the process by providing a gradient down which active rills can wash remaining fines from the coarse borders; this enhances the sorting process by positive feedback, because the coarse borders are then more efficient freezing centers to push any remaining fines into the fine stripes, which is usually wider than the coarse one.

19

The polygonal form is only gradually lost as the slope gradient increases. On fairly gentle slopes the polygonal pattern is elongated to form a stone garland, which is an elongated polygon rimmed at its lower margin by stones, which are usually arranged on edge parallel to the outline of the garland. These features are well developed, for example, on the hill slopes of moderate gradient in Snowdonia in north Wales.

Pingos

The term "pingo" is one of the two Eskimo words that have been adopted in geomorphology. It was introduced by Porsild in 1938 to describe the dome-shaped hills that are common in parts of the Arctic regions. There are many pingos in the Mackenzie delta area, and they occur in Greenland and Siberia. They also extend into the discontinuous permafrost zone of the subarctic boreal forest in North America. Their size ranges up to a diameter of 300 m and a height of over 60 m. Most pingos have cracks across their length, giving them the appearance of having a crater. They are composed of an intrusive ice core covered by bedded sediments dipping parallel to their sides. They occur in greatest profusion in the Mackenzie delta, where there are about 1400 in the 25,000-km² delta area. These pingos, often referred to as the Mackenzie type, differ from the other main type, the East Greenland type.

The Mackenzie type of pingo has been studied in detail, especially by Mackay (Paper 11), according to whom this type of pingo is formed in association with the shallow, round lakes that are very common in the permafrost areas. As sediment gradually fills in the lake it becomes shallower, so that eventually it is able to freeze solid in the winter. Under these conditions the winter cold can penetrate and permafrost gradually extends downward under the lake, reducing the volume of thawed ground that previously occurred beneath the lake when it was too deep to freeze solid in winter. As the talik, the unfrozen patch, is gradually reduced in volume, ice is forced up under pressure beneath the lake silts, and this gives rise to the intrusive ice that creates the pingo mound on the surface. The upheaval leads to cracking of the silt blanket over the ice mound; under these conditions of exposure to the air the ice can melt in summer to eventually destroy the pingo form, leaving a ring of sediment around the central depression. Pingos grow slowly as the permafrost extends beneath them, and rates between 0.3 and 0.5 m per 1000 years have been suggested.

In contrast, the East Greenland type of pingo is thought to be formed by hydraulic forces. Downward freezing of the soil disturbs the hydraulic flow of groundwater and sets up pressure in the water, which under extreme conditions can cause a rupture of the ground and an outburst of water and debris. These are called "open-system" pingos because they require flowing water in the aquifer beneath the permafrost. Some of the pingos of central Alaska have been interpreted as being of the East Greenland type, and it has even been suggested that some of the Mackenzie delta pingos could have formed by this method. The argument is that the area is undergoing subsidence, which causes thawing at the base of the permafrost as the ground subsides. The water so created moves up under pressure, melting its way through the overlying permafrost, until it causes the upbulging near the surface to form the pingo mound.

Pingos seem to undergo melting and degradation as soon as their ice core is exposed by the rupture of the overlying sediments. They also occur in degraded and fossil form in areas that are now far from the periglacial zone. When they can be recognized in fossil form, they provide useful evidence of earlier permafrost conditions. In these conditions they usually occur as a low circular rampart, with a gap in the ridge being common. Pissart (1963) has identified fossil pingos in Wales and Belgium, where they are up to 120 m in diameter. They have also been recognized in Germany, and Watson has described fossil pingos from Wales and the Isle of Man. Paper 12, by G. F. Mitchell, discusses the type of fossil pingo that has been identified in Ireland. Other examples have been located in the Netherlands, Scandinavia, and East Anglia. The pingos of southern Ireland all lie to the south of the limit of the last glaciation. They, therefore, were probably formed when the area concerned lay fairly close to the ice front and had a periglacial climate.

Palsas and Related Features

Pingos are the largest of a range of features that form mounds in the periglacial environment. J. Lundqvist in Paper 13 discusses the four different sizes in which these mounds appear. The largest are the pingos, with diameters of up to 300 m or more and heights of 100 m; next in size are palsas, which can reach a

21

maximum height of 10 m. Smaller than the palsas are earth hummocks, which can reach a height of about 1 m; the final type are frozen peat hummocks, which have a maximum height of about 1 m. The term "palsa," according to Seppala (1972), has been used for a number of different features. The term was originally used by the Lapps and northern Finns for a hummock rising out of a peat bog and having a core of ice. They occur in permafrost areas and are always associated with peat. The high insulating properties of peat have preserved palsas in areas of discontinuous and sporadic permafrost. There is an intermediate form in Fennoscandia, Alaska, and Canada between pingos and palsas, in which the peat is underlain by a layer of mineral soil. Pingos should only contain mineral material and not peat. Where palsas occur the permafrost is usually restricted to them, owing to their exceptional insulating qualities. Palsas sometimes have a low, winding, ridge-like form; others are oval hummocks. Palsas near the Great Whale River in Canada are 5 to 7 m high, 15 to 40 m long, and 13 to 16 m broad. It appears that palsas may go through a cycle of development and decay that is not dependent on the climate. As soon as the mound forms, cracking is liable to occur; this exposes the ice core beneath the peat, resulting in the gradual melting of the ice and the decay of the feature. Thus at any one time some palsas may be devleoping while others are degrading. In general, they develop where the mean annual temperature is between −1.2 and −1.9°C and summer temperatures are +5.2 to 5.5°C.

Palsas are distinct from thurfurs, which are small frost mounds, more closely related to earth hummocks. Other features that have an ice core, called "hydrolaccoliths," were described by French (1971) and occur in the western Canadian Arctic. They are found in the low central area of ice-wedge polygons on poorly drained meadow tundra soils. They appear to be similar in formation to the Mackenzie-type pingos. The mounds are smaller than pingos, however; none exceeds 50 cm in height and 2 to 5 m in diameter. All had cracks in the covering tundra mat, revealing a clear ice core. They form within a closed system inside the polygon and thus resemble the Mackenzie-type pingo, which is also a closed-system form. Like palsas, they exist in all stages of formation and decay, and appear to be micropingos.

A later stage of the same type of feature has been described by Svensson (1969) from examples in northern Norway. They have the characteristics of both pingo mounds and palsas, and are in

22

the form of small, round lakes surrounded by a low, uniform wall of material. They appear to be a collapse structure, in a late stage of the development of a collapse structure that originated as a pingo or palsa. Fossil circular forms have also been found in which the ridges are only rarely more than 1½ m high. The outer slope is steeper than the inner one, and fissures occur, which indicate slipping of the inner part of the ring. In some instances the rings are so flattened that they only show by a change of vegetation. They are normally composed of fine-grained material. The smaller features of somewhat similar type are called "earth hummocks" in English or *buttes gazonnées* in French; the term "thurfur" is also used. They occur in the Yukon on fairly gently sloping surfaces and can reach a height of 0.6 m and a diameter of 1.5 m. They appear to be due to differential pressure resulting from uneven freezing. The areas between the hummocks freeze first, and material is forced from these freezing centers into the hummocks, causing them to form bulges. They tend to occur in high-altitude areas where freeze–thaw cycles are frequent. Another intriguing type of periglacial mound is the mima mounds of Washington state near Puget Sound. These low mounds have been associated with periglacial action, although the theory that they are formed by pocket gophers has not yet been disproved.

Thermokarst and Thaw Lakes

Paper 13, by J. Lundqvist, attempts to provide a terminology for the many somewhat similar features considered in this and the preceding section, and papers by R. F. Black (Paper 14) and J. C. F. Tedrow (1969) deal mainly with one of the most conspicuous features on aerial photographs of some periglacial areas. These are the round or oval shallow lakes that are very common on low ground, such as the Great Plain of the Koukjuak in central Baffin Island. These thaw lakes are one of the manifestations of the process known as "thermokarst." Just as the solution of limestone in true karst areas gives rise to distinctive landforms, so the melting of ground ice can give rise to somewhat similar landforms in an area of degrading permafrost.

The term "thermokarst" was introduced by M. M. Ermolaev in Russia in 1932. The main result of thermokarst processes resulting from melting of the permafrost is one of collapse. The resulting features, many of which have their equivalent in a true karst landscape, include surface pits, basins, funnel-shaped sinks, and

23

dry valleys or ravines. Caves may occur for a time before collapse takes place. Conical hills also occur, sometimes called "baydjarakhs," in the areas between the melting-out ice wedges. This type of hill may be similar to the mima mounds mentioned previously. Thermokarst features develop best where the thickness of segregated ice in the upper layers is considerable. Thermokarst is actively developing in the southern areas of the permafrost zone of the USSR at present, where the ice is melting out of the ground. Thermokarst development may be hastened by human interference at the surface of a delicately balanced area in periglacial zones, when the ground ice is exposed beneath peat or vegetation by excavations. Such a disturbance can lead to the development of thaw lakes. These lakes can also form where ground ice thaws in a patchy way, or where its occurrence is irregular. Where permafrost remains at depth, the impermeable nature of the ground causes the melt water to remain on the surface in areas of low relief and disorganized drainage, such as occur widely in northern North America, especially in previously glaciated areas. Once the shallow lakes have started to form, they will become larger, as a result of wind action, which drives the water against the banks, facilitating further melting. The alignment of the long axis of oval lakes can often be related to the prevailing wind direction. As the lakes gradually grow in size they may merge; though drainage may develop, which will drain some of the lakes, and such drained lakes can often be identified on aerial photographs.

The area studied by Black was shown to contain between 30 and 90 percent of ice by volume in the upper 5 to 10 m of fine-grained material. The thaw lakes are often associated with the melting out of ice wedges. Where the ice wedges are up to 5 m wide, they contain a great deal of ice, so relatively large thaw lakes can form when they melt. Where the ice wedges are 1 to 3 m wide and have a 10 to 30 percent ice content, the thaw lakes could be between 3 and 6 m deep. The lakes increase in size by bank collapse due to melting, which causes water to be blown against the lake side and to undercut it. The rate of enlargement has been recorded at between 6 and 19 cm per year in eastern Alaska. Four stages of development have been recognized as the lakes grow and then decay. Those in the Seward Peninsula of Alaska are normally up to 300 m across and 0.3 to 2 m deep. The age of thaw lakes in aeolian silts south of Umiak in Alaska, which are now mainly drained, was determined from organic matter in the silts of the lake floor. The ages given by Tedrow were 9325 and 9130 years.

Apart from wind, a number of other possibilities have been suggested to account for the alignment of the thaw lakes. One suggestion is orientation determined by the maximum insolation at noon, and another is control by the pattern of ice wedges, although one or other of the wind direction controls appears more likely. The thaw lakes of Alaska appear to vary in age, some being 14,000 years old while others are less than 8000 years old; yet others have formed in the last 150 years.

Fossil thaw lakes have been identified on the coastal plain of New Jersey and in the Paris basin.

1

Reprinted from *Jour. Geol.*, **38**, 303–317 (1930)

THE MECHANICS OF FROST HEAVING

STEPHEN TABER

University of South Carolina

ABSTRACT

The old theory that frost heaving is due to change in volume of water frozen was based on experiments with closed systems. Field observations and recent experiments indicate that soils, when subjected to freezing under normal conditions, usually behave as open systems. When the freezing of saturated soils results in little or no heaving, part of the water is forced through the soil voids below the zone of freezing, compressing or expelling air. Excessive heaving results when water is pulled through the soil to build up layers of segregated ice. These ice layers grow in thickness because water molecules are pulled into the thin film that separates the growing ice crystals from underlying soil particles. Since heavy surface loads may be heaved and much force is required to pull water through impervious clay, the water is put under high tension.

Heaving is limited by the tensile stress that may be developed in the water and by downward growth of ice crystals in soil voids. These two factors also probably explain the rhythmic banding due to alternating layers of ice and clay.

In well-consolidated clays the surface uplift equals the total thickness of the ice layers, the water content of the clay between the ice layers remaining approximately constant; but heaving is continuous and regular instead of intermittent. Clay is soft near the lowest ice layer because much of the water is unfrozen, the hardness increasing higher up where the temperature is lower and freezing has gone on for a longer time. Additional evidence has been obtained by freezing in open systems other liquids than water.

It has been generally assumed that frost heaving, as well as other pressure effects accompanying the freezing of water, is due to increase in volume of the water frozen. This assumption was based on experiments in which water was frozen in closed systems, i.e., systems from which the water present could not escape and into which additional water could not enter. The experiments of early investigators were limited to such systems.

In an investigation of frost heaving in 1914, the writer experimented with open systems on nights when the temperature was sufficiently low. The work was resumed in March, 1927, with an electric refrigerator, which made close control of temperature possible.[1]

Simple apparatus for freezing water in an open system is shown in Figure 1. A cylindrical container having a perforated bottom is

[1] The refrigerator was made available through the courtesy of the Frigidaire Corporation. Since November, 1927, the U.S. Bureau of Public Roads has given financial co-operation.

tightly packed with the soil to be tested, and is stood in a vessel containing sand, which can be kept saturated with water. This apparatus is placed in the bottom of the refrigerator and buried to the top in dry sand so that the soil is cooled from the top downward. A collar, fitting snugly around the container and resting on top of the vessel, prevents the entrance of dry sand. Container and vessel are held rigidly in place by a strong frame. A heavy lead disk resting on the soil in the container carries a recording pen which is in contact with a drum turned by a clock. This gives a graphic record of the amount and rate of movement of the soil surface during an experiment. Temperature is recorded on the same drum by another pen. Freezing extends gradually downward from the surface, and the interstitial spaces below the zone of freezing permit the movement of water either upward or downward according to the forces to which it may be subjected. This movement is without interference from outside

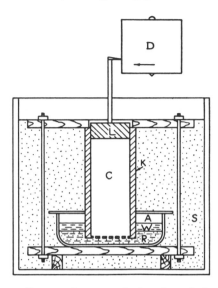

Fig. 1.—Apparatus for freezing soils in a system that is open with respect to water. C, soil; K, container; A, air; W, water; R, sand saturated with water; S, dry sand; L, lead disk with recording pen; and D, drum turned by a clock.

forces, since air may readily enter the vessel, or be expelled, through the dry sand.

If a definite amount of water is cooled under atmospheric pressure, it freezes at $0°$ C., with an expansion in volume of about 10 per cent. Greater pressures tend to prevent solidification, because the distance between the water molecules increases when ice crystals form; therefore, with increasing pressure the freezing point is lowered. If the water is confined so as to prevent expansion, freezing results in high pressure; but the maximum pressure obtainable is 2,115 kilograms per square centimeter, for at this pressure and

—22° C. ice III, which is denser than water, begins to form.[1] In such a system pressure develops because openings through which water can escape are absent.

In experiments with open systems, it was found that some soils, when kept saturated with water, freeze with no uplift of the surface, while others give uplifts ranging up to more than 60 per cent of the depth of freezing. An uplift of even 100 per cent was obtained in some tests through the formation of a layer of ice at the surface. When no heaving occurs, some of the water must be pushed downward by the growing ice crystals and expelled from the soil container; when the surface uplift is in excess of that which could result from the expansion in volume of the water frozen, it is obvious that additional water must have entered the container as a result of the freezing.

When liquids that solidify with decrease in volume are substituted for water in freezing experiments, similar pressure effects are obtained in open systems, although these liquids could develop no pressure as a result of freezing in closed systems.

Numerous observations made out of doors indicate that soils, when subjected to freezing under normal conditions, usually behave as open systems rather than closed systems. Frost heaving is often too great to be explained by change in volume of water present in the voids, and in other instances it is too little. Assuming the water content of soil to be 50 per cent—and it is usually much less—the change in volume, if all of it froze *in situ*, could account for an uplift of 5 per cent of the depth of freezing. The depth of freezing in the northern part of the United States seldom exceeds 2 or 3 feet; but a surface heaving of 6 inches is not uncommon, and an uplift of "a couple of feet" has been reported from Minnesota. On the other hand, the freezing of saturated soils is accompanied in many places by no appreciable uplift.

The entrance of additional water where soil heaving is excessive, and the partial expulsion of water from the soil voids where freezing results in little or no heaving, may be compensated for, in the one case, by the expansion and, in the other, by the compression of imprisoned air, or by the entrance and expulsion of air through pores

[1] P. W. Brigman, *Proc. Amer. Acad.*, Vol. XLVII (1912), pp. 441–558.

and cracks. Most soils probably contain sufficient air to permit the free movement of water during freezing.

Soil may behave as a closed system with respect to water when the water-table is flat over large areas and practically coincides with the surface so as to exclude air; but even under these conditions, lenses and layers of segregated ice can form if the soil texture is favorable, and differential heaving may be produced if there is sufficient variation in the conductivity of soil cover so that freezing does not begin everywhere at the same time. The distinction between closed and open systems in the ground is not sharply defined. As the resistance to the movement of ground water increases, an open system tends to grade into a closed system.

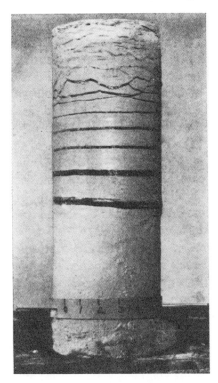

FIG. 2.—Cylinder of frozen clay. The surface uplift equals the total thickness of the ice layers. Scale in centimeters.

In laboratory experiments excessive heaving is always accompanied by the introduction of additional water and its segregation to form layers or masses of more or less pure ice (see Fig. 2). Similar layers of segregated ice have been formed in the ground where excessive heaving has been observed. An excellent illustration of such ice layers (see Fig. 3) was sent to me by Mr. F. C. Lang, of the Minnesota Department of Highways. The lump of clay shown in the photograph was excavated in the spring of 1929 from a street where the heaving was "a couple of feet." It is obvious that these ice layers could not have been formed by the freezing of water *in situ*.

An investigation of the factors involved in segregation and excessive heaving has been made.[1] The chief factors are: size and shape of soil particle, amount of water available, size and percentage of voids, rate of cooling, and the surface load or resistance to heaving. Segregation occurs readily if the particle diameter is less than a micron, and under favorable conditions if the particles are slightly larger. Tabular particles, such as are found in clay, probably give results

FIG. 3.—Clay containing ice layers from under a badly heaved street in St. Peter, Minnesota.

similar to those obtained with somewhat smaller particles that are more nearly isodiametric. Water occupying very small voids in soil does not freeze readily and may be undercooled in the immediate vicinity of ice crystals. The water remaining unfrozen at low temperatures is probably adsorbed water. Rapid cooling, when the temperature is near the freezing-point, prevents segregation in some soils but has little effect if the soil particles are sufficiently small and the other factors favorable. Moderate loads have less influence on the heaving of clay than the heat conductivity of the material used in applying the pressure; heavier loads tend to reduce heaving, but in

[1] Stephen Taber, "Frost Heaving," *Jour. Geol.*, Vol. XXXVII (1929), pp. 428-61.

laboratory tests with pure clay, surface loads of nearly 15 kilograms per square centimeter were required to prevent heaving. Moderate loads prevent the heaving of relatively coarse material.

It is commonly stated that frost heaving is upward because of less resistance to expansion in that direction, but this hypothesis is not supported by the facts. Experiments prove that the pressure effects accompanying freezing are due to the growth of ice crystals. The pressure is developed only in the direction of crystal growth, which is determined chiefly by the direction of heat conduction and the availability of water. Differential pressure is a minor factor, of little importance when soils freeze under normal conditions.

A tentative hypothesis has been advanced to explain the mechanics of the process by which ice crystals growing in open systems are able to exert pressure and overcome resistance.[1] A continuation of the investigation has led to the discovery of additional facts which shed more light on the problem.

A growing ice crystal is in contact with a thin film of water similar to the adsorbed layer which forms on many other solids that are in contact with water. As a molecule in the film is oriented and attached to the crystal, it is replaced by another from the adjacent liquid, thus maintaining the integrity of the film. If resistance to growth is offered by a solid body such as a soil particle, the pressure is exerted through the thin film of water that separates them. This film may consist of a single layer of molecules; but I am inclined to think that it is somewhat thicker, for the molecules in it possess considerable mobility. A unimolecular layer could not be expelled by pressure alone; thicker layers could be reduced through expulsion of some of the molecules under pressure, though this is probably resisted by the strong attractive forces. After the available water has been exhausted, the film may be frozen; but it does not freeze readily.

The orientation and attachment of a molecule to the crystal is accompanied by a slight repulsion, proportional to the change in volume; but this cannot be considered an essential factor in the process, for liquids that freeze with decrease in volume give pressure effects in open systems similar to those obtained with water.

[1] *Ibid.*, pp. 458-60.

Cohesion is greater between the molecules in the film and between these molecules and the ice, than it is between water molecules that are not similarly located close to ice crystals. Since no outside force is competent to push molecules of water into the film between the growing ice crystal and the resisting solid, the crystal must be displaced relative to the adjacent solid because water is pulled in between them; and this is made possible by the high intermolecular attraction, which prevents the separation of the water molecules.

During the growth of an ice layer in soils, water is pulled up through the interstitial capillary passages; but the upward flow cannot be attributed to capillarity, for there is no free surface or meniscus. Since ice crystals growing in an open system are able to overcome a resistance many times as great as atmospheric pressure, the water, in such cases, is actually pulled into the nourishing film under high tension.

The ultimate tensile strength of water has been estimated, from the energy required to separate the molecules when it is converted into gas, as over 1,300 atmospheres. A water column of large cross-section is placed under tension with difficulty, for the water is superheated when under negative pressure, even at temperatures below 0° C., and the formation of a gas phase immediately breaks the column. Extremely slender columns, or filaments, and thin films are more easily maintained under tension than larger ones; and this is the form in which water is present in clays.

Only rough estimates can be made of the maximum tensional stress developed in water during the freezing of soil. The surface load lifted can be measured in laboratory experiments; but to this it is necessary to add the frictional resistance of the frozen soil in contact with the container, the force necessary to separate the clay cylinder, and the force required to pull water through impervious clay to the growing ice crystals. Moreover, an ice layer must begin to form at some favorable point and then spread outward rather than form across the entire cross-section of the cylinder instantaneously.

Since the ratio of surface uplift to depth of freezing varies greatly with different soils under similar conditions of saturation, and with the same soils under different conditions of loading, the tensile stress that can be developed in the water is not the only factor limiting soil heaving. If tensile strength were the only factor, the amount of uplift would equal the depth of freezing or be zero. The principal additional factors are the rate of freezing and the downward growth of ice crystals due to the freezing of interstitial water *in situ*. The gradual formation of ice in the soil yoids below a growing ice layer probably explains the rhythmic banding due to development of successive ice layers as freezing extends downward in clays (see Figs. 2, 3, and 4).

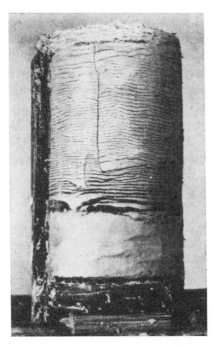

FIG. 4.—Clay frozen under pressure of 15 kilograms per square centimeter. Ice is clear and sharply separated from clay.

In indurated clay, or in clay that has been thoroughly consolidated artificially, the layers of segregated ice are clear, for the most part, and very sharply separated from the frozen clay. The total thickness of these layers, as close as can be measured, is the same as the amount of surface uplift. The ice layers, because of slower cooling, tend to become thicker and more widely spaced with distance from the cooling surface. In unconsolidated clay the ice layers are not so sharply defined, and it is not usually possible to estimate accurately the thickness of the segregated ice. In tests with open systems where water can enter, moisture determinations show that the frozen clay between the ice layers contains the same percentage of water as clay below the depth of freezing. This observation confirms the conclusion that heaving in these tests is due to the formation of the ice layers and that the freezing of interstitial water causes practically no uplift. However, in spite of this, the rate of heaving, as shown by repeated experiments, is continuous under constant temperature, and never intermittent.

Using the apparatus shown in Figure 1, tests were made on indurated clay cut in the form of smooth, true cylinders and set in paraffin so as to equalize friction on all sides, and thus, as far as possible, prevent development of irregular ice layers. Temperature and heaving graphs for one of these tests, redrawn so that they start from the same vertical axis, are shown in Figure 5; and the distribution of the ice layers at the close of the experiment is shown in Figure 2. A sample of clay taken just below the frozen material and three samples taken between successive ice layers gave 22.0, 22.1, 21.8, and 22.0 per cent water respectively. When tested with the point of a knife,

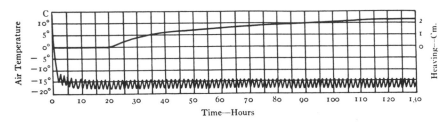

F IG. 5.—Graphs showing air temperature and rate of heaving during the freezing of the clay cylinder shown in Figure 2.

the clay between the lowest ice layers was found to be soft, as compared with that higher up, the hardness tending to increase toward the surface where the temperature was lower and freezing had gone on for a longer time.

The hypothesis which best explains the facts here outlined is that during the growth of an ice layer the voids in the adjacent underlying clay, beginning with the larger ones, tend to fill gradually with ice and thus increase the work that must be performed in supplying water to the ice layer. When resistance becomes too great, the flow of water to this ice layer stops and a new layer of ice begins to form near the bottom of the zone of frost penetration. More of the water in the clay between the ice layers freezes as the temperature gradually becomes lower, some of it possibly migrating to the ice layers; but previous tests with this clay have shown that 6 parts of water per 100 parts of dry clay do not freeze on prolonged exposure to low temperatures.

An ice prism in one of the layers continues to grow at the base where it is exerting pressure against an adjacent soil particle as long as water molecules can enter the separating film and be attached to the crystal. If the soil particle is very small, the molecules travel only a short distance through the film to reach the points where they are attached. If the soil particle is larger, it takes longer for the molecules to reach their points of attachment; and meanwhile freezing may extend downward around the particle so as to gradually inclose it in ice. As soon as crystal growth is checked at any point, the temperature of the adjacent soil particle begins to fall, for heat is no longer liberated at this point by the conversion of water into ice; and, since water is a poorer conductor of heat and has a higher specific heat than the minerals present in soils, the temperature of the bottom of the soil particle will reach the freezing point sooner than the water with which it is in contact. This helps to bring about the inclusion of the soil particle in the ice. The formation of layers of segregated ice is possible only when the soil particles are excluded by the growing crystals, a process which is aided by slow freezing, for then the molecules have more time in which to penetrate between crystal and soil particle.

In carefully sized material having a particle diameter of about 2 microns, segregated ice forms under conditions of slow freezing, but not when freezing is rapid. Even with slow freezing segregation may be prevented by applying a little pressure to the surface. As the diameter of the soil particles diminishes, higher pressures are required to prevent heaving. The effect of pressure in reducing and preventing heaving in open systems seems to be due in part to a decrease in the mobility of molecules in films that are under pressure.

If heat is conducted upward from the bottom of a growing ice layer just fast enough to remove both the heat set free when water is converted into ice and the heat brought up by the water, the ice layer will continue to grow in thickness; but if heat is conducted away more rapidly, the freezing isotherm will move gradually downward, causing ice to form in soil voids below. As freezing extends downward in clay, cooling is slower and there is less resistance to the upward movement of water; therefore, the ice layers tend to become thicker. Conditions most favorable for slow freezing are a relatively steep temperature gradient below the freezing-isotherm and a relatively gentle temperature gradient above.

Separation of the air dissolved in water, as freezing progresses, is a factor in interrupting the supply of water to the growing ice layers, but it is of relatively little importance. Examination of the ice with a strong magnifying glass shows that it contains small filiform cavities filled with air and oriented in the direction of crystal growth. These cavities, together with the parallel orientation of the ice prisms, give the ice a satiny luster. The cavities begin a short distance below the top of an ice layer and become more numerous lower down, but there is no difference in their abundance between the central and lower portions of thick ice layers. When boiled distilled water is used, with protective oil layers to prevent the absorption of air, the results obtained on freezing are not appreciably different. Also, when liquids which do not absorb air are substituted for water in freezing experiments, the same rhythmic banding occurs.

In order to prove that the pressure developed through the growth of ice crystals in an open system is limited by the failure of the water supply, and not by inability of the crystals to grow under greater pressure, the following experiment was performed.

The upper part of a strong fiber container, water-proofed with paraffin, was filled with clay and the lower part with water, the two being separated by a perforated disk of hard rubber with filter paper resting on top. The disk was held in place by a stiff steel spring that rested on the bottom of the container. Before filling the lower part of the container with water, air was removed from the clay by displacement with water and the clay thoroughly compacted for several days under a pressure of 14 kilograms per square centimeter. After introducing the water, the lower part of the container was stood in a can of impervious cement, and the upper part was strengthened by surrounding it with wood strips held in place by steel bands.

The container was placed in the refrigerator and buried to the top in dry sand. With the apparatus set up as described, and with the clay placed under a pressure of 14 kilograms per square centimeter by means of a piston inserted in the top of the container, the water in the lower part, because of the supporting action of the spring, was under a pressure of only about 10 kilograms per square centimeter.

As freezing progressed, the pressure on the clay increased; and at the close of the experiment it was about 15 kilograms per square centimeter.

The walls of the container were sawed through and one-half was removed to expose the cylinder of frozen clay (see Fig. 4). The rubber disk in the photograph is slightly displaced. The total thickness of the ice layers was between 2 and 3 centimeters, whereas almost no heaving could be obtained in experiments with this clay under similar pressures but with the water under no outside pressure. The spring in the reservoir was completely compressed, for the water had been pulled into the clay to build up ice layers in spite of the high pressure on the clay.

There is some evidence that rupturing of the upward-moving filaments and films of water under tension is one way by which water supply is interrupted and growth of ice layers stopped. The inability of molecules to enter films that are under pressure, or to enter fast enough to prevent the downward freezing of interstitial water, is probably another way in which the growth of ice layers is stopped. The data now available are insufficient to determine which of these methods is the more important.

When water is under tension because of the growth of an ice layer in clay, the stress is ordinarily greatest right at the points where water molecules are being pulled into the film which separates the growing ice crystals from the clay below, and it is therefore at this place that any break would occur. The breaking of a water filament, or film which is being drawn toward the ice layer, means that the molecules are separated beyond their range of attraction; or, in other words, a gas phase is formed between the liquid water and the ice. The gas phase, however, could not extend far below the ice layer, for the tensile stress in water a little below the ice is due to frictional resistance to the passage of the water through the small soil voids, and this stops as soon as the upward movement is checked.

On feeezing clays that are extremely impermeable because of high colloid content and that contain a high percentage of water, tensional cracks of polygonal pattern develop and extend downward in advance of freezing. As freezing progresses, the cracks are gradually filled with ice.[1] These results were obtained with South Carolina

[1] *Ibid.*, pp. 457–58.

Cretaceous white clay to which from 5 to 95 per cent bentonite had been added.

The hypothesis that the growth of ice crystals stops because of interruption of the water supply by the formation of a gas phase receives some support from freezing experiments with open systems in which liquids having different boiling-points were substituted for water. The ratio of uplift to depth of freezing for nitrobenzene, which boils at 210.85° C., is greater than for water under the same surface loads and conditions of freezing; and this is true in spite of the fact that greater force is required in pulling nitrobenzene through the clay to the growing crystals, since its viscosity at low temperatures is about 1.7 times that of water. Benzene, with a boiling-point of 79.6° C. and a viscosity of 0.5 that of water, on freezing heaves much less than does water under similar conditions. Nitrobenzene and benzene both freeze with decrease in volume at temperatures between 5° and 6° C.

It would be interesting to know the maximum tensile stress that could be developed in water as a result of freezing in an open system. By using finer material and cooling with extreme slowness, it should be possible to obtain somewhat higher tensile stress in the water, and therefore lift heavier loads, than was possible in the writer's experiments. In an ideal system it is theoretically possible to obtain a tensile stress of about 1,350 kilograms per square centimeter, but it is improbable that this stress could be approached in any attainable system.

The energy for frost heaving is supplied, of course, by the removal of heat. The greater the pressure on an ice layer the lower the temperature must be if growth is to continue. Therefore, if the films of water are not ruptured under tension, the load lifted is determined by the amount of undercooling possible in the clay immediately below an ice layer. The presence, between the base of an ice layer and the underlying soil, of a separating film connected with water below makes frost heaving a reversible action from the standpoint of thermodynamics.

When liquids that solidify with decrease in volume are substituted for water in freezing experiments with open systems, surface load, or pressure, reduces heaving, although removal of heat is the only source of energy; and the freezing point of such liquids is raised by hydrostatic pressure acting on both liquid and solid phases.

This anomaly might be explained if we assume that the base of a crystal where growth takes place is subjected to non-uniform pressure because of adjacent soil particles, for the effect of unequal pressure would be to lower, locally, the melting-point; but this assumption seems improbable, for crystal growth would tend to equalize the pressure, and as long as the pressure is transmitted across a separating film it should be practically hydrostatic.

An alternative hypothesis is that pressure tends to reduce the thickness of the nourishing film by expelling some of the molecules and, since the expulsive forces are increased, the attractive forces must likewise be increased, by lowering the temperature, else molecules cannot enter the film. Pressure decreases molecular mobility in the film and retards crystal growth.

Growth of ice crystals under resisting pressure in open systems is analogous to the growth of crystals from solutions under similar conditions.[1] One is possible because water occupying very small voids can be undercooled; the other, because solutions in such voids can be supersaturated. Water molecules are pulled toward a growing ice crystal to replace those that become attached to it, while molecules in solutions reach their points of attachment because of osmotic pressure. If water at low temperatures is considered to be a saturated solution of trihydrol, or ice molecules in dihydrol, the analogy would be even closer; but it would be unwise to draw conclusions until more data are available. Benzene and nitrobenzene have usually been classed as non-associated liquids; but, according to Schames, molecules of $(C_6H_6)_4$ begin to appear near the freezing point in liquid benzene, which corresponds to $(C_6H_6)_2$; and Dutoit and Moijou find nitrobenzene slightly associated.[2] Tension in water should favor the formation of trihydrol molecules, and therefore of ice, in soil voids below a growing ice layer; but tension should op-

[1] Stephen Taber, "The Growth of Crystals under External Pressure," *Amer. Jour. of Sci.*, 4th Ser., Vol. XLI (1916), pp. 532–56; "Pressure Phenomena Accompanying the Growth of Crystals," *Proc. Nat. Acad. Sci.*, Vol. III (1917), pp. 297–302.

[2] Quoted in *Molecular Association* by W. E. S. Turner (London, 1915), pp. 87 and 101.

pose the formation of associated molecules in benzene and nitro-benzene, and thus help to make undercooling possible.

For one whose conceptions of frost heaving have been derived from freezing experiments with closed systems, the results obtained in this investigation seem paradoxical. (1) Heaving on clay is greater than on sand, although part of the water does not freeze in clays, whereas practically all of it, within the zone of frost penetration, freezes in sand. (2) The pressure developed during heaving in an open system is limited by the tensile stress that can be developed in the water. (3) The boiling-point of a liquid seems to be more important than the freezing-point in determining the pressure developed by freezing in open systems. (4) While water expands on freezing, the freezing of saturated clays may be accompanied by the formation of shrinkage cracks, owing to withdrawal of water to build ice layers. (5) Heaving is upward because that is the direction of heat conduction rather than because it is the direction of least resistance.

It cannot be expected that the explanation of frost heaving here given is complete and without error in all of its details. It is a first attempt and will doubtless need modification as additional data become available. A better knowledge is needed of the constitution of water and the nature of the changes occurring as the temperature is lowered and freezing takes place. Also, more must be learned about the properties of thin films of water in soils. And, it is highly desirable that accurate field observations be made of the conditions under which soil freezing is accompanied by excessive heaving in some places and no heaving in others.

2

Reprinted from *Science*, **142**, 499–501 (Oct. 1963)

Particle Sorting by Repeated Freezing and Thawing

Abstract. *If a hetrogeneous mixture of particles of various sizes is frozen and thawed repeatedly, the particles are sorted into relatively uniform groups by size. The movement of particles depends on the amount of water between the ice-water interface and the particle, the rate of freezing, the distribution of the particles by size, and the orientation of the freeze-thaw plane.*

In regions of cold winters, stones often rise to the surface of the ground, fence posts move upwards in their holes, plants are often uprooted, and soil particles are sorted into pockets or layers of relatively uniform size by the action of frost. The upward movement of stones is well-documented, but there is no agreement on the causes of the movement (1, 2). This report describes laboratory experiments which demonstrate the effects of direction of freezing, rate of freezing, and shape and size of particles on their movement during repeated freezing and thawing, and the effect of a moving plane of freezing upon the migration of individual particles.

The basic test material was a noncohesive mixture of rounded sand grains and gravel, ranging in size from 0.074 to 7.0 mm in diameter, and a fine fraction (<0.074 mm) composed of crushed quartz. Freezing cabinets were placed in cold rooms maintained at −5°, −10°, and −20°C. The rate of freezing or thawing was regulated with a heating tape situated opposite the freezing front. Freezing rates were determined by thermocouple measurements at the contact between the sample and the cabinet wall. After the sample had been subjected to the required number of freeze-thaw cycles, horizontal layers 15 mm thick were examined, and the number of particles finer than 0.074 mm in each layer was determined by sieve analysis. The difference in concentration, by weight, of these particles in successive layers indicated the amount of particle migration produced.

One set of experiments was designed to simulate conditions in regions of the earth where the soil both freezes and thaws from the surface. The thermally controlled cabinet of Schmertmann (3), provided these conditions. Freezing rates ranged between 0.6 and 1.4 mm per hour; the thawing rate was 6.0 mm per hour. Preliminary experiments with tetrahedrons, 2.5 cm in height, and glass marbles, 2.5 cm in diameter, showed that coarse particles moved upward, whether they were buried just beneath the sand surface or 12 cm below. The tetrahedrons moved upward faster than the marbles. Analysis of the grain size of the sand showed that 20 freeze-thaw cycles increased both the number of fine particles in the lower part of the sample and the number of coarse particles near the surface. The term "vertical sorting" is proposed for this type of movement in the freeze-thaw layer (4). The rate of downward movement of sand particles finer than 0.074 mm was 0.2 to 0.3 percent (by weight) per cycle.

Samples which exhibited vertical sorting also showed changes in volume which were related to the percentage by weight of the different grain sizes, and were inversely proportional to the rate of freezing. The more unsorted the mixture, the larger the volume change due to freezing and thawing.

In extremely cold regions, the layer above the perennially frozen ground freezes from the bottom as well as from the top, and thaws exclusively from the surface. The cabinet designed to study this type of freezing was composed of two Lucite cylinders of equal height (15 cm) and 12.5 and 15.0 cm in diameter, respectively, placed one inside the other, and attached at the bottom to an aluminum plate 2.5 cm thick. The air space between the cylinders insulated the sample, and the cylinders were covered with a wooden box containing a heating tape. A sample of sand, 7 cm high, was placed in the inner cylinder, which was then filled with water to a depth of 5 cm above the surface of the sample.

When the freezing line advanced upward from the bottom, the differential movement of particles created by ice growth at the ice-water interface affected the soil particles below the freeze-thaw plane as well as above it (Fig. 1). This mechanical "shaking" seemed to be responsible for the upward migration of coarse particles and the downward migration of the finer particles in the unfrozen layer, for which the term "mechanical sorting" is proposed. Sieve analysis showed that particles finer than 0.014 mm moved downward 0.1 percent (by weight) per cycle. Under the same conditions, glass beads showed a downward migration of 1.0 percent per cycle, suggesting that an important factor in this type of sorting is the shape of the particles. It is also possible that this phenomenon is related to the thickness of the sample used.

The downward migration of fine particles (<0.074 mm) produced a layer 5 mm thick when the freezing plane moved at a rate of 0.1 mm per hour for 1 week (Fig. 1). Hence, layers can be produced in the ground by freeze-thaw action as well as by irregularities of the original deposition. The same type of vertical sorting can be observed in the freeze-thaw layer above perennially frozen ground, as shown in Fig. 2. Gravels deposited by man in Alaska show sorting after 15 to 20 cycles of annual freezing and thawing. Vertical sorting is not necessarily uniform throughout the layer; because particles migrate at different rates, areas of both fine and coarse particles may appear at the surface.

In regions of intensive seasonal frost or perennially frozen ground, the soil freezes and thaws laterally in places such as riverbanks, retaining walls, culverts, ditches, and terraces. The cabinet designed to simulate these conditions consisted of a rectangular plywood box waterproofed with shellac, measuring 15.0 by 12.5 cm, and 12.5 cm in height, with one side replaced by an aluminum plate which acted as a cooling front. Heating tapes in the opposite side and in the cover were used to regulate the rate of movement of the freeze-thaw plane between 30 and 42 mm per hour. Thermocouple readings in the soil indicated that a vertical freeze-thaw plane was produced. The sample was the same as used in

Fig. 1. Side view of cabinet, after seven freeze-thaw cycles, in which freezing was effected from the bottom upward. The sample was frozen at a rate of 0.6 mm per hour, and thawed at approximately 6.0 mm per hour.

the previous experiments, containing 14 percent by weight of fine particles (<0.074 mm). Water was added to saturate the sample.

After 22 cycles of freezing and thawing, the sample was divided into 9 horizontal sections. Sieve analysis showed that particles migrated in front of the freezing plane and that the percentage of finer particles, by weight, increased from the top downward as well as away from the freezing plane. The term "horizontal-vertical sorting" is proposed for this; it occurs in naturally deposited materials in the freeze-thaw layer (4) and should also be found in seasonally frozen soils. The shape of the path followed by the particles seemed to be determined by the rate of freezing and by the gravitational force acting on the particles; but this point requires further experimental verification. As in the previous experiments, the volume change was inversely proportional to the rate of freezing.

The actual migration of particles in front of a moving freezing plane was experimentally demonstrated in the cabinet designed for studying freezing from the bottom. Freezing was effected at the contact between the inner cylinder, in which distilled water had been placed, and the aluminum base plate. As the horizontal freezing plane was raised (between 0.2 and 7.0 mm/

hr). 20 grams of particles of different sizes were placed on the interface. After freezing the mixture upward 1 cm, those particles that had not been engulfed by the ice were removed by siphoning; the percentage of particles carried was expressed on the basis of weight. The fine particles were carried to the top leaving streams of bubbles trapped in the ice beneath them, while the coarser particles were engulfed by the ice close to the seeding position. Evidently, a moving interface of ice and water can exclude particles located in its path. Various materials were tested, including particles of SiO_2, of the same size but different shapes (glass beads, broken glass, and quartz), and particles of different materials (calcite, rutile, quartz, shale, and mica) which also differed in shape. Some of the results are shown in Fig. 3.

Side freezing is produced in nature around large particles or concentrations of particles which migrate upward and become slightly exposed at the surface. Such particles are better heat conductors than the surrounding finer materials, and initiate freezing from the sides (2, 6). The effect of repeated side freezing and thawing was demonstrated by placing sorted layers of particles in aluminum pans and alternately placing them in cold (−20°C) and warm (+20°C) rooms so that the soil froze

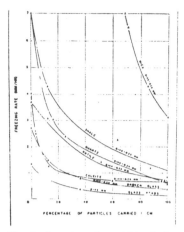

Fig. 3. Percentage, by weight, of glass beads (0.14 mm in diameter), calcite, broken glass, rutile, quartz, shale, and mica particles (0.14 to 0.29 mm) carried 1 cm at different rates of freezing.

mainly from the sides. Vertical sorting was changed into horizontal-vertical sorting regardless of whether the layers were dome-shaped or flat. This can be explained by the fact that the freezing front pushes the soil particles toward the center of the pan, forming a mound. Coarse particles in the upper layers roll from the top of the mound toward the edge of the pan, thus creating sorting.

The results of these experiments indicate that particle migration in nature is dependent on certain conditions.

1) The rate of freezing. Horizontal-vertical sorting is produced experimentally with freezing rates of 30 to 42 mm/hr. Natural freezing rates range up to 40 mm/hr. Vertical sorting has been produced with freezing rates of 0.6 and 1.4 mm/hr and a thawing rate of 6.0 mm/hr.

2) The size, shape, and distribution of particles. Fine particles (<0.1 to 0.5 mm in diameter) migrate when the freezing rate is either high or low; coarse particles (>1.0 mm) migrate only if the freezing rate is low. An important factor may be the area of contact or the pressure of the particle on the interface.

3) The direction of freezing. It is evident that particles become sorted by the cyclic effect of a freeze-thaw plane; the particles move away from the freezing point. For vertical sorting to be changed to horizontal-vertical sorting, either large particles must be present and partially exposed at the

Fig. 2. Profile of vertical sorting in flat terrain in which a "boulder field" is forming (Thule area). The scale rests at 25 cm below the level of perennially frozen ground.

surface, so that the adjacent soil will freeze and thaw in a non-horizontal plane, or non-uniform vertical sorting must exist.

4) The amount of moisture at the freezing plane. In materials deposited by man, the amount of sorting is directly proportional to the moisture content of the soil (7). The actual presence of a layer of water between the ice front and the particle was proposed by Taber (8) but still remains to be demonstrated. This water would have to be continuously replenished as the ice advanced, moving the particle in front of it. If the water layer froze faster than it was replenished, the particles would be trapped by the ice. It is reasonable to assume that the layer of water would be more easily and rapidly replenished in the presence of small, rather than large, particles. This would account for the fact that coarse particles move only when the rate of freezing is low.

The results of these experiments may be important in determining the susceptibility of materials to frost over long periods of time. Evidently, the chances of frost-heaving or formation of ice lenses will increase in proportion to the concentration of fine particles. On the other hand, it may be predicted that the soil adjacent to a retaining wall will have a tendency to become less susceptible to the formation of ice lenses as fine particles move away from it.

Arturo E. Corte*
*Cold Regions Research and Engineering Laboratory,
Hanover, New Hampshire*

References and Notes

1. C. Troll. *Geol. Rundschau* **34**, 545 (1944); L. Washbur. *Geol. Soc. Am. Bull.* **67**, 823 (1956); A. Hamberg, *Geol. Foren. Stockholm Forh.* **37**, 583 (1915); G. Lundqvist, *Geogr. Ann. Stockholm* **31**, 335 (1949); ———, *Geol. Foren. Stockholm Forh.* **73**, 505 (1951).
2. A. Cailleux and G. Taylor, *Cryopédologie Etude des Sols Gelés* (Herman, Paris, 1954), p. 58.
3. J. H. Schmertmann, U.S. Army Snow, Ice and Permafrost Establishment, Corps of Engineers, *Tech. Rept. 50* (1958).
4. A. E. Corte, U.S. Army Cold Regions Research and Engineering Laboratory, Corps of Engineers, *Research Rept. 85*, part I (1961), part II (1962); also *Highway Res. Board. Bull. No. 317* (1962), pp. 9–34; *No. 331* (1962), pp. 46–66.
5. ———, *J. Geophys. Res.* **67**, 1085 (1962).
6. S. Taylor, thesis, Univ. of Minnesota (1956), p. 189.
7. A. E. Corte, unpublished data.
8. S. Taber, *J. Geol.* **38**, 303 (1930).
9. The author thanks W. Brulle, J. Gallippi, E. Murabayashi, R. Ramseier, A. Tice, and R. Tien for able assistance in the field and laboratory investigations. Dr. Jerry Brown was very helpful in the editing stage.
* Present address: Universidad del Sud, Bahía Blanca, Argentina.

26 August 1963

3

Reprinted from *Science*, **149**(3691), 1520–1521 (1965)

STONE MIGRATION BY FREEZING OF SOIL

Chester W. Kaplar

*U. S. Army Cold Regions Research and Engineering
Laboratory
Hanover, New Hampshire*

I have read with great interest D. R. Inglis's explanation of the uplifting of a large object such as a stone or fence post during freezing of soil [*Science* **148**, 1616 (1965)]. As early as 1958, using time-lapse color photography, I made observations of this phenomenon at the U.S. Army Arctic Construction and Frost Effects Laboratory, Waltham, Massachusetts (now merged with U.S. Army Cold Regions Research and Engineering Laboratory, Hanover, New Hampshire). A 15-minute sound and color film entitled "Frost action in soils" was presented at the International Permafrost Conference at Purdue University in November 1963, and has been in continued circulation since. It has been shown at various gatherings of the American Society of Civil Engineers. It has also been shown in Japan and Sweden.

This film demonstrates, among other things, the upward movement of a stone and simulated piles within a silt soil by frost action, when freezing is from the top down in an open-system test. The narration accompanying the film is in very close agreement (with some exceptions) with the explanation set forth by Inglis. The film shows that adhesion of ice to the top of the stone was not the lifting force, although it might possibly be under certain conditions of object shape and relative position. What actually happened in the frost-susceptible material studied was that the soil directly over a stone was lifted above the stone, leaving a void. A stone rises only when the adfreeze force around it is greater than the forces holding it in place. The total movement of the stone from its initial position depends upon the heaving rate of the soil and the time required for freezing to penetrate down to a level below the cavity formed under the stone. In saturated non-frost-susceptible soils in a closed system, uplift by the expansion of water upon freezing, as proposed by Inglis, is possible. Figures 1–4 illustrate the sequence of events.

The film is available on loan from the Commanding Officer, U.S. Army Cold Regions Research and Engineering Laboratory, Hanover, New Hampshire. Further study of the freezing of soils and related phenomena is being carried on in that laboratory.

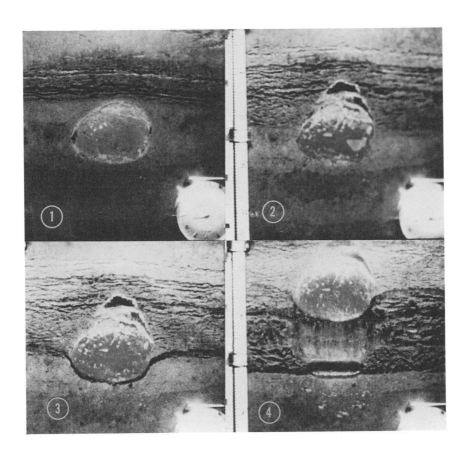

Figs. 1–4 Movement of a stone (actual size, about 3 cm diameter) in freezing soil. (The object partly visible at the lower right is a wrist watch.) Fig. 1: frost line approaching the top of the stone. Fig. 2 (about 30 hours later): the soil above the stone has heaved, leaving a void. Fig. 3 (about 12 hours later): the stone is being lifted by the grip of the frozen soil; there is now a void under the stone. Fig. 4 (about 79 hours later): the stone has moved up a considerable distance, and the cavity below has become slightly narrower and filled with water, which has frozen.

4

Reprinted from *U.S. Geol. Survey Prof. Paper 109*, 205–214 (1919)

OBSERVATIONS ON THE NORTH SHORE OF ALASKA

E. de K. Leffingwell

GROUND ICE WEDGES.

THEORY OF FORMATION.

The theory of the origin of ground ice here presented was formed in July, 1914, after nine summers spent upon the north shore of Alaska. An article was written in December of the same year and afterward published.[1] This article has been expanded and embodied in the present discussion.

On reexamining the literature, with which the writer already had a fair acquaintance, it was found that in 1883 Bunge had propounded a similar theory. Bunge, however, did not give detailed descriptions of exposures, nor describe the stages of growth of the ice wedges with sufficient clearness to make his exposition plain to those who followed him. In fact, by citing others as supporting him he greatly weakened his own case, for the men whom he cited held theories that differed greatly from his own.

Hooper, in 1881, also suggested a probability of the growth of ice in cracks in the ground, under successive changes from freezing to thawing.

The writer here endeavors to present his theory with such detail that it may be grasped and to support it with diagrams and photographs showing the stages of the formation of ground ice according to this view.

He went into the field in the summer of 1906 with the idea that the coastal ground ice occurred in horizontal sheets and did not learn its real distribution until 1914. During the first eight summers, although the ground ice was examined at every opportunity, little insight was gained into the method of its formation.

The usual theory advanced in the literature is that bodies of snow or ice were buried by peat or wash material and thus preserved. The writer sought to interpret the Alaskan coastal ground ice in this way, but could neither postulate a satisfactory source for the ice nor find any workable hypothesis to account for its preservation. Not until the summer of 1914 did he discover that most of the ice was formed in place in the ground. A vertical wedge of ice within a peat bed first drew his attention to the fact; for such a dike of ice could not have stood up in the air for the hundreds of years that were necessary for the formation of the peat.

Several dozen photographs of the ice were made, but most of them were later damaged by water, so that the writer has to depend chiefly upon sketches which were often hastily made. Fortunately P. S. Smith, of the United States Geological Survey, some years ago had secured photographs of ground ice on Noatak River. One of these photographs illustrates wedge-form ice of the kind that is the subject of this chapter. (See Pl. XXVIII, *A*.)

FROST CRACKS.

During the Arctic winter frequently loud reports are heard, coming apparently from the ground. Often the sound is accompanied by a distinct shock, which is in fact an earthquake of sufficient intensity to rattle dishes. The writer has spent six winters in the region under discussion, living most of the time upon the tundra, which is chiefly underlain by muck. Many camps have been made upon other formations, such as sands and silts, and his impression is that the sound of the cracking ground was heard everywhere. This impression has been confirmed by a prospector who has lived nearly 30 years in the country.

It was at first thought that the reports were caused by the cracking of hard snowdrifts, but the cracks in these drifts were seen to run into the ground below. When the snow melts in the summer fresh open cracks can be seen cutting across all the tundra formations, even across mud and growing moss beds, and dividing the surface into polygonal blocks, which resemble mud-crack blocks but are of a much larger size. These blocks have an estimated average diameter of about 16 yards and have a tendency toward the hexagonal form, although rectangles and pentagons are commonly developed.

Occasionally a crack is seen to run across a flat surface, with no associated features (see Pl. XXIX, *A*), but usually it is accompanied by other modifications of the surface. Either the cracks lie in a gentle depression which surrounds elevated polygonal blocks or they run

[*Editor's Note:* Certain plates have been omitted owing to limitations of space.]

between two parallel ridges which surround depressed blocks. These features do not vary from block to block, but each is locally developed over a considerable area. Few of the elevated blocks have a relief of more than 1 foot, but that of the depressed blocks may be twice as much.

The parallel ridges form shallow reservoirs, very similar to those of the block system of irrigation, especially when they take a rectangular form, as many of them do. They may contain ponds and are invariably swampy, so that when crossing such an area dry footing can only be had on the ridges. (See Pls. XXIX, B; XXX, A.) On Flaxman Island the tendency of the depressed blocks is toward a rectangular arrangement, and the frost cracks extend in a straight line for a distance sufficient to include several polygons. One double ridge was paced for a hundred yards in nearly a straight line. On either side similar ridges ran parallel to it about 75 feet away. These ridges were perpendicularly intersected by four others equally spaced, thus forming blocks about 25 yards square.

In an area of elevated blocks near by there were many places where these cracks radiated symmetrically from a point, so that fairly regular hexagons were formed.

FORMATION OF ICE WEDGES.

The open frost cracks are in a favorable position for being filled with water from the melting snow, as most of them lie in depressions upon a flat surface. Those that by chance should get no water may become filled with ice crystals deposited by the damp air by internal "breathing." The crack is thus filled with solid ice from the freezing of the water, or contains much ice in the form of frost crystals, so that a narrow vein of true ground ice is formed in the portion which lies below the depth reached by the annual thawing. When the frozen ground expands under the summer's heat, the readjustment to the strain may take place in four ways: (1) The pressure may melt the ice, so that the crack is closed again; (2) the ground may be sufficiently elastic to absorb the strain, so that no deformation occurs; (3) the ground may be deformed and bulged up, either as a whole or locally along the edge of

the ice wedge; (4) the ice itself may be deformed.

If the summer's strain has been relieved by readjustment of the material within the polygonal block, the next winter will again bring about the conditions which caused the first cracking of the ground. The first crack contains ice and probably is a plane of weakness for tensile strains, especially if the crack has been only partly filled. If it is a plane of weakness, new cracks will annually open at the same place and a constantly growing body of ice will be formed there. That this condition is common in tundra formations is shown by the constant association of ice wedges with definite centers of frost cracks.

Thus the growth of the wedge goes on from year to year, possibly failing during mild winters, when all the cracks may not be forced open in order to relieve the strain. The writer's observations show that the ice may increase until it underlies about one-fifth of the area of the block, but there is no theoretical reason for stopping at this stage. The wedges might grow until the amount of ice would exceed the amount of earth, so that instead of intrusions of ice into the earth an exposure would apparently show intrusions of earth into a heavy body of ground ice. Although nothing approaching this possible stage of development has been seen by the writer, the descriptions of localities in Siberia by Maydell and Toll, which show intrusions of earth scattered through a heavy body of ice, lead the writer to believe that there we possibly have ice wedges in an advanced stage of growth.

APPEARANCE OF THE WEDGES.

Cracks have been seen accompanied by no other surface manifestations (Pl. XXIX, A) and with no visible ice below them. The smallest wedge that has come under observation was about a foot wide. No doubt there are smaller wedges, especially in areas such as recently drained lake bottoms, where the process is just beginning. The thin veins have nearly parallel sides and flat tops, as shown in figure 16. As the ice increases in size it approaches more and more to the wedge shape, as the growth is greatest near the top, where the crack opens widest. There is a tendency in the large

A. FROST CRACKS, PARALLEL RIDGES, AND BLOCK PONDS ON FLAXMAN ISLAND.

B. EROSION OF A POLYGON FIELD NEAR THE 141ST MERIDIAN.

wedges to spread out under the surface of the ground. This tendency is exaggerated in oblique sections. (See fig. 17 and Pl. XXVIII, *B*.)

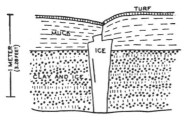

FIGURE 16.—Narrow ice wedge in a deposit of mixed clay and ice granules. There is an open crack within the wedge.

The bottom of an ice wedge has never been observed by the writer. Most of the bluffs on the north shore of Alaska are less than 10 feet high, and their bottoms are concealed by slumping. The maximum vertical dimension observed is about 10 feet, but the wedge had a thickness sufficient to have carried it two or three times that distance before pinching out. The ultimate depth must depend upon the

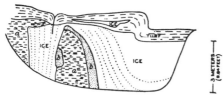

FIGURE 17.—Two joining ice wedges; the one on the right is cut obliquely. The dotted lines represent lines of air bubbles within the ice. *a*, Muck and clay, much disturbed; *b*, sand. (See Pl. XXVIII, *B*.)

depth of annual change in the ground temperature in the region. In the Schergin shaft (p. 184) the annual change is not noticeable at 50 feet. It is not probable that the cracks extend to the extreme limit, but once having been started by the great tension at the surface they may be expected to approach that depth. In discussing the question it seems safe to assume 30 feet as a working basis.

In muck formations the upper surface of the ice is usually less than 2 feet under the ground, about the limit to which the summer's thawing penetrates. This surface is usually horizontal or undulates with the surface of the ground. One or two exposures showed a dome-shaped surface and another a

central projection above the general surface (fig. 16). Some more complicated exposures are shown in figures 19 and 20. The overlying material, as a rule, is muck capped by a few inches of turf. Occasionally it is peat capped by growing sphagnum (?) moss.

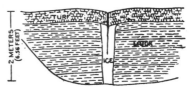

FIGURE 18.—Narrow ice wedge in muck beds. An open frost crack runs through the turf and into the ice.

STRUCTURE OF THE ICE.

A fresh transverse section of an ice wedge shows a face of whitish ice with numerous parallel vertical markings. These markings are usually formed by whiter ice, which con-

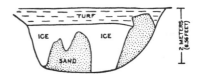

FIGURE 19.—Complicated exposure of ice in sand.

tains a large amount of air bubbles. In many specimens it is visibly granular, yet it shows a general vertical structure. As a rule the granules are less than half an inch in diameter, the average being about a quarter of an inch. Some specimens of ice when allowed to lie in the shade on a cool day broke up into short, irregular pieces, the greatest dimensions of

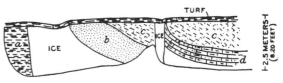

FIGURE 20.—Complicated exposure of ground ice. *a*, Clay; *b*, sand; *c*, peaty detritus, no structure visible; *d*, peaty detritus with upturned beds.

which were vertical. These pieces were an inch long and half an inch in diameter.

From the method of formation the crystallization should be irregular. Where water freezes in an open crack, short prisms might develop with the axes perpendicular to the freezing surface, which in these places is vertical. The axes of the ice crystals in the walls would have a disturbing effect, as would also any frost crystals within the crack. Every new crack runs across the crystals indiscriminately, and the new intrusion of ice adds to the complication. A careful study of the structure of wedge ice would probably develop criteria by which it may be separated from névé ice.

The writer's notes upon the air bubbles bring out nothing as to their shape and size. The chief characteristics are their abundance and orientation into vertical rows. In content of air wedge ice may closely resemble névé ice, the

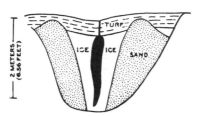

FIGURE 21.—Ice wedge in sand. A tunnel has been cut in the ice by drainage of surface water through the frost crack. The sand on either side is apparently bulged up.

difference lying in the distribution of the air bubbles. In recent névé ice examined by the writer the bubbles were scattered haphazard throughout the ice. Any concentration is more apt to occur along horizontal lines than along vertical ones.

At the sides of the wedge the markings of the ice are inclined from the vertical and approach parallelism with the sides. As the growth seems to occur near the center of the wedge the older lines, though originally vertical, are spread apart at the top. Oblique sections of wedges give exaggerated angles of inclination or even curves (fig. 17).

In several places cracks were seen running down a few feet into the ice as prolongations of an open frost crack in the tundra above. Once or twice isolated open cracks were seen within the body of the ice, so that a thin sheath knife could be shoved in for several inches (fig. 16). Near the edge of a bank these open cracks may become drainage lines for surface water, so that a tunnel is developed within the ice (fig. 21). As the tunnel widens the roof caves in,

and a deep gully is formed in the bank. These gullies work back and around the polygonal blocks and make walking rather difficult in the neighborhood of an old bank (Pl. XXX, B).

Some earth is likely to get down into the open frost crack while it is being filled with water, especially when the overlying material is mud or clay. Under growing peat the chance is not so great. The exposures of ice are usually so dirty from the mud which has dropped off from above that vertical earth veins might not be noticeable.

The possibility also exists that the pressure exerted by the growing wedge should fracture it so that earth inclusions might make their way into the sides of a wedge. Such an occurrence has never been seen by the writer, although it is reported from Siberia (fig. 30, p. 217).

STRUCTURE OF INCLOSED BLOCK OF EARTH.

The typical formation associated with ice wedges in the region under discussion is muck, a black mud containing much vegetable matter more or less decomposed. It varies from a peaty detritus that shows signs of having been water-laid to sand or mud mixed with varying amounts of decaying vegetation. Undisturbed sections of this muck usually show horizontal bedding. Occasionally sand or a

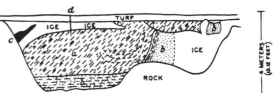

FIGURE 22.—Complicated exposure of ground ice. a, Disturbed muck and clay; b, clay; c, peat; d, a frost crack through the turf and ice.

slimy clay was seen under the muck at points where a good exposure revealed the lower strata (fig. 22). As the ice wedge grows in thickness and presses against the edges of the cleaved muck and sand beds they may become upturned and in time bent to the vertical or even beyond, causing the ridges which run along either side of the frost crack in the area of "depressed blocks." In areas of "elevated blocks" the process is not so easily understood. It may be that the block as a whole has been bulged up sufficiently to bring its surface above the general level, or else a central depression has been filled in and capped by turf.

The upturned beds of muck on either side of a wedge are a striking feature, and almost invariably occur in good exposures (figs. 23 and 24). Wherever the muck is underlain by sand, the sand is apparently forced up along the edge of the wedge, so that it lies between the ice and the upturned beds of muck. Some

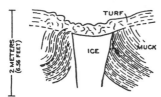

FIGURE 23.—Ice wedge in a muck bed, showing upturned strata.

of the illustrations show the sand apparently forced higher than the ice, or nearly through the layer of turf (fig. 21). The bedding of peat, on the other hand, does not seem to be disturbed by the growth of wedges.

In figure 25 is shown a hypothetical section of ice wedges and a depressed polygon block.

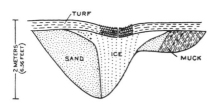

FIGURE 24.—Large ice wedge which spreads out under the surface of the ground. The vertical lines indicate rows of whiter ice full of air bubbles. The material on both sides is sand. To the right are upturned muck beds.

The writer was unable to ascertain the factors which control the character of block development. The "elevated blocks" are characteristic of drained areas and seem to be nearly constant features near banks. The "depressed blocks" are associated in the writer's mind with flat, marshy country. This, however, may be the effect rather than the cause of the difference in character of the blocks. The network of depressions will drain the elevated blocks, but the ridges form dams which interfere with surface drainage.

As the growing vein of ice becomes more wedge-shaped the pressure from its walls exerts a vertical component against the sides of the wedge. This tends to force the wedge upward. If an upward movement should occur, the ice would carry its protective covering with it and be able to exist level with or even somewhat above the general surface of the block. Since a bulging of the block by the growing wedge seems necessary, some upward

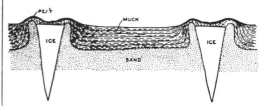

FIGURE 25.—Hypothetical section of ice wedges and depressed polygonal block.

motion of the wedge may take place without bringing the top of the wedge up to the general level. In the depressed blocks (Pls. XXIX, B, p. 205, and XXX, A, p. 206) the surface of the ground between the parallel ridges (probably underlain by ice) is higher than that of the blocks on either side.

THICKNESS OF PROTECTIVE MANTLE.

The usual covering of the ice is muck capped by turf, or peat capped by growing sphagnum (?) moss. As the thickness of this mantle increases by surface growth, the limit of the summer's thawing should rise, thus allowing a constant upward extension of the surface of the ice wedge at the locus of growth. Apparent upward growth of the surface was seen at only two or three places. In one place (see fig. 16, p. 207) there is an upward projection of ice above the general surface of the wedge, indicating a sudden change of the limit of thawing. In the other places the surface was dome-shaped, indicating a gradual change. As most exposures show the surface of the wedges to be nearly parallel with the surface of the ground, it seems that a balance must be maintained between the thickness of the covering and the increase in area to be covered, as the wedge becomes wider.

The rate of growth of turf must be very slow in this region, for there are many half-buried boulders on the surface of the tundra, which have been there since glacial time, or at least since the coastal plain emerged from the sea. Many bare spots exist where the turf has not been able to obtain a footing.

Wherever the protective covering is of muck, creeping of the soil will tend to close up the open frost crack. This will thin the covering, and if the rate of surface growth of turf is not sufficient to counteract the resulting decrease in thickness the upper surface of the ice will be lowered by melting. The increased slopes will cause more material to creep down over the ice, thus keeping the protective mantle up

CHARACTER OF EXPOSURES OF ICE WEDGES.

When a bank is undercut by wave or river action large masses of tundra commonly break off. As the frost cracks are planes of weakness, the break is likely to be along the ice wedges (Pl. XXXI, A and B), especially in high banks, where whole blocks break out, leaving many reentrant angles in the resulting bank, along which a nearly continuous exposure of ice may be seen. There appears to be a heavy horizontal sheet of ice of great thickness, whereas, in fact, little of it extends back more than a few feet from the face of the exposure.

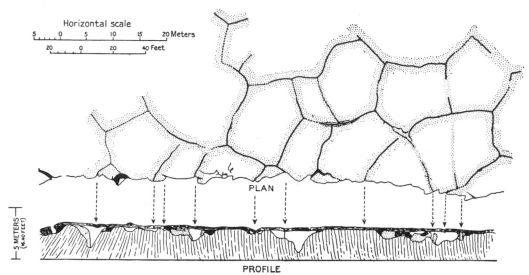

FIGURE 26.—Map of frost cracks on the tundra, with a sketch of the exposures of ground ice in the bank at one edge of the mapped area. Heavy lines represent frost cracks open in July. The dotted lines show evident loci of frost cracks. Stipple marks show areas probably underlain by ground ice. In the section below the map white areas represent ground ice; dotted areas, sand; lined areas, upturned muck beds. The rest of the exposure has slumped.

to the required thickness. A shallow depression will thus be formed, the slopes being sufficiently steep to cause the proper amount of creeping.

If the covering is sphagnum (?) moss the conditions may be somewhat similar. The moss and subjacent peat may also close the crack by creeping. At the same time the bed will become thinner, but growth of the moss will soon cause it to become thicker. If the moss grows too rapidly, the depression will be filled, and the conditions of moisture favorable to growth will cease. Thus it is possible for the growing ice wedge to maintain a peat covering of constant thickness.

On a 250-mile boat trip from Flaxman Island to Point Barrow in the summer of 1914 exposures of muck banks invariably revealed ice. Many miles were examined closely on foot or from the boat, and very little ice was observed which was not definitely in the form of vertical wedges, associated with frost cracks on the surface of the tundra.

DETAILS OF A FROST-CRACK AREA.

In figure 26 is shown a plane-table map of an area of frost cracks and below it a sketch of the exposures of ground ice in the bank. The polygonal blocks were of the elevated type. The relief was very faint, being somewhat ob-

A. TUNDRA BLOCK BROKEN OFF FROM FLAXMAN ISLAND.

Ice wedge at left.

B. TUNDRA BLOCK BROKEN FROM THE NORTH SIDE OF FLAXMAN ISLAND.

scured by sand which had drifted up from a sand spit on the left. The exposure was somewhat slumped but gave sufficient details to be illustrated. The heavy lines on the map show open frost cracks; the dotted lines, the evident locations of cracks. Where there was no surface indication, no lines were drawn. The shaded areas are supposedly underlain by ice wedges.

The average diameter of 11 blocks shown on the map is about 36 feet. The largest block is about 36 by 49 feet and the smallest 16 by 26 feet. The largest wedges of ice are about 8 feet wide at the top, and this width has been used in estimating the areas probably underlain by ice, except where surface indications pointed to new centers of development. About 20 per cent of the tundra is probably underlain by ice.

ASSOCIATED FORMATIONS.

Wedges of ground ice, with their accompanying surface features, are typically associated with muck formations. None were seen elsewhere. River silts, elevated deposits of sand and gravel, and soft shales have been carefully examined. The only ice found in such places was evidently of another form and of a different origin. Straight lines leading across the gentle surface undulations of sand spits have frequently been observed, and they can be explained only as frost cracks. No polygonal forms have been seen in such places. The writer is unable to say whether ice wedges develop in such sands, for the exposures made by fresh wave cutting are seldom more than 2 or 3 feet deep, which is less than the depth reached by the summer's thawing of sand.

RATE OF GROWTH OF ICE WEDGES.

Fresh ice-filled cracks 8 to 10 millimeters wide have been observed in the ground immediately above the ice wedges. This may be put as the maximum width of the crack. Open cracks about 5 millimeters wide have been found in the ice itself near the upper surface. The width, of course, diminishes downward. If 5 millimeters is assumed as the width at the top it would require only 600 years to build up a wedge 3 meters wide, which is about the maximum width seen in the region. If the cracks do not all open every winter this period must be multiplied by some factor. The writer frequently observed open cracks during the early years of his residence in Arctic Alaska, but as he did not realize their bearing he did not keep any record of their abundance.

About 1,000 years seems to be the age of the largest wedges. Unless some unknown cause prevents a greater growth, the temperature could not have been sufficiently low to bring them into existence at an earlier date or else the coastal plain has not been elevated above sea level for a longer period.

It would be well worth while to endeavor to measure the actual rate of growth of ice wedges over a series of years. In this way considerable insight as to the age of existing wedges might be gained. By measuring wedges of different sizes some law might be found by which the rate of growth could be extrapolated beyond the limit of size found in the locality. If the frost-crack theory is found to apply in areas where the wedges are more greatly developed than in Alaska a fairly definite time limit could be set for the existence of the Alaskan coastal plain above sea level. In areas of greater development a minimum might be set to the duration of the cold climate which was the cause of their development.

A series of posts in pairs might be set well into the ground on both sides of the ice wedge. Careful measurements made over a period of years would show any growth of the wedge. Any irregularities from possible disturbances of the posts would be eliminated by having a sufficient number of pairs of posts.

UPWARD BULGE OF SURFACE OF GROUND.

If we assume that the elevated blocks are bulged up by the growing ice, the amount of general elevation of the surface of the tundra can readily be calculated. If 20 per cent is taken for the surface compression of the block, as has been done for the north shore of Alaska, the average compression will be 10 per cent. An average block 11 meters in diameter will be compressed horizontally 1.1 meters to an assumed depth of 8 to 10 meters. This will cause an increase in a vertical direction of about 1 meter.

In the depressed blocks the adjustment is concentrated into the surrounding ridges, and the conditions are different. The depressions are being continually filled with growing vegetation, as well as by wind-blown material. In this way a much greater general elevation of the surface of the tundra is possible.

ULTIMATE STAGES OF GROWTH OF ICE WEDGE.

It seems to the writer that the wedges should bulge up the inclosed polygonal block more and more as they encroach upon it, so that a stage would be reached where a central mound of earth would rise above the surface of the block, which is generally underlain by ice. The earth block should now be pyramidal, with the apex pointing upward, and the wedges of ice should be in contact along the greater part of their upper edges. A perpendicular section through the middle of the inclusion of earth would show a section of a pyramid, but one which was taken eccentrically through the inclusion would invariably show earth overlain by ice. In this way the inclusions of earth overlain by ice which appear in the photographs in Von Toll's work (Pls. XXXII, XXXIII, XXXIV) might find an explanation, but the fact that some of the inclusions are apparently cylindrical is hard to reconcile with the writer's understanding of the theory.

As the wedges grow into contact along their upper edges, the deformatory pressure of the summer's expansion of the ground is supported more and more by the ice itself, until at last the included earth is in such subordinate amount that new conditions are set up. That the annual fissuring of the ground must go on even after it is chiefly underlain by ice is shown by the numerous contraction fissures in all ice covering bodies of water, though it is not certain that the original polygonal network of fissures would be maintained. If these fissures are filled with new ice the old ice must either be bulged up or caused to flow horizontally, unless the compression can be absorbed by the elasticity of the ice. It is possible that the pressure might melt enough of the ice to relieve the strain, but because of the low temperature of the ground and the plasticity of ice under strains, it is more likely that it will be deformed.

Consequently, if bodily flowage is not possible, the surface of the ice must gradually bulge upward, of course carrying its protecting mantle with it. The great thickness of the New Siberian ice may be thus explained, if it is to come under the frost-crack theory.

The conditions seen by the writer in Alaska are widely different from those described in Siberia. He does not maintain that the ground ice of both localities can be explained by the same theory, but as no other theory explains all the facts in the Siberian field the frost-crack theory may be entertained as a working hypothesis. Detailed study is necessary to show the intermediate stages. If they are found, the theory can be expanded so as to fit them.

If the ice of New Siberia is to be accounted for under this theory, the surface must have grown upward many feet. If 30 feet is assumed as the depth to which the fissures are opened by the annual changes in temperature, then the ice under discussion, which is at least 60 feet thick, must have increased 30 feet in thickness by upward growth.

MISCELLANEOUS FORMS OF GROUND ICE.

In many places along the coast there are exposures of a mixture of nearly equal amounts of clay and ice granules. The ice granules are clear and from one-half to three-quarters of an inch in diameter. There are also horizontal patches of clear ice an inch thick and 4 or 5 inches long. The clay granules are about the same size as the ice granules. They are roughly concentrated into horizontal rows, and this arrangement, with the intervening ribbons of clear ice, gives a stratified appearance to the formation. The clear ice in many places has fine vertical lines of air bubbles, which suggests a prismatic structure. Figure 16 (p. 207) illustrates an exposure of this ice. The clay granules shown in the illustration have rootlets in them, but elsewhere they are of a slimy clay similar to that found on the bottoms of shallow ponds.

The origin of this ice is not clear. Concentration out of saturated earth, as described by Hesselmann,[1] seems the best hypothesis. It is

[1] Holmsen, Gunnar, Spitzbergens jordbundsis : Norske Geog. Selskaps Aarbok, vol. 24, p. 42, 1912–13.

possible that the granules might increase in size in the manner suggested by Chamberlin [1] for glacial ice.

Another minor form of ground ice is found in ribbons or thin sheets of clear ice alternating with layers of earth. Only one actual exposure was observed by the writer, but as the details of such an exposure would be quickly masked by melting it is possible that this form is fairly common. In figure 22 (p. 208) parallel curved lines are drawn to indicate such ribbons of ice, which in the place sketched were 3 to 6 millimeters thick. In July, 1914, when the sketch was made, no ice was seen, but only the marks left on the surface where the ice had melted out. When examined during the previous October, a short time after a block had broken off and revealed the exposure, the ice was seen to be clear and near the outside to contain fine lines of air bubbles running perpendicularly to the plane of the ribbon. This orientation still occurred where the ribbons were bent upward. These alternating bands of ice and earth occurred both in the slimy clay and in the muck which lay above. The phenomenon at first suggested conchoidal cleavage of the tundra under the winter's contraction and subsequent filling of the cracks with surface water, but later insight into the working of frost wedges forced the abandonment of this hypothesis. In the clearness of the ribbons as well as in the apparent prismatic structure, it is very different from the finely granular and air-filled ice of the wedges. There was no way to determine whether the ice and earth bands had been horizontal at first and had been disturbed afterward or had been formed in their present attitude.

This same kind of ice has been noted by Quackenbush at Eschscholtz Bay, by Holmsen in Spitzbergen, and by several of the Siberian observers. The presence of thin bands of ice, which increased in amount toward the bottom of the Schergin shaft, is mentioned by Middendorff.

The writer is at a loss to explain this banded form of ice, where it has evidently been formed after the ground formations had been deposited. There are objections to the hypothesis that the ice has concentrated out of the mois-

ture in the freezing ground; the same objections are raised against Lieut. Belcher's hypothesis as elaborated by Holmsen. (See p. 229.) Nor does it seem possible that small fractures in the frozen ground, caused by gradual contractions as the increasing cold penetrated downward, could have become filled with water. The ground circulation would have been previously shut off by the freezing.

It may be that in the readjustment of the polygonal blocks, owing to the pressure of the growing ice wedges, conchoidal fractures are formed in the frozen ground. These spaces may be filled with surface water, but it is difficult to explain the solidity and clearness of the ice in comparison with the ice in the wedges, which is whitish.

The only notable exposure of ground ice upon the coast that did not show ice in the form of wedges was examined in July, 1912, before the writer had worked out this theory. The exposure is 3 or 4 miles east of Collinson Point, at a place where the ocean is cutting a 40-foot bank into the gentle seaward slope of the Anaktuvuk Plateau. In many respects this exposure resembles some of the exposures described by Quackenbush at Eschscholtz Bay.

This exposure was in a crescentic slump scarp which showed about 100 feet of ice that had a maximum vertical thickness of 7 feet. The upper surface of the ice was parallel with the surface of the ground and lay about 18 inches below it. The hillside here sloped about 15°. Immediately over the ice was an apparently undisturbed layer of yellow and black clays in thin alternating beds. Over this was about a foot of turf mingled with clay, which was slumping down over the lower clay and ice. The ice immediately below the soil had a perpendicular face, but 3 or 4 feet lower the slope decreased until one could stand upon the surface of the exposure. On this floor mud and blocks of turf had lodged, so that the lower part of the ice was concealed.

Where the ice was clean on the outside it was unusually clear and blue. The greater part was nearly free from air bubbles. In many places a network of lines over its surface showed that its structure was granular. The average size of the granules was 10 to 20 millimeters, a larger size than that of those in other ex-

[1] Chamberlin, T. C., and Salisbury, R. D., Geology, 2d ed., vol. 1, pp. 309–310, 1905.

posures, where later investigations have shown the ice to be in the form of wedges. A block of this clear ice that was left in the shade for a few hours disintegrated into small fragments. Where the nearly horizontal foot of the exposure could be seen it was crisscrossed with many closely spaced straight lines, which were perhaps the locations of as many fractures. In one place there was a roughly **V**-shaped mass of discolored ice, which contained many vertically elongated air bubbles. Close inspection showed this discolored ice to be in part composed of thin alternating bands of clear and yellowish ice, which curved parallel to the line of demarcation between it and the body of the exposure, much as if it had been frozen while flowing down the side of a depression in the clear ice. Several irregular masses of clay were inclosed in the main body of the ice, as is shown in figure 27.

could it be told from the needle-shaped crystals of river ice over which the observer had been recently walking. This ice seems clearly to be aufeis that has been buried by gravels.

Much ice was observed in the 4 to 6 foot banks of the flat between the two distributaries of the same river. The ice lay far back under the overhanging turf, so that its structure could not be made out from the canoe in which the observer was traveling.

On the east bank of the east mouth of Katakturuk River, about a mile from the coast, were two exposures of ground ice, each about 50 feet long. They lay under gentle dome-like elevations which rose 3 or 4 feet above the general delta level. One of these exposures, when carefully examined, showed about 3 feet of ice under 1 to 1½ feet of turf. Where the exposure was quite fresh the ice was solid, blue, and very clear. Where weathered, it was white and showed a vertical prismatic structure. Individual crystals could be traced for 12 to 18 inches, and by running the point of a knife along the surface of the exposure a mass of needles 8 inches long and three-quarters of an inch in diameter could be stripped

FIGURE 27.—Exposure and section of ground ice near Collinson Point. A, Front view ; B, profile.

As no glaciers came down to the coast in this neighborhood, and as no boulders of the Flaxman formation are found within several miles, this ice is probably not glacial. A recent slump could hardly have buried a snowdrift under a former steep bank, for the size of the granules points to a great age for the ice. Still the inclosed masses of clay, the granulation, and the locality make a buried snowdrift the probable source of this ice.

Very little ground ice was seen inland, probably because no search was made in favorable localities. In winter the snow obscured most of the river banks, and in summer the packing routes usually lay along the tops of the banks. On Canning River, near the end of the Shublik Mountains, a fresh cut against a 12-foot flat showed 4 to 5 feet of clear ice overlain by 6 inches of gravel and as much muck and moss. Small willows were growing at the top. The ice was definitely prismatic and in no way

off. These domes can hardly be remnants of a former generally higher level of the delta flat. No gravels or sand were seen above the ice but only turf. The burial of flood ice by wash material is excluded, for the river here is transporting material as large as small cobblestones. The most probable hypothesis is that these domed exposures of ice were formed where hydraulic pressure had bulged up the frozen turf. Water then solidified in the cavity. According to Tyrrell, horizontal sheets of ice of great extent are thus formed.

[Editor's Note: Material has been omitted at this point.]

5

Reprinted from *Geol. Soc. America Spec. Paper 70*, 3–6, 55–61, 63–65 (1962)

MECHANICS OF THERMAL CONTRACTION CRACKS AND ICE-WEDGE POLYGONS IN PERMAFROST

Arthur H. Lachenbruch

THROUGHOUT many thousands of square miles of Arctic terrain the most conspicuous surface feature is a striking pattern of markings known as "ice-wedge polygons." The markings are shallow troughs that intersect to form irregular polygons[1] ranging from 10 to 100 or more feet in diameter. Superficially these polygons resemble those produced by desiccation cracks in mud, although the scale is generally much larger and the medium is solidly frozen permafrost (Pl. 1, figs. 2, 3). The troughs that delineate such polygons are generally underlain by vertical wedge-shaped masses of relatively clear ice, commonly 10–20 or more feet deep and several feet wide at the top (Pl. 1, fig. 1). "Ice wedges," as these masses are called, have mechanical and thermal manifestations that can profoundly affect the geomorphic and biological processes that mold the Arctic landscape; they play a central role in engineering problems that arise when this landscape is modified. Although ice wedges and ice-wedge polygons have been discussed widely in the literature, the details of their origin are still poorly understood. Even as recently as 1950, relatively few American scientists and engineers concerned with the Arctic were aware of the existence of ice wedges beneath the polygonal surface markings.

A more detailed understanding of the origin of ice wedges can be expected to yield greater insight into the natural history of the Arctic tundra and into other superficially unrelated problems. The polygonal surface manifestations of ice wedges are closely related to polygonal surface fractures of cooling lava, drying mud, and hardening concrete. As the details of all of these phenomena are poorly understood, it will be profitable to keep the more general problem in mind.

---

[1] "Polygon" as used throughout, will refer to a closed figure bounded by several sides, some or all of which may be curved. This is common usage in discussions of the subject although it departs from the geometric definition.

[*Editor's Note:* Certain plates and figures have been omitted owing to limitations of space.]

Geomorphic aspects of the origin of ice-wedge polygons will be considered only to explain the assumptions upon which the theoretical model is based. The chief effort is devoted to explaining the phenomena observed in terms of physical theory. The geomorphic starting points are based largely upon observations made over the past 15 years by the writer in Alaska. Most observations were made between the crest of the Brooks Range and the Arctic Ocean (about 68° to 71° north latitude) where mean annual air temperatures range between −6°C and −12°C and permafrost is normally on the order of 1000 feet thick. The theory is general, however, and with appropriate choice of parameters, should apply equally well to other permafrost areas. Much of the discussion applies also to "frost-crack polygons" which occur in the deep seasonal frost in subarctic regions where no permafrost occurs (Hopkins, and others, 1955, p. 138). They have no ice wedges but are delineated by slumping or aeolian deposition in the seasonal contraction cracks.

Contraction Theory of Origin

THE contraction theory of ice-wedge polygons as set forth by Leffingwell in 1915 is given briefly as follows:

During the Arctic winter, vertical fractures on the order of one-tenth of an inch wide and several feet deep are known to form in the frozen tundra—a process generally accompanied by loud reports (Fig. 1A). They are assumed to be the results of tension caused by thermal contraction of the tundra surface. In early spring it is supposed that water from the melting snow freezes in these cracks and, with accumulating hoarfrost, produces a vertical vein of ice that penetrates permafrost (Fig. 1B). Horizontal compression caused by re-expansion of the permafrost during the following summer results in the upturning of permafrost strata by plastic deformation. In the winter that follows, renewed thermal tension supposedly reopens the vertical ice-cemented crack which is presumed to be a zone of weakness. Another increment of ice is added when the spring meltwater enters the renewed crack and freezes. Such a cycle, it is argued, acting over centuries, would produce the vertical wedge-shaped masses of ice (Fig. 1C, D). The polygonal configuration is generally thought to be a natural consequence of a contraction origin.

With minor variations this general qualitative theory is subscribed to by most investigators who have firsthand knowledge of the phenomenon (Bunge, 1884, 1902; Leffingwell, 1915, 1919; Black, 1952b.

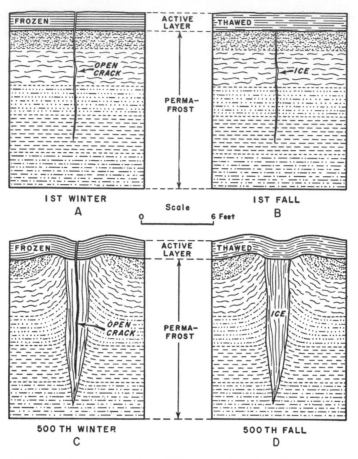

Figure 1. Schematic representation of the evolution of an ice wedge according to the contraction-crack theory (width of crack exaggerated for illustrative purposes)

1954; Popov, 1955; Hopkins, 1955; Shoumsky, Shvetsov, and Dostovalov, 1955; Washburn, 1956; Britton, 1957; Péwé, 1959; Lachenbruch, 1959, 1961); only a few workers (Taber, 1943, Irwin Schenk, oral communication) doubt its general validity. With the exception of the simplified mathematical treatment by Dostovalov (1957) and the petrographic work by Black (1954), these authors have generally dealt with the problem from a qualitative point of view; the thermal and mechanical processes responsible for the formation and growth of ice wedges have yet to be identified and studied from the standpoint of physics. Not only is the physical point of view useful for testing a qualitative theory, but it inevitably leads to greater insight into the phenomenon itself. Before proceeding with the analysis it is necessary to develop a few points regarding stress and strength implied by qualitative field observations.

[*Editor's Note:* Material has been omitted at this point.]

Summary

ICE-WEDGE polygons in permafrost are a special case of a broad class of phenomena that could be called contraction-crack polygons, which form in response to tensions resulting from decrease in volume such as could be caused by thermal contraction, desiccation, chemical reaction, or phase change. They occur in a wide variety of media on a wide variety of scales ranging from a fraction of an inch in crackle-finish ceramic ware to 100 feet or more in permafrost. They include desiccation cracks in mud, columnar jointing in cooling basalt, and shrinkage cracks in hardening concrete. The size and configuration of contraction-crack polygons depend upon the rheological behavior of the medium and the nature of the induced volume change. It is of interest to consider how these factors interrelate to control the diverse polygonal cracking phenomena. As in most complicated problems in deformation the best we can hope, from a theoretical approach, is that pertinent, simple, idealized models can be selected and that studying them will yield insight into the natural processes. One approach to the problem is to break it into the following parts:

(1) The relationship between temperature (or desiccation, etc.) and stress in the unfractured medium

(2) The mechanics of the fracture process and prediction of crack depth

(3) The effect of a single crack on the general stress distribution in the medium

(4) Multiple cracking and the evolution of polygonal form. These four problems were investigated for the case of ice-wedge polygons to the extent that existing empirical data warrant. Some of the results are fairly general and should apply with little or no modification to other cases of contraction-crack polygons.

Ice-wedge polygons evidently form in response to thermal tension set up in the frozen ground by its tendency to contract during the cold Arctic winter. The polygonal tension cracks, formed in winter, penetrate permafrost to depths on the order of 10 or 20 feet and are sealed by the freezing of water which enters when the surface thaws in early summer. The resulting vertical ice veins are zones of weakness subject to recurrent fracture and growth by repetition of the cycle in succeeding winters. Differential thawing of the thickened veins or ice wedges produces the shallow troughs by which ice-wedge polygons are often delineated on the tundra surface.

The stresses that lead to rupture of the ground must be accounted for by a viscoelastic response to thermal contraction—not an elastic one; a formal calculation shows that thermal stresses developed in an elastic medium under the thermal regime of permafrost would be too large by an order of magnitude to satisfy requirements of the frost-crack theory. A simple linear relation between rate of cooling and stress is also unsatisfactory, for it predicts that the stress falls off too rapidly with depth to account for observed details of the fracture process.

A nonlinear relation between the principal components of distortional stress, (P_i), and distortional strain, (E_i), of the form.

$$\frac{1}{\bar{\eta}} P_i^r + \frac{dP_i}{dt} = \mu \frac{dE_i}{dt}, \qquad i = 1, 2, 3 \tag{47}$$

yields stress distributions of a form compatible with field inferences regarding the fracture process. Numerical values required for the exponent, n, elastic shear modulus, μ, and viscoelastic parameter, $\bar{\eta}$, are compatible with pertinent, although fragmentary, laboratory results. Equation (47) leads to the following relation between horizontal stress, $\tau(x, t)$, and ground temperature $\Theta(x, t)$:

$$\frac{d\tau}{dt} + \frac{Y}{1-\nu}\frac{1}{3^n 2\bar{\eta}}\tau^n = \frac{-Y}{1-\nu}\frac{d\Theta}{dt}. \tag{48}$$

Inferences drawn from this equation and from experimental results on polycrystalline ice suggest that the large thermal stresses that crack frozen ground are caused by rapid cooling at low temperatures.

When the tension near the surface is approaching the tensile strength, horizontal stresses at depths greater than 10 or 20 feet are often compressive, owing to the combined effects of the time lag in the penetration of the seasonal thermal wave and compression caused by the weight of overburden. When the tensile strength is exceeded near the surface, a tension crack forms and propagates downward. The depth at which it stops depends primarily upon the effect of compression at depth and the dissipation of strain energy by plastic deformation near the leading edge of the crack. If the medium is relatively nonplastic under rapid deformation, the tension crack will penetrate deep into the compressional zone; if not it may stop near the neutral horizon or even within the surficial tensile zone. Subsequent viscoelastic effects may produce a slow closing of part of the crack from the bottom.

The crack depth is dependent upon the actual stress distribution at the time of cracking. Other things being equal, deeper cooling generally produces deeper tension and deeper cracks. The depth of a crack can, of course, change as the stress regime changes with the progressing season. Numerical calculations based upon temperatures measured in Alaskan permafrost, equation (48), and a mathematical formulation of the present considerations of fracture mechanics lead to crack depths of the same order of magnitude as those observed in the field.

The formation of a crack causes a local relief of tension in the surficial materials. The component of stress normal to the crack vanishes at the crack walls but passes asymptotically to the precracking value at horizontal distances that are large relative to the crack depth. Each crack is, therefore, surrounded by a band in which cracking has caused appreciable reduction of horizontal tension—the "zone of stress relief." The stress configuration within this zone depends primarily upon the stress distribution with depth that would exist in the ground if it were not fractured, *i.e.*, upon the thermal stress. It has been calculated approximately by applying Muskhelishvili's method of complex stress functions to the problem of a crack in a semi-infinite elastic medium with a nonuniform stress field. The thermal stress **upon which** this solution is superimposed is calculated from the viscoelastic relation (eq. 48), and, hence, most inelastic effects are accounted for (the superposition error due to nonlinearity is considered separately). In general, increased depth of thermal tension and/or increased crack depth result in wider zones of stress relief and, hence, more widely spaced cracks and larger polygons. The theory yields numerical results compatible in order of magnitude with observed diameters of ice-wedge polygons.

The component of thermal tension at the ground surface in the direction parallel to the crack is relieved only slightly by the cracking and, thus, large horizontal stress differences occur within the zone of stress relief. A second crack entering this zone tends to align itself perpendicular to the direction of greatest tension and, hence, tends to intersect the first crack at right angles. Conversely, the occurrence of an orthogonal intersection generally implies that one of the cracks predated the other. This suggests a useful scheme for classifying contraction-crack polygons as follows:

1. *"Orthogonal systems"* of polygons are those that have a preponderance of orthogonal intersections. They are evidently characteristic of somewhat inhomogeneous or plastic media in which the stress builds up gradually with cracks forming first at loci of low strength or high stress concentration. With increasing thermal stress, polygons are subdivided by cracks initiating near polygon centers or, under certain circumstances, at pre-existing cracks. The polygons attain such a size that zones of stress relief of neighboring cracks are superimposed at the polygon centers so as to keep the stress there below the tensile strength. The polygon size depends upon the magnitude of the thermal stress and configuration of the zones of stress relief of individual cracks. The cracks do not all form simultaneously, and hence each new crack tends to join pre-existing ones at orthogonal intersections. Orthogonal systems can be conveniently divided into two subgroups:

a. *"Random orthogonal systems"* in which the cracks have no preferred directional orientation (Pl. 1, fig. 2) and,

b. *"Oriented orthogonal systems"* in which they do. Most oriented orthogonal systems in permafrost are probably caused by horizontal stress differences generated by horizontal thermal gradients near edges of gradually receding bodies of water (*e.g.*, slowly draining lakes and shifting river channels). Under such conditions the induced polygonal system has one set of cracks parallel to the position of the water's edge and a second set normal to it (Pl. 1, fig. 3). Oriented orthogonal systems can result also from topographic effects or from anisotropy of horizontal tensile strength as in steeply dipping shales.

2. "*Nonorthogonal systems*" of polygons are those that have a preponderance of tri-radial intersections, commonly forming three obtuse angles of about 120°. It is suggested that they result from the uniform cooling of very homogeneous, relatively nonplastic media. Judging from experimental and theoretical results from the field of fracture mechanics, it is likely that under these conditions cracks propagate laterally until they reach a limiting velocity (on the order of half that of elastic shear waves) and then branch at obtuse angles. The branches then accelerate to the limiting velocity and bifurcate again. Unlike the orthogonal system, all elements of nonorthogonal intersections are generated virtually simultaneously.

Although many ice-wedge intersections appear to be of the nonorthogonal type, the surface expression of a wedge is so wide that it is generally difficult to distinguish between a nonorthogonal intersection and an orthogonal intersection at a convexity in the primary fracture (Fig. 13).

As this geometric classification seems to have genetic significance, it might be useful when applied to contraction-crack polygons in general. Crackle-finish ceramic ware generally displays a random orthogonal pattern, except locally where topography of the object (ash tray, vase, etc.) may produce an oriented orthogonal system. Some of the classic occurrences of columnar-basalt joints seem to be of the nonorthogonal type; the less regular ones commonly have a preponderance of orthogonal intersections. According to the present point of view the nonorthogonal ones presumably formed under conditions of greater thermal and mechanical homogeneity. Perhaps also the medium was less plastic at the time of cracking. (For a discussion of the depth and spacing of cooling joints in lava *see* Lachenbruch, 1961.)

Desiccation polygons in mud and shrinkage polygons in concrete

65

seem generally to be of the orthogonal type although in both systems nonorthogonal intersections of the tri-radial, $(o/o/o)$, type occur. As cracks in these media are often irregular, they have many "convexities" at which orthogonal intersections (Fig. 13b) could be confused with nonorthogonal ones (Fig. 13a) as the cracks widened. Limited laboratory experiments by the writer confirm that, in general, mud cracks propagate slowly, do not bifurcate, and tend to form orthogonal intersections. [Since this paper was written, an extensive series of mud-crack experiments has been reported (Corte and Higashi, 1960)].

Conclusion

IN VIEW of the limited data pertinent to the mechanics of ice-wedge polygons, this paper is necessarily speculative. However, the attempt to provide an explicit mechanical and thermal model focuses attention on the areas of ignorance and provides a framework within which to orient further study. The validity of conditions (1) and (2), upon which part of this paper is based, can be tested by careful field observations. The related problem of the tensile strength of wedge ice normal to the plane of foliation can be resolved by laboratory tests. The strength and rheological properties of frozen soils and their temperature dependence also play a central role in this problem, and inferences made can be verified, refined, or rejected on the basis of their measurement.

As the theory presented must be considered tentative, it is premature to proceed to a discussion of its detailed geomorphic and engineering implications. However, a few interesting problems will be mentioned.

If the recurrent contraction cracks initiate at the top of permafrost ice-wedge polygons could form and grow under shallow lakes; if the cracks initiate at the surface, they could not. Thus, careful observations of shallow lakes could provide verification of this fundamental assumption.

Whether lakes in partially filled basins are growing or shrinking slowly can probably be determined in many cases by whether the associated oriented orthogonal pattern is discordant or concordant with the shore line. Both types occur on the Arctic coastal plain, and the distinction between them is important to the study of lake cycles in permafrost. Oriented orthogonal polygons can also be used to reconstruct past positions of migrating river channels.

When winter stresses at a wedge top cease to exceed the strength of wedge ice, cracking and growth stops, and eventually the wedge gets buried by normal aggradation of the tundra. Hence, the tops of buried wedges often mark horizons at which ancient climatic or geomorphic events acted to change the stress regime. Buried wedges can, therefore, provide a useful stratigraphic tool if their mechanics is understood well enough to reconstruct the event that led to extinction. On the basis of the theory presented, extinction of wedges can be brought about by three main causes: (1) the increase in minimum winter temperatures (causing decrease in $\bar{\eta}$, equation (12), (2) the decrease in maximum winter cooling rates at the top of permafrost (decrease in $-\dot{\Theta}_{max}$), and (3) the development of an adjacent competing wedge or similar mechanical events—increase in S, equation (44). Causes (1) and (2) can both be brought about by climatic change, the formation of lakes, and, more commonly, by increased winter snow accumulation in deepening interpolygonal troughs. Cause (2) can also be brought about by convergent solifluction or any factor that tends to thicken the active layer. Cause (3), which could be called "wedge piracy," probably goes on continuously in some ice-wedge communities. Other sources of these changes can be recognized in the field, and a careful study of them should lead to a better understanding of the history and mechanics of permafrost.

References Cited

BLACK, R. F., 1952a, Growth of ice-wedge polygons in permafrost near Barrow, Alaska (Abstract): Geol. Soc. America Bull., v. 63, p. 1235

— 1952b, Polygonal patterns and ground conditions from aerial photographs: Photogrammetric Eng., v. 18, no. 1, p. 123–134

— 1954, Permafrost—a review: Geol. Soc. America Bull., v. 65, p. 839–855

BRITTON, M. E., 1957, Vegetation of the Arctic tundra, p. 26–61 in Arctic biology: Oreg. State Colloquim, 18th Ann. Biology Colloquim Proc., 134 p.

BUNGE, Alexander, 1884, Naturhistorische Boebachtungen und Fahrten im Lena-Delta: Acad. Imp. Sci. St. Petersbourg Bull., v. 29, p. 422–476

— 1902, Einige Worte zur Bodeneisfrage: Russian K. Min. Gesell. Verh., Zweite Ser., v. 40, p. 203–209

CORTE, ARTURO AND HIGASHI, AKIRA, 1960, Experimental research on desiccation cracks in soil: SIPRE Rept., no. 66, 48 p.

DOSTOVALOV, B. N., 1957, Izmenenie obshchema rykhlykh gornykh porod pri promerzanin i obrazovanie morozoboynykh treshchin [Change in volume of unconsolidated ground during the freezing and formation of frost cracks], p. 41–51 *in* Materially po laboratornym issledovaniyam merzlykh gruntov, Sbornik 3: Akad. Nauk. SSSR, 323 p.

HOPKINS, D. M., KARLSTROM, T. N. V. AND OTHERS, 1955, Permafrost and ground water in Alaska: U. S. Geol. Survey Prof. Paper 264–F, p. 113–146

LACHENBRUCH, A. H., 1959, Contraction theory of ice-wedge polygons (Abstract): Geol. Soc. America Bull., v. 70, p. 1796

— 1961, Depth and spacing of tension cracks: Jour. Geophys. Research, v. 66, no. 12

LEFFINGWELL, E. DEK., 1915, Ground-ice wedges; the dominant form of ground-ice on the north coast of Alaska: Jour. Geology, v. 23, p. 635–654

— 1919, The Canning River region, northern Alaska: U. S. Geol. Survey Prof. Paper 109, 251 p.

PÉWÉ, T. L., 1959, Sand-wedge polygons (tesselations) in the McMurdo Sound region, Antarctica—a progress report: Am. Jour. Sci., v. 257, p. 545–551

POPOV, A. I., 1955, Materialy k osnovam ucheniya o merzlykh zonakh zemnoi kory [The origin and evolution of fossil ice]: Inst. Merzlotovedeniya, Akad. Nauk, SSSR, no. 2, p. 5–25

SHOUMSKY, P. A., SHVETZOV, P. F., AND DOSTOVALOV, B N., 1955, Ice veins, *in* Special application of engineering geology in reconnaissance of underground

TABER, STEPHEN, 1943, Perennially frozen ground in Alaska, its origin and history: Geol. Soc. America Bull., v. 54, p. 1433–1548

WASHBURN, A. L., 1956, Classification of patterned ground and review of suggested origins: Geol. Soc. America Bull., v. 67, p. 823–866

6

Reprinted from *Arctic*, 25(2), 142–150 (1972)

Thermal Contraction Cracks in an Arctic Tundra Environment

DENIS E. KERFOOT[1]

ABSTRACT. Field observations in the Mackenzie Delta area largely substantiate Lachenbruch's theoretical considerations of thermal contraction crack development. Frost crack patterns, representing the incipient stage of tundra polygons, were observed on both bare and vegetated surfaces of low alluvial flats and sandspits of three islands. Individual polygons, where developed, ranged in size from 20 to 30 metres diameter on bare surfaces to 2 to 3 metres on sedge-covered areas, and 80 per cent of the angular intersections measured were of the orthogonal type. Most cracks exhibited random orientations, except in close proximity to water bodies where tendencies toward normal and subparallel orientations occurred.

RÉSUMÉ. *Fentes de contraction thermique dans un milieu de toundra arctique.* Des observations sur le terrain, dans la région du delta du Mackenzie, appuient largement les considérations théoriques de Lachenbruch sur le développement des fentes de contraction thermique. Des figurés de fentes de gel, représentant le stade initial de polygones de toundra, ont été observés à la fois sur des surfaces nues et sur des surfaces recouvertes de végétation, sur des alluvions et des flèches sableuses de trois îles. Les polygones individuellement développés mesurent de 20 à 30 mètres de diamètre sur les surfaces nues, et de 2 à 3 mètres sur les surfaces peuplées de laîches; 80 pour cent des intersections angulaires mesurées sont de type orthogonal. La plupart des fentes sont orientées au hasard, sauf au voisinage de nappes d'eau, où apparaissent des tendances à des orientations normales et subparallèles.

РЕЗЮМЕ. *Термические контракционные трещины в условиях арктической тундры.* Полевые наблюдения, выполненные в дельте реки Макензи, в значительной степени подтверждают теорию развития контракционных термических трещин, предложенную Лахенбрухом. Структура мерзлотных трещин, представляющих начальную стадию тундровых полигонов, была исследована как на лишенных, так и обладающих растительностью низких аллювиальных береговых участках и песчаных косах трех островов. Размеры отдельных сформировавшихся полигонов колеблются от 20 до 30 метров в диаметре на голых, и от 2 до 3 метров на заросших камышом участках почвы. 80% всех обмеренных полигонов оказались ортогонального типа. За исключением случаев тесной близости к воде, когда наблюдается тенденция к параллельному или перпендикулярному расположению, большинство трещин обладает случайной ориентацией.

INTRODUCTION

Patterned ground in the Mackenzie Delta area is restricted primarily to non-sorted types (Washburn 1956). Although other factors may be involved, the absence of the sorted forms can largely be attributed to the fact that the mantle frequently lacks a sufficient concentration of stones to exhibit marked frost sorting (Mackay 1963, cf. Mackay 1967). Of the non-sorted forms, tundra or ice-wedge polygons constitute one of the most widespread types of patterned ground in the delta area, and this paper describes some aspects of thermal contraction cracks, which represent the incipient stage of tundra polygon development.

---

[1]Department of Geography, Brock University, St. Catharines, Ontario, Canada.

Leffingwell (1915), working on the coastal plain of northern Alaska, was among the first to postulate that the networks of tundra polygons were generated by contraction cracks in the frozen ground, produced by intense stresses created as a result of pronounced seasonal changes in the ground temperature. His "thermal contraction" theory was outlined as follows (Leffingwell 1915):

> The permanently frozen ground contracts in the cold Arctic winter and cracks are formed which divide the surface into polygonal blocks. In the spring these frost cracks become filled with surface water which immediately freezes. In the expansion of the frozen ground as the temperature rises in summer, the vein of ice becomes more rigid than the country formation, and the readjustment takes place in the latter. The result is to bulge up the inclosed block either bodily or else locally along the sides of the ice. During the next winter's cold wave a new crack forms at the same locus so that a continually growing wedge of ground ice is formed. Thus the tundra becomes underlain by a network of ice-wedges, which inclose bodies of the original formation.

The general principles of Leffingwell's contraction theory have been accepted by most of the subsequent research workers investigating tundra polygons. Despite the voluminous literature on this subject, however, the precise details of their origin are still imperfectly understood. Black (1952) and Lachenbruch (1962) attribute some of this ignorance to an absence of quantitative, rather than qualitative, data but it may also reflect the paucity of observations describing the initial development of the frost crack patterns.

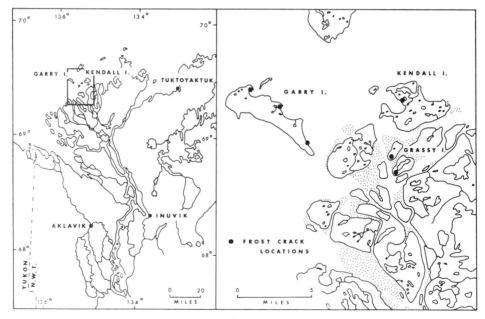

FIG. 1.   Location Map.

INCIPIENT FROST CRACK PATTERNS

An examination of the literature pertaining to tundra polygons reveals abundant references to, and descriptions of, the polygonal ground in relatively advanced stages of development, but surprisingly little relating to the formation of the initial frost crack patterns. Leffingwell (1919) shows illustrations of incipient cracks on the coastal plain of northern Alaska, and Black (1952) also makes brief reference to similar features. Lachenbruch (1962) has made a theoretical study of frost crack patterns in his examination of Leffingwell's thermal contraction hypothesis from the point of view of mechanics, but he does not cite any field evidence to corroborate his conclusions. Washburn *et al.* (1963), to my knowledge, have produced one of the few papers describing the occurrence of frost cracks, albeit in a non-arctic environment.

Observations in the Mackenzie Delta area during the summers of 1964 and 1965 revealed 3 locations where incipient frost cracks had developed on the ground surface. The locations, each characterized by an absence of any major relief features and elevations of less than 1 metre (3 feet) above mean sea level, included lake-strewn, alluvial flats on Kendall and Grassy Islands and the flats bordering the lagoons enclosed behind sandspits on Garry Island (Fig. 1). All of these sites are frequently inundated during periods of high water, especially during storm surges. The frost cracks were found to be equally well developed on bare

FIG. 2. Orthogonal intersections of frost cracks on bare alluvial surface of Grassy Island, Mackenzie Delta, N.W.T.

ground surfaces (Fig. 2), and on flats supporting a dense cover of grasses and sedges 20 to 40 cms. (8.0 to 15.5 inches) tall (Fig. 3). In these latter areas, the vegetation was flattened, presumably by a combination of prevailing winds, snowfall and flood surges in the preceding fall. The vegetation was cut by sharp, knife-like fractures and was probably frozen to the ground surface at the time of the frost cracking to produce the clean break. In many places, the cracks were observed to extend beneath the surfaces of shallow lakes, though none was traced entirely across the lake floor. This indicates that the water, at least in the shallower parts of the lakes, was frozen to the bottom although water in the central

FIG. 3.   Frost cracks
developed through a layer of
vegetation on Kendall Island,
N.W.T.

sections of the lakes may have remained unfrozen at depth. The fact that the
fissures on Kendall Island cut through the vegetation into the underlying mineral
soil, would appear to substantiate the view that the cracks were produced by
intensive frost action. The cracks on the bare ground surfaces were interpreted
as having originated in a similar manner, although the possibility that they were
produced by desiccation of the soil cannot definitely be excluded.

Excavations in one of the sandspit-lagoon areas on Garry Island revealed that
the frost cracks were best developed in organic-rich silts and silty loams. Mechanical analyses of the soils showed that they were composed of 59.2 per cent
silt, 21.3 per cent sand, and 19.5 per cent clay (Wentworth classification categories). The soils were mantled by a thin layer of organic material, and the organic
content of the soil samples averaged 5.5 per cent by weight of the dried sample.
The surficial organic layer and the organic matter at depth, often in the form of
thin intercalations, probably represent washed peat derived through erosion of
the adjacent coastal bluffs. The soils also possessed a high moisture content, with
frozen samples having an average ice content of 91.7 per cent (expressed as weight
of ice to dry soil). Mackay (1965) has already documented the granulometric
composition of the soils underlying the sedge-covered flats on Kendall Island,
where the proportions of clay and sand were slightly lower (silt 79 per cent, sand
13 per cent and clay 8 per cent).

Most of the frost cracks exhibited little or no topographic expression at the
ground surface, but some of the larger fissures traversing the unvegetated areas
were marked by the presence of shallow troughs 30 to 70 cms. (12.0-27.5 inches)
across and 10-20 cms. (4-8 inches) deep. When examined in early August 1964,
the frost cracks on Garry Island exhibited a wedge-like form extending down
through the active layer and into the frozen ground at depths of 50 cms. (19.5
inches) below the ground surface. At the surface, the open fissures were up to
4 cms. (1.5 inches) across, and they remained open to depths of 15 to 25 cms.
(6 to 10 inches) but narrowed to only a few millimetres in width at the level of
the frost table. In many localities, the cracks were infilled at depth by sand-size

material that had probably been blown or washed in from the ground surface. These miniature sand-wedges are similar in form to the larger scale features described by Péwé (1959) in the McMurdo Sound area of Antarctica. Below the level of the frost table, the cracks were occupied by small veins of ice, approximately one millimetre across, which could be traced to a depth of 76 cms. (30 inches) below the ground surface.

Rarely did the frost cracks in the sandspit-lagoon areas of Garry Island reveal any arrangement into a definite polygonal network. The majority of the fissures appeared to be randomly distributed over the ground surface, and showed no preferred directional orientation except for a weak tendency to occur along lines developed at right angles and parallel to the margins of the lagoons. The only other salient feature of the distribution was the contrast in the density of the pattern between the silty loam and coarser sand areas of the sandspit. Although the frost cracks were present in both of these areas, the density was much higher in the silty loam sections.

The distribution of the frost cracks on Kendall and Grassy Islands showed a much greater tendency to be organized into crude polygonal patterns. Fig. 4 shows the spatial arrangement of the frost cracks on a part of the sedge-covered flat on Kendall Island. As may be seen, the ground surface was subdivided into a number of highly irregular polygons of variable size but averaging 2 to 3 metres (6.5 to 10.0 feet) across. The majority of these irregularly-shaped polygons were 4- or 5-sided, and hexagonal forms were notably conspicuous by their absence. Most of the fissures on Kendall Island also exhibited little preferred directional orientation except in the vicinity of the larger water bodies.

The spatial distribution of the frost cracks on Grassy Island demonstrated the existence of a much more regular polygonal network, although on a considerably larger scale. On these bare alluvial flats the individual polygons averaged 20 to 30 metres (65 to 100 feet) across, and tetragonal forms were predominant. The pattern of these frost cracks moreover revealed much stronger trends in their preferred directional orientation. The larger fissures, up to 5 cms. (2 inches) wide and located in the floors of shallow troughs in the ground surface, were oriented at right angles to the bank of a distributary of the Mackenzie River. The smaller cracks, less open and having almost no topographic expression on the ground surface, on the other hand, were aligned more or less parallel to the same river bank. These preferred orientations were especially noticeable within distances of approximately 50 to 60 metres (165 to 200 feet) from the edge of the channel, but became less distinct with increasing distance from the bank.

ANGULAR INTERSECTION PATTERNS

It is generally agreed that frost cracks originate as a result of large thermal stresses created by a sudden cooling of the ground (Lachenbruch 1962):

> When the tensile strength (of the ground) is exceeded near the surface, a tension crack forms and propagates downward. . . . The formation of a crack causes a local relief of tension in the surficial materials. . . . Each

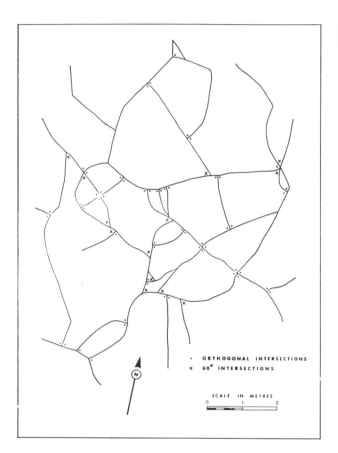

FIG. 4. Incipient frost crack pattern on Kendall Island, N.W.T.

crack is, therefore, surrounded by a band in which cracking has caused appreciable reduction of horizontal tension — the "zone of stress relief". . . . The component of thermal tension at the ground surface in the direction parallel to the crack is relieved only slightly by the cracking and, thus, large horizontal stress differences occur within the zone of stress relief. A second crack entering this zone tends to align itself perpendicular to the direction of greatest tension, and, hence, tends to intersect the first crack at right angles. Conversely, the occurrence of an orthogonal intersection generally implies that one of the cracks predated the other.

Lachenbruch's conclusion that the angular intersections of a polygonal network of frost cracks will exhibit a preferred tendency towards an orthogonal pattern, contrasts with many descriptions of polygonal ground in which authors have expressed a tendency for hexagonal forms and angular intersections of 120 degrees to predominate (Leffingwell 1915; Conrad 1946; Black 1952). The implications of the hexagonal pattern, and angular intersections of 120 degrees, are that the frost cracks originated at a series of points and each crack developed more or less simultaneously. In an attempt to determine the validity of Lachenbruch's conclusion, particular attention was paid to the nature of the angular

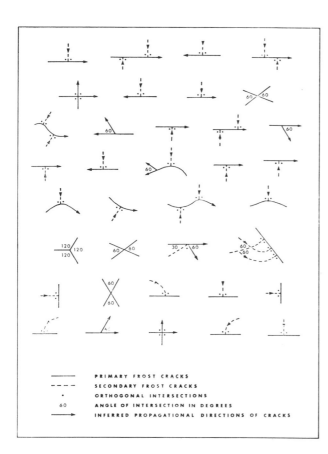

FIG. 5. Diagrammatic sketches of frost crack intersection patterns.

intersections of the frost crack patterns on Garry and adjacent islands. A total of 101 angular measurements was recorded including those shown in Fig. 4. An additional fifty intersection patterns are illustrated diagrammatically in Fig. 5, where an attempt has also been made to indicate the relative order of occurrence and propagational direction of each of the fissures. Primary frost cracks, usually the larger, are defined as those which originated first ·at any location, and their propagational directions were inferred, wherever possible, from their orientation with respect to the water bodies. (i.e. cracks which were oriented at right angles to, and propagated outward from, the body of water). Secondary frost cracks are defined as those which developed later at each location, and these cracks terminate at, and propagate towards, pre-existing primary fissures.

Of the 101 angular measurements recorded, no fewer than 79, or 80 per cent, were of the orthogonal type. As Figs. 4 and 5 indicate, most of these orthogonal intersections were formed by the junction of a primary and secondary frost crack, and only rarely were two primary frost cracks observed to intersect one another. The influence of a zone of stress relief is also manifest in the manner in which many of the secondary fissures curve to intersect the primary cracks at right angles. Where the primary cracks were sinuous, the most favoured loci for the intersection points of secondary frost cracks were located on the convex sides of the curves. This is in accordance with the distributional pattern of stress relief on a curved section of a frost crack as described by Lachenbruch (1962).

Angular intersection values of 60 degrees were by far the most common of the non-orthogonal intersection patterns. This angle was most frequently developed as a result of the bifurcation of a primary frost crack, and at points of intersection where 2 frost cracks of the same order approached one another obliquely. A great majority of the angles measured were either 90 degrees or 60 degrees, and only 2 examples were found of tri-radial intersections, forming 3 obtuse angles of about 120 degrees, suggesting that the frost cracks originated at a point (Fig. 6).

FIG. 6. Frost crack intersection pattern forming three obtuse angles of about 120 degrees, Kendall Island, Delta, N.W.T.

SUMMARY

The field evidence collected in the outer Mackenzie Delta area thus appears to substantiate the conclusions of Lachenbruch's theoretical study. Primary frost cracks were developed, essentially in a random pattern, across the ground surface, and the junctions of secondary frost cracks with these primary fissures showed a definite preferred tendency towards an orthogonal intersection pattern. According to Lachenbruch's classification scheme, the resultant crude polygonal network would therefore be classified as a 'random orthogonal system' (Lachenbruch 1962). Only in the vicinity of large bodies of water did preferred directional orientations become sufficiently pronounced to be classified as 'oriented orthogonal systems'.

The apparent dichotomy between this evidence and the fact that most tundra polygons appear to be of a non-orthogonal type has been explained by Lachenbruch to be the result of an obscuring of the intersection angles by the growth of large ice-wedges (Lachenbruch 1962). The incipient frost crack pattern, therefore, is practically the only stage in the development of a network of polygonal ground in which the angular intersection patterns can be determined with any real degree of accuracy. Moreover, these determinations can only be made in the field through ground inspection, since the frost cracks are generally too small to be identified from aerial photographs.

The transformation of an initial pattern of frost cracks into a network of tundra or ice-wedge polygons requires that recurrent fracturing takes place at the same loci. The evidence collected on Garry Island suggests that once a fracture is formed, it tends to persist as a permanent line of weakness in the mantle. Most of the larger cracks remained as open fissures throughout the summer months, possibly indicating that the ground was sufficiently elastic to absorb the strain produced by its expansion under the summer's heat and consequently no deformation took place (Leffingwell 1915). Where the ground surface was covered by a thin layer of washed peat, it is also possible that the maintenance of the open fissure may have been aided locally by a slight desiccation and shrinkage of this layer. Even where the cracks were closed, this probably did not take place before some material had infiltrated from the ground surface, and the miniature sand-wedges, produced in this manner, also assist in the preservation of the lines of weakness.

ACKNOWLEDGEMENTS

I gratefully acknowledge financial support of the field work from the Department of Energy, Mines and Resources, Ottawa; the Department of Indian Affairs and Northern Development, Ottawa; and research funds of the University of British Columbia. I am also deeply indebted to Dr. J. Ross Mackay, University of British Columbia, for his assistance in the preparation of the manuscript.

REFERENCES

BLACK, R. F. 1952.   Polygonal patterns and ground conditions from aerial photographs. *Photogrammetric Engineering*, 18: 123-34.

CONRAD, V. 1946.   Polygon nets and their physical development. *American Journal of Science*, 244: 277-96.

LACHENBRUCH, A. H. 1962.   Mechanics of thermal contraction cracks and ice-wedge polygons in permafrost. *Geological Society of America, Special Paper*, 70: 69 pp.

LEFFINGWELL, E. de K. 1915.   Ground Ice-Wedges. The dominant form of ground ice on the north coast of Alaska. *Journal of Geology*, 23: 635-54.

————. 1919.   The Canning River Region, North Alaska. *United States Geological Survey, Professional Paper*. 109: 251 pp.

MACKAY, J. R. 1963.   The Mackenzie Delta Area, N.W.T. Canada, Department of Mines and Technical Surveys, Geographical Branch, *Memoir*, 8: 202 pp.

————. 1965.   Gas-Domed mounds in permafrost, Kendall Island, N.W.T. *Geographical Bulletin*, 7: 105-15.

————. 1967.   Underwater patterned ground in artificially drained lakes, Garry Island, N.W.T. *Geographical Bulletin*, 9: 33-44.

PÉWÉ, T. L. 1959.   Sand-Wedge Polygons (Tesselations) in the McMurdo Sound Region, Antarctica — A Progress Report. *American Journal of Science*, 257: 545-52.

WASHBURN, A. L. 1956.   Classification of patterned ground and review of suggested origins. *Geological Society of America, Bulletin*, 67: 823-66.

WASHBURN, A. L., SMITH, D. D. and GODDARD, R. H. 1963.   Frost-Cracking in a Middle-Latitude Climate. *Biuletyn Peryglacjalny*, 12: 175-89.

Reprinted from *Geoforum*, **15**, 15–26 (1973)

Ice Wedge Casts and Past Permafrost Distribution in North America

Fossile Eiskeile und Dauerfrost in der Vergangenheit in Nordamerika

Pseudomorphoses des coins de glace et distribution du permafrost au passé en Amérique du Nord

Troy L. PÉWÉ, Tempe, Ariz.*

Abstract: Ice wedge casts are the most accurate and widespread indicators of past permafrost. Many ice wedge casts exist in Alaska, some in areas of existing ice wedges. In addition to indicating paleotemperature conditions and a wider distribution of permafrost in Wisconsinan time than now, casts in Alaska also indicate permafrost in Illinoian and pre-Illinoian time. Hundreds of ice wedge casts are now known in temperate North America and are described from about 22 widespread localities coast to coast in Canada and United States. Permafrost existed in late Wisconsinan time, 20,000 to 10,000 years ago, along the glacial border in temperate United States. Later permafrost formed north of the glacial border as the continental ice sheet withdrew exposing drift to the rigorous periglacial climate. Ice wedge casts indicate that the – 7 °C mean annual air isotherm was about 2 000 km farther south in late Wisconsinan time than now.

Zusammenfassung: Fossile Eiskeile sind die zuverlässigsten und am weitesten verbreiteten Merkmale einstigen Dauerfrosts. In Alaska gibt es viele Beispiele dafür. In einigen Gebieten sind die Eiskeile noch rezent. Sie markieren nicht nur das Dauerfrostareal des Wisconsin (Würm), sondern lassen sich auch für das Illinois (Riss) und noch früher datieren. Aus dem jetzt gemäßigten Klimabereich Nordamerikas sind hunderte von fossilen Eiskeilen bekannt. Sie werden von 22 verschiedenen Lokalitäten in Canada und USA beschrieben. Dauerfrost herrschte im Spät-Wisconsin vor 20–10 000 Jahren entlang dem Inlandeisrand in USA. Mit dem Eisrückzug breitete sich der Dauerfrost über die nun dem strengen Periglazialklima ausgesetzten Grundmoränen aus. Für das Spät-Wisconsin lassen die fossilen Eiskeile auf eine Südwärtsverschiebung der – 7 °C mittleren Jahresisotherme um ca. 2 000 km schließen.

Introduction

Permafrost was more widespread in North America at different times in the past than at present; undoubtedly, at times, it was also less extensive than now. The distribution of permafrost in subarctic and arctic North America has changed little since Pleistocene time, but in temperate latitudes the permafrost distribution has changed dramatically in late Quaternary time.

Permafrost is defined as a thickness of soil or other superficial deposits, or even bedrock, which has had a temperature below 0 °C for two or more years; it is defined exclusively on the basis of temperature, irrespective of texture, degree of induration, water content, or lithologic character (MULLER, 1945, p. 3). It has long been held that about 20 % of the land area of the world is underlain by permafrost (MULLER, 1945, pl. 1; BLACK, 1954, p. 842; PÉWÉ, 1966b, 1969, Fig. 1; FERRIANS, KACHADOORIAN, and GREEN, 1969).

Permafrost is widespread in North America today (Fig. 1) underlying 82 % of the land surface of Alaska (PÉWÉ, in press b) and about 50 % of Canada (BROWN, 1970; BROWN and PÉWÉ, 1973). The perennially frozen ground in North America has been arbitrarily divided into two zones: the continuous and the discontinuous. In the continuous zone permafrost is almost everywhere present and extends to a depth of 650 m at Prudhoe Bay in Alaska and considerably more than 500 m in northern Canada (BROWN and PÉWÉ, 1973).

Southward, in the discontinuous zone of permafrost, the thickness of frozen ground decreases and unfrozen areas are more and more abundant until near the southern boundary where only rare patches of permafrost exist today. Permafrost is also reported to exist offshore in the Arctic Ocean in both Canada and Alaska (MACKAY, 1972; SHEARER, *et al.*, 1971; LEWELLEN, 1973).

Present knowledge of the southern limit of permafrost indicates that it coincides roughly with the − 1 °C mean annual air isotherm in Canada between James Bay and the Western Cordillera (BROWN, 1967). The southern limit of

---

* Prof. Dr. Troy L. PÉWÉ, Department of Geology, Arizona State University, Tempe, Ariz. 85821, USA

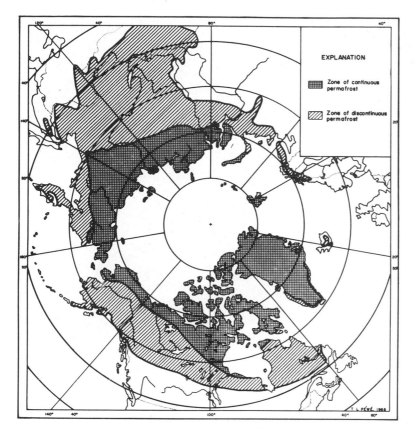

Fig. 1

● Distribution of permafrost in the
   Northern Hemisphere

● Verbreitung des Dauerfrost-
   bodens auf der Nordhalbkugel

● Distribution du permafrost sur
   l'hémisphère nord

permafrost appears to extend south of the 0 °C isotherm in Alaska due to the presence of numerous large bodies of relict permafrost not in equilibrium with the present climate, which have not been similarly encountered on such a large scale in Canada. The southern boundary of the continuous permafrost zone appears to lie in the neighborhood of the $-7°$ to -8 °C mean annual air isotherm in North America which corresponds to the mean annual ground temperature at the level of zero amplitude of -5 °C (GOLD and LACHENBRUCH, 1973).

Criteria for Former Extent of Permafrost

Although floral and faunal studies may indicate that a more rigorous environment was present in temperate latitudes in the past than now, they do not in themselves prove the former presence of permafrost. The same may be said for the presence of a continental ice sheet. Solifluction, soil involutions, and sorted patterned ground features are best developed in areas of permafrost but do not require perennially frozen ground; therefore, the former presence of permafrost cannot be inferred from these features.

Past existence of permafrost can be told only from the existence of former ice wedges and sand wedges and associated large scale polygons, pingos, rock glaciers, and perhaps altiplanation (cryoplanation) terraces. These features are common in the polar areas today (PÉWÉ, in press a) and have received considerable study in the last two decades.

Inactive rock glaciers and altiplanation terraces are limited to highlands or mountainous areas and exist in temperate parts of North America today. However, they are relatively small in geographical extent and are not discussed in this report.

Pingos are conical ice cored hills or mounds, round to oval in plan, 20—400 m in diameter, and 10—70 m high, that form where thick layers of massive ice grow in permafrost near the surface. Pingos are of two distinct genetic types: the closed system and the open system. The closed system type forms in more or less level areas when unfrozen ground water migrates under pressure to a site where a

layer of thin near-surface permafrost can be domed up to
form a mound. Closed system forms are the larger of the
two types of pingos. The open system type forms on sloping terrain where water beneath, or within, the permafrost
penetrates the permafrost under high hydraulic pressure
and a hydrolaccolith forms in the upper part of the permafrost and freezes. Hydraulic pressure together with the
crystallization pressure of the growing ice lens heaves the
overlying material to form a mound.

Thousands of pingos are known in northwestern North
America and by far most of them are in the unglaciated
part of the country (HOLMES, *et al.*, 1968; PÉWÉ, in press
b; MACKAY, 1972, in press; HUGHES, 1969). Radiometric
dating indicates that most pingos in North America are less
than 4,000–7,000 years old. MACKAY (in press) shows by
careful field measurements of four closed system pingos
over several years that many are probably less than
1,000 years old and some less than 25 years old.

When the ice core melts the pingo collapses commonly
leaving a raised ring of ground surrounding a central pond.
The presence of pingo remains in temperate areas as an
excellent indicator of the former distribution of permafrost. Although collapsed pingos have been reported from
widespread localities in Europe (PISSART, 1963, 1965;
WATSON, 1972; and SVENSSON, 1969, 1971), no such
features have been reported from the temperate areas of
North America. However, from the nonpermafrost areas
of northern British Columbia and southern Alberta and
Saskatchewan, as well as from northcentral Illinois, hundreds of pingo-like mounds are reported (BIK, 1968, 1969;
MATTHEWS, 1963; FLEMAL, HINKLEY and HESSLER,
1969; FLEMAL, 1972). Although it is apparent that permafrost existed in late Quaternary time in these particular
areas based on ice wedge casts and perhaps relationship to
glacial fronts, these mounds in themselves do not necessarily indicate permafrost or even pingos.

Ice wedges and the polygonal patterns they reflect on the
surface of the ground are a characteristic feature of permafrost. Ice wedges are characterized by parallel or subparallel foliated structures. Most such ice masses occur as wedge-
shaped, vertical, or inclined sheets or dikes 1 cm–3 m
wide and 1–10 m high where seen in transverse cross
sections (Fig. 2). Some masses exposed in the face of
frozen cliffs may appear as horizontal bodies a few centimeters to 3 m in thickness and 0.5–15 m long. The true
shape of these ice wedges can be seen only in three dimensions. Ice wedges are parts of polygonal networks of ice
enclosing polygons or cells of frozen ground 3–30 m or
more in diameter.

The network of ice wedges in the ground generally causes
a microrelief pattern on the surface of the ground, called
polygonal ground or tundra polygons (Fig. 3). Troughs
that delineate polygons are generally underlain by ice

Photograph No. 474 by T. L. PÉWÉ, September 19, 1949

Fig. 2

● Foliated ground ice mass (ice wedge) in organic-rich silt exposed in
  placer gold mining operations on Wilber Creek near Livengood, Alaska

● Bei der Goldförderung entblößter Eiskeil in humusreichem Silt am
  Wilber Creek bei Livengood, Alaska

● Coin de glace mis à nu au cours de l'extraction aurifère au Wilber
  Creek près de Livengood, Alaska

wedges 1–2 m wide at the top. These polygons are 3–30 m
in diameter and should not be confused with the small
scale polygons or patterned ground produced by seasonal
frost sorting.

Existing ice wedges are classed as active or inactive. The
area of active ice wedges in North America roughly coincides with the continuous permafrost zone; this is especially
true in Alaska and northwestern Canada (Fig. 4). They
occur widely in the northern reaches of the Arctic
Archipelago (BROWN and PÉWÉ, 1973) being reported in
various areas including Prince Patrick Island (PISSART,
1967a, b, 1968) and as far north as Ellesmere Island
(HOGDSON, 1972). It appears that ice wedges crack and
continue to grow when the maximum annual thermal

Photograph by R. I. LEWELLEN, August 11, 1966

Fig. 3

● Oblique summer aerial view of raised-edge polygons on the northern Alaskan seacoast near Barrow, Alaska. Diameters of polygons are 7- 15 m

● Sommerliches Schrägluftbild der Eispolygone mit konkaven Kernen an der Nordküste von Alaska bei Barrow. Polygondurchmesser 7—15 m

● Photo aérienne oblique des polygones négatifs sur la côte nord de l'Alaska près de Barrow. Diamètres des polygones 7—15 m

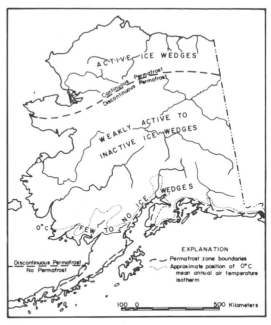

Fig. 4

● Distribution of ice wedges and permafrost in Alaska

● Vorkommen von Eiskeilen und Dauerfrostboden in Alaska

● Distribution des coins de glace et du permafrost en Alaska

tension at the top of the permafrost is the same order of the magnitude but exceeds the strength of the ground ice. For the thermal tension to reach this magnitude the ground must cool at a certain rate for a certain period of time in a winter cold snap. LACHENBRUCH (oral commun., 1962) states that active cracking of ice wedges might be expected in many permafrost materials when the minimum winter temperature at the top of the permafrost is colder than − 15° to − 20 °C. This area is almost limited to the tundra and to the continuous permafrost zone. It occurs in areas where the mean annual air temperature is approximately − 6° to − 8 °C or colder (Fig. 4).

In the discontinuous permafrost zone, especially in north-western North America, the wedges are essentially inactive

and a line dividing the zones of active and inactive ice wedges is arbitrarily placed at a position where low centered or raised edged polygons are uncommon and where it is thought most wedges do not crack frequently (Fig. 4). When more data become available concerning the tempe-rature at the top of permafrost, this arbitrary line may be more accurately placed.

In the area of inactive ice wedges of Alaska (Fig. 4), the mean annual air temperature ranges from about − 2 °C in the south to about − 6° to − 8 °C in the north. Minimum winter temperatures at the top of permafrost in central Alaska range from about − 3° to − 4 °C (PÉWÉ, in press b). Such ground temperatures probably rarely permit thermal cracking of the ice wedges; therefore, no or rare ice in-crements are added to existing ice wedges and they can be considered dormant, relict, or inactive.

The writer has examined ice wedges in Alaska, Canada, Scandinavia, Siberia, and Antarctica and considers all wed-ges he has seen to be Wisconsinan (Würmian) or Holocene in age. He is unaware of any proven existing ice wedges of pre-Wisconsinan age. Wedges in northwestern North America are of several ages, but none appear older than the last major cold period — the Wisconsinan. Those formed prior to that time either melted or are so rare or deeply buried that they are no longer exposed or penetrated in drilling.

Ice Wedge Casts and Permafrost Distribution

In many parts of the world, mainly temperate latitudes, former permafrost is indicated by ice wedge casts. Such casts form when ice wedges slowly melt as the permafrost table is lowered, generally the result of a warming climate. As the top of the ice wedge is melted down a pronounced polygonal pattern of shallow trenches generally appears on the ground surface. Runoff water is channeled into these furrows. The water percolates down into the melting ice wedges carrying sediments downward. As the ice wedge surface is further lowered and the surrounding sediments thawed, losing their cementing material (ice), they tend to collapse toward and into the area formerly occupied by the ice wedge. This is especially true in coarse, sandy gravel. In loess, the walls may stand and the fill originate only from above. Collapse of the bordering material plus downwashing of fines results in a mixture of sediments from above and the side.

Although a voluminous literature has developed in Europe on the subject of ice wedge casts (see PÉWÉ, CHURCH, and ANDRESEN, 1969) only a few examples are known in temperate North America prior to the last decade. Most ice wedge casts are replacements of ice wedges that formed in Wisconsinan time, however, in Alaska there are evidences of ice wedges of a much more ancient time.

Pre-Wisconsinan Ice Wedge Casts

Although it is well known that ice wedges are common in arctic and subarctic North America, it is not generally realized that ice wedge casts also exist there. More references have been made from ice wedge casts of different ages in the subarctic, especially Alaska, than in temperate latitudes. Perhaps the earliest evidence of permafrost in North America is indicated from central and western Alaska. In central Alaska on the campus of the University of Alaska ice wedge casts occur in solifluction layers (Table 1; Fig. 5). These ice wedge casts are thought to be at least a million years old inasmuch as they represent the older of two permafrost periods, both of which predate the Illinoian glaciation. Ice wedge pseudomorphs are also reported from Cape Deceit on the south shore of Kotzebue Sound (GUTHRIE and MATTHEWS, 1971) and HOPKINS (in press) believes they represent a permafrost period of

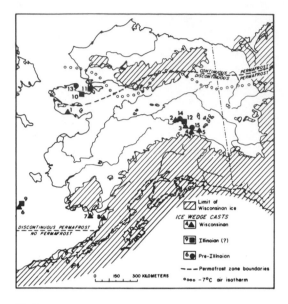

Fig. 5

- Distribution and age of ice wedge casts in Alaska
- Verbreitung und Alter der fossilen Eiskeile in Alaska
- Distribution et l'age des pseudomorphoses des coins de glace en Alaska

1.0 ± 0.5 million years ago. The presence of ice wedges at that time indicates that the mean annual air temperature was at least as cold as − 6° to − 8 °C.

Abundant evidence in Alaska indicates the presence of permafrost and ice wedges during the Illinoian time. Ice wedge casts are present near Fairbanks (PÉWÉ, 1966b) and ice wedge casts in loess of Illinoian age have also been reported from central and western Alaska (McCULLOCH, TAYLOR and RUBIN, 1965; McCULLOCH and HOPKINS, 1966). HOPKINS and EINARSSON (1966) also record ancient ice wedges on the Pribilof Islands (Table 1, Fig. 5).

Wisconsinan ice wedge casts

Alaska. Ice wedge casts of Wisconsinan age in outwash gravel and loess deposits are known in central Alaska (PÉWÉ, 1965b, c; BLACKWELL, 1965; AGER, 1972). The most thorough work on ice wedge casts in North America is probably that of PÉWÉ, CHURCH and ANDRESEN (1969) near Big Delta (Fig. 5, locality No. 4) in central Alaska (Fig. 6). In this area ice wedge casts are reflected on the

Map location number	Author	Date	Locality	Time of ice wedge formation
1	Hopkins, et al.	1960	Nome	Wisconsinan
2	Péwé	(unpub. data)	Fairbanks	Wisconsinan
3	Blackwell	1965	Tanana River	Wisconsinan
4	Péwé, et al.	1969	Big Delta	Wisconsinan
5	Ager	1972	Healy Lake	Wisconsinan
6	Hopkins	1972 (oral commun.)	Pribilof Island	Wisconsinan
7	Hopkins	1972 (oral commun.)	Kvichak Peninsula	Wisconsinan (?)
8	Hopkins, et al.	1955 p. 139	Bristol Bay	Wisconsinan
9	Hopkins and Einarsson	1966	Pribilof Island	Illinoian (?)
10	McCulloch, et al.	1966	NW Alaska	Illinoian (?)
11	McCulloch, et al.	1965 p. 449	Baldwin Peninsula NW Alaska	Illinoian (?)
12	Péwé	1966b	Fairbanks	Illinoian (?)
13	Guthrie and Matthews	1971	Northern Seward Peninsula	Pre-Illinoian
14	Péwé	1965a p. 17	Fairbanks	Pre-Illinoian
15	Péwé	1965b Fig. 4—22	Shaw Creek	Pre-Illinoian

surface by large scale polygons (Fig. 7). Based on a comparison of areas where ice wedges are actively growing today, the mean annual air temperature of the Big Delta area at the time of ice wedge growth was at least $-6\ °C$, in contrast with the mean annual air temperature of $-2.8\ °C$ there today. After the formation of the wedges the climate warmed to $0\ °C$ or above and the ice wedges as well as most of the permafrost in the outwash gravel disappeared. Only relict permafrost exists at great depths in the gravel today. This interval of warmer climate was evidently rather short, inasmuch as permafrost and ice wedges in the adjacent regions were only partially thawed.

Study of ice wedge casts in central Alaska near Big Delta has demonstrated that extensive areas of ice wedge cast-polygons can occur in coarse-grained sediments in regions where permafrost is actively growing, such as in central Alaska, and where large ice wedges are still present in fine-grained sediments. Such an association supports the

suggestion that the permafrost and ice wedges thaw more rapidly in coarse-grained sediments and ice-rich fine-grained sediments. This is especially true where percolating water modifies the thermal regime of the ground, such as in outwash plains, alluvial fans, and river floodplains. In fact, speed or progress of thawing basicly depends on the transfer of heat by groundwater circulation.

Casts of ice wedges of Wisconsinan age occur in southwestern Alaska near or outside the glacial border and near or south of the present southern boundary of permafrost. HOPKINS (oral commun., December 4, 1972) reports such casts on Pribilof Island where permafrost no longer exists. Only relict permafrost exists around the Bristol Bay area where ice wedge casts and large scale polygonal markings are present (HOPKINS and KARLSTROM, et al., 1955; HOPKINS, oral commun. December 4, 1972).

Temperate North America. The distribution of late Wisconsinan permafrost in periglacial areas adjacent to the

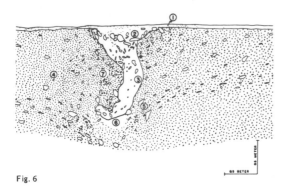

Fig. 6

- Ice wedge cast in Wisconsinan outwash gravel near Big Delta, Alaska. Circled numbers indicate sediment type: (1) silt mantle, (2) fill of upper part of wedge, (3) fill of middle and lower part of wedge, (4) undisturbed outwash gravel, (5) disturbed outwash gravel, (6) sand, and (7) iron-stained sediments

- Fossiler Eiskeil in würmzeitlichem Fluvioglazialschutt in der Nähe von Big Delta, Alaska. (1) Deckenlehm, (2) und (3) Füllung, (4) ungestörter Sanderschotter, (5) umgelagerter Sanderschotter, (6) Sand, (7) eisenverfestigte Ablagerungen (Ortstein)

- Pseudomorphose d'un coin de glace dans le wisconsien près de Big Delta, Alaska. (1) glaises de couverture, (2) et (3) comblement, (4) cailloutis non détruits, (5) cailloutis détruits, (6) sable, (7) sédiments durcis par fer (Ortstein)

continental ice sheet of temperate North America is known only in the most sketchy manner. Unlike Europe, no permafrost belt has heretofore been mapped in temperate North America; however, inasmuch as information has been accumulated in the last 10 to 20 years on the location and age of ice wedge casts (Fig. 8; Table 2) the minimum extent of past permafrost can be visualized.

It is assumed that permafrost formed as refrigerating conditions developed in front of the expanding continental glaciers which covered all of northern North America except western Yukon Territory and parts of Alaska. It has been postulated that permafrost existed beneath those parts of the ice sheets where the bottom temperature was colder than 0 °C either in a pressure melting situation ($-2°$ to -1 °C) or polar glacier conditions when the ice was frozen to the underlying ground and bottom temperatures are lower (BROWN and PÉWÉ, 1973). The thickness of permafrost depended on the ice bottom temperatures. Perhaps some permafrost was thawed by running water under the temperate glaciers, but reformed as glaciers withdrew. The permafrost adjusted to the ensuing cold periglacial conditions with the formation of ice wedges and other features as the ice sheet retreated (Fig. 8).

Fig. 7

- Distribution and age of ice wedge casts in North America (exclusive of Alaska) in relation to position of Wisconsinan glacial ice fronts. See Table 2 for source of data and age of ice wedges. Glacier ice borders generalized from PREST 1969

- Alter und Verbreitung fossiler Eiskeile in Nordamerika (ohne Alaska) in Beziehung zum Wisconsin-Inlandeis. Für Altersangaben siehe Table 2. Die Ausdehnung des Inlandeises wurde nach PREST (1969) generalisiert

- Age et distribution des pseudomorphoses de coins de glace en Amérique du Nord (sauf Alaska) en relation avec la position du front glaciaire wisconsien. Pour les données sur l'age voir Table 2. Les limites glaciaires sont généralisées d'après PREST (1969)

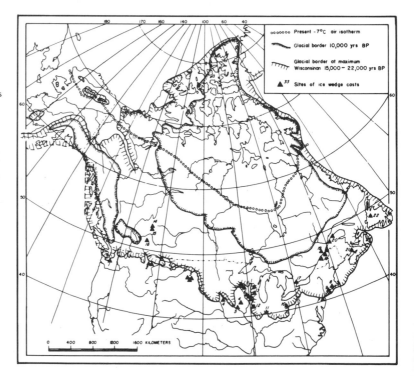

Fig. 8

● Ice wedge cast near Wisconsinan glacial border at River Falls,
St. Croix County, Wisconsin. Cast filling consists of clean,
medium-grained sand enclosed by poorly-sorted, silty, gravelly
sand kame deposit (from BLACK 1965a)

● Fossiler Eiskeil in der Nähe der Grenze des Wisconsin-Inlandeises
bei River Falls, Wisconsin. Die Füllung besteht aus lockerem mittel-
körnigem schlecht sortiertem Sand umschlossen von Kames-Ab-
lagerungen (aus BLACK 1965a)

● Pseudomorphose d'un coin de glace près de la limite glaciaire
wisconsien près de River Falls, Wisconsin. Le comblement est com-
posé de sable meuble à grains moyens contenant des sédiments
de kames mal triés (de BLACK 1965a)

Large areas of central Canada were inundated by post-
glacial lakes (Lake Agassiz and others) beneath which existing permafrost dissipated and did not reform until these
bodies of water disappeared. Much of the Quaternary
record in Canada was greatly disturbed or destroyed by
Pleistocene glaciation and disturbed by post-glacial sub-
mergence making the use of ancient deposits for interpreting part of permafrost conditions difficult (BROWN
and PÉWÉ, 1973).

One of the most logical places to find ice wedge casts in
temperate North America is near the border of the Wisconsinan ice sheet. This was the latest of the major glaciations and features formed at that time are the most likely
to be preserved. The area adjacent to and south of the ice
sheet for a distance of 50 to 150 km was subject to rigorous climate of the periglacial environment, a climate
favorable for the growth of permafrost and ice wedges.
Although permafrost may form when the mean annual air
temperature is 0 °C or slightly colder, ice wedges will not
form, unless, as mentioned earlier, the top of the permafrost becomes very cold and contracts overcoming the
tensile strength of the frozen ground. The presence of ice
wedge casts around the Wisconsinan ice sheet in the United
States indicates that the mean annual air temperature
isotherm of $-6°$ to -8 °C in North America was probably
shifted at least 2,000 km to the south during Wisconsinan
maximum (Fig. 8).

Many ice wedge casts have been reported outside of and
adjacent to the southern border of the continental ice
sheet (SCHAFER, 1949; CLAYTON and BAILEY, 1970;
WILSON, 1958; RUHE, 1969; BLACK, 1964, 1965a, b,
1969). The writer believes that localities south of the
glacial border (Numbers 2–5) undoubtedly reflect true
ice wedges in a cold continental environment, however,
questions are still unanswered concerning reported ice
wedge casts along the Atlantic and Pacific coasts in temperate latitudes just in front of the ice sheet. At locality
Number 1, between Seattle and Tacoma, are the widely
discussed Mima Mounds of Thurston County. Even though
PEWÉ (1948), NEWCOMB (1952) and RITCHIE (1953)
believe the mounds are the result of erosion of ice wedge
polygons, no typical ice wedge casts have ever been reported with these polygonal surface structures. In northcentral
United States at locality Number 6 (Table 2) WOLFE
(1953) reports various periglacial features and alludes to
ice wedge casts. The writer believes these are not true ice
wedge casts. In most instances, students of ice wedge casts
in temperate areas, especially prior to the last 10 to 15
years, were not familiar with modern ice wedges in the
field and were hindered therefore by lack of knowledge
of what an ice wedge or ice wedge cast should look like.

Most of the ice wedges south of the border have been
reported from southwestern North Dakota where more
than 100 are known to exist. RUHE (1969) and WILSON
(1958) mention several localities in Iowa. The most
thorough study of ice wedge casts south of the glacial
border has been by BLACK (1965a) in Wisconsin. The
wedges that existed south of the glacial border are thought
to have formed about 14,000 to perhaps 22,000 years
ago, although some could have formed earlier.

When the glacial ice sheet withdrew to the north from its
maximum position the drift in the recently uncovered

Table 2

● Ice wedge cast localities in temperate North America

● Fossile Eiskeilvorkommen in gemäßigten Nordamerika

Map location number	Author	Date	Locality	Time of ice wedge formation	Distance from max. late Wisc. border
USA — Outside of late Wisconsinan glacial border					
1	Péwé	1948	Thurston Co., Washington	Wisconsinan ~14,000	At border
1	Newcomb	1952	Thurston Co., Washington	Wisconsinan ~14,000	At border
1	Ritchie	1953	Thurston Co., Washington	Wisconsinan ~14,000	At border
2	Schafer	1949 p. 165	near Vaugn, Montana	Wisconsinan	16 km
3	Clayton and Bailey	1970 p. 382	SW North Dakota	Early Wisconsinan (?)	10—100 (?) km
4	Ruhe	1969 p. 179	NW Iowa	Wisconsinan	10—100 (?) km
4	Ruhe	1969 p. 179	Tama Co., Iowa	Wisconsinan	40 (?) km
5	Wilson	1958	Central Iowa	Wisconsinan	70 km
6	Black	1965a	SW Wisconsin	~20,000	0.5—100 km
7	Wolfe	1953	New Jersey*	Wisconsinan	20 km
USA — Inside of late Wisconsinan glacial border					
8	Birman	1952	Rhode Island*	Late Wisconsinan	~70 km
9	Denny	1936	S Connecticut*	Late Wisconsinan	~50 km
10	Totten	1972	Richland Co., Ohio	<14,000	5 km
11	Wayne	1965, p. 33, 1967	W Central Indiana	<14,500	75 km
12	Horberg	1949	Bureau Co., N Illinois	~14,000	30 km
12	Frye and Willman	1958	Woodford Co., N Illinois	Late Wisconsinan	30 km
13	Black	1965a	Outagamie Co., and Columbia Co., Wisconsin	<12,000	50 km
Southern Canada					
14	Westgate and Bayrock	1964	Edmonton, Alberta	>31,000	600 km
15	Berg	1969	Edmonton, Alberta**	Early Wisconsinan	600 km
16	Morgan	1969	Calgary, Alberta	Pre-late Wisconsinan	~300 km
17	Morgan	1972	Toronto, Ontario	~13,000	~300 km
18	Dionne	1971	15 km NW Québec City, Québec	12,000 to 11,5000	~600 km
19	Dionne	1969	S shore of St. Lawrence River	~12,000	~600 km
20	Lagarec	1973	SE Québec	~12,000	~600 km
21	Borns	1965	N Nova Scotia	~12.000	~200 km
22	Brookes	1971	Newfoundland	~11,200	~100 km

*   Doubtful ice wedge casts

** Extinct sand wedge, not ice wedge cast

areas became perennially frozen. In many regions the climatic conditions were especially rigorous and ice wedges formed. As indicated on Fig. 8 (Table 2) ice wedge casts have been reported from glacial deposits just north of the glacial border in the United States, this is especially true in the central part of the country (BLACK, 1965a; HORBER, 1949; FRYE and WILLMAN, 1958; WAYNE, 1965, 1967; TOTTEN, 1973. These wedges formed perhaps 12,000 to about 14,000 years ago.

As the continental ice sheet continued to withdraw northward the wedges south of the maximum extent undoubtedly became inactive. As new areas were uncovered by the glacier, ice wedges formed in the perennially frozen ground. Many casts are recorded in southern Canada in the west (MORGAN, 1969) and especially in the St. Lawrence Valley in the east (MORGAN, 1972; DIONNE, 1967, 1969, 1970, 1971; LAGAREC, 1973). Hundreds of wedges are known on the south side of the St. Lawrence River in southern Quebec.

In addition to the ice wedge casts of the continental interior, they have been reported along the eastern coast of North America. Several years ago possible ice wedge casts were reported by DENNY (1936) in Connecticut, and in Rhode Island by BIRMAN (1952). From the nature of the descriptions the writer believes these are not ice wedge casts, and DENNY (oral commun. December 4, 1972) concurs that his description of periglacial features did not include ice wedge casts. Robert F. BLACK (oral commun. December 4, 1972) states that examination of wedge forms he has seen in Connecticut and Massachusetts indicate they are not true ice wedge casts, and he further believes that perhaps the climate was too mild for ice wedge formation as the late Wisconsinan glacier withdrew in this area. Farther north along the Atlantic coast (Fig. 8), typical ice wedge casts have been reported from Nova Scotia (BORNS, 1965) and Newfoundland (BROOKES, 1971).

Conclusions

Ice wedge casts are the most accurate and widespread indicators of past permafrost. Many ice wedge casts exist in Alaska, some in areas of existing ice wedges. In addition to indicating paleotemperature conditions and a wider distribution of permafrost in Wisconsinan time than now, casts in Alaska also indicate permafrost in Illinoian and pre-Illinoian time.

Hundreds of ice wedge casts are now known in temperate North America and are described from about 22 widespread localities coast to coast in Canada und United States. Permafrost existed in late Wisconsinan time, 20,000 to 10,000 years ago, along the glacial border in temperate United States. Later permafrost formed north of the glacial

border as the continental ice sheet withdrew exposing drift to the rigorous periglacial climate. Ice wedge casts indicate that the $-7\,^\circ$C mean annual air isotherm was about 2000 km farther south in late Wisconsinan time than now.

References

AGER, T. A. (1972): *Surficial Geology and Quaternary History of the Healy Lake Area, Alaska.* Unpub. M.S. Thes. Univ. Alaska.

BERG, T. E. (1969): Fossil sand wedges at Edmonton, Alberta, Canada; *Bull. Peryglac.* **19**, 325−333.

BIK, M. J. J. (1968): Morphological observations on prairie mounds; *Z. Geomorph.*, N.F. **12**, 4, 409−469.

BIK, M. J. J. (1969): The origin and age of the prairie mounds of southern Alberta, Canada; *Bull. Peryglac.* **19**, 85−130.

BIRMAN, J. H. (1952): Pleistocene clastic dikes in weathered granite-gneiss, Rhode Island; *Am. J. Sci.* **250**, 721−734.

BLACK, R. F. (1954): Permafrost: a review; *Bull. geol. Soc. Am.* **65**, 839−855.

BLACK, R. F. (1964): Periglacial phenomena of Wisconsin, north central United States; *VI Conf. int. Ass. quatern. Res.* **4**, 21−28.

BLACK, R. F. (1965a): Ice-wedge casts of Wisconsin; *Trans. Wis. Acad. Sci. Arts Lett.* **54**, 187−222.

BLACK, R. F. (1965b): *Paleoclimatologic Implications of Ice-Wedge Casts of Wisconsin* (Abs.); INQUA Abs. V., 7th Congr., Boulder.

BLACK, R. F. (1969): Climatically significant fossil periglacial phenomena in north central United States; *Bull. Peryglac.* **20**, 225−238.

BLACKWELL, M. F. (1965): *Surficial Geology and Geomorphology of the Harding Lake Area, Big Delta Quadrangle, Alaska.* M.S. Thes. Univ. Alaska.

BORNS, H. W., Jr. (1965): Late glacial ice-wedge casts in northern Nova Scotia Canada; *Science*, **148**, 1223−1225.

BROOKES, I. A. (1971): Fossil ice wedge casts in western Newfoundland; *Maritime Sediments*, **7**, 3, 118−122.

BROWN, R. J. E. (1967): *Permafrost in Canada;* Map pub. by Div. of Bldg. Research, Natn. Res. Coun. (NRC 9769) and Geol. Surv. Can. (Map 1246 A).

BROWN, R. J. E. (1970): *Permafrost in Canada.* Toronto Canada: Univ. Toronto Press.

BROWN, R. J. E. and T. L. PÉWÉ (1973): Distribution of permafrost in North America and its relationship to the environment: 1963−1973− a review; *Proc. of Second Internat. Permafrost Conf., Yakutsk.* US Nat. Acad. Sci.

CLAYTON, L. and P. K. BAILEY (1970): Tundra polygons in the northern Great Plains; *Geol. Soc. Am. Abstr. with Programs*, North-Central Sec., **2**, 6.

DENNY, C. S. (1936): Periglacial phenomena in southern Connecticut; *Am. J. Sci.* **32**, 322−342.

DIONNE J. C. (1967): Formes de cryoturbation fossiles dans le sud-est du Québec; *Bull. Géogr.* **9**, 4, 13−24.

DIONNE, J. C. (1969): Nouvelles observations de fentes de gel fossiles sur la côté sud du Saint-Laurent; *Revue Géogr. Montreal* **23**, 3, 307−316.

DIONNE, J. C. (1970): Fentes en coin fossile dans la région de Québec; *Revue Géogr. Montreal* 24, 3, 313—318.

DIONNE, J. C. (1971): Fentes de cryoturbation tardiglaciaires dans la région de Québec; *Revue Géogr. Montreal* 25, 3, 245—264.

FERRIANS, O. J. R. KACHADOORIAN, and G. W. GREENE (1969): Permafrost and related engineering problems in Alaska; *Prof. Pap. U. S. geol. Surv.* 678.

FLEMAL, R. C. (1972): Ice injection origin of the DeKalb Mounds, north-central Illinois, United States; *Proc. Intern. geol. Congr. Sec.* 12, 130—135.

FLEMAL, R. C. K. HINKLEY and J. L. HESSLER (1969): Fossil pingo field in north-central Illinois; *Geol. Soc. Am. Abstr. with Programs*, North-Central Sec. 6.

FRYE, J. C. and H. B. WILLMAN (1958): Permafrost features near the Wisconsin glacial margin in Illinois; *Am. J. Sci.*, 256, 518—524.

GOLD, L. W. and A. H. LACHENBRUCH (1973): Thermal conditions in permafrost; *Proc. of Second Internat. Permafrost Conf., Yakutsk.* US Nat. Acad. Sci.

GUTHRIE, R. D. and J. V. MATTHEWS, Jr. (1971): The Cape Deceit fauna — early Pleistocene mammalian assemblage from the Alaskan Arctic; *Quatern. Res.* 1, 4, 474—510.

HOGDSON, D. A. (1972): Surficial geology and geomorphology of central Ellesmere Island; *Geol. Surv. Can. Pap.* 73—1 (Rep. Actic., April to Oct.).

HOLMES, G. W., D. M. HOPKINS, and H. L. FOSTER (1968): Pingos in central Alaska; *Bull. U. S. geol. Surv.* 1241-H.

HOPKINS, D. M. (in press): The paleogeography and climatic history of Beringia during late Cenozoic time; *Internord* 12.

HOPKINS, D. M. and Th. EINARSSON (1966): Pleistocene glaciation on St. George, Pribilof Islands; *Science* 152, 343—345.

HOPKINS, D. M., T. N. V. KARLSTROM et al. (1955): Permafrost and ground water in Alaska; *Prof. Pap. U. S. geol. Surv.* 264-F, 113—145.

HOPKINS, D. M., F. S. MacNEIL and E. B. LEOPOLD (1960): The coastal plain at Nome; Alaska; a late-Cenozoic type section for the Bering Strait region; *Proc. 21st int. geol. Congr. Copenhagen*, pt. 4, 46—57.

HORBERG, L. (1949): A possible fossil ice wedge in Bureau County, Illinois; *J. Geol.* 57, 2, 132—136.

HUGHES, O. L. (1969): Distribution of open-system pingos in central Yukon territory with respect to glacial limit; *Geol. Surv. Pap. Can.* 69—34.

LAGAREC, D. (1973): Postglacial permafrost features in eastern Canada; *Proc. of Second Internat. Permafrost Conf., Yakutsk.* US Nat. Acad. Sci.

LEWELLEN, R. I. (1973): The currents and characteristics of nearshore permafrost, northern Alaska; *Proc. of Second Internat. Permafrost Conf., Yakutsk.* US Nat. Acad. Sci.

MacKAY, J. R. (1972): The world of underground ice; *Ann. Ass. Am. Geogr.* 62, 1, 1—22.

MacKAY, J. R. (in press) The growth of pingos, western arctic coast, Canada; *Can. J. Earth Sci.*

MATTHEWS, W. H. (1963): Quaternary stratigraphy and geomorphology of the Fort St. Johns area, northeastern British Columbia; *Br. Columbia Mines Petroleum Resour.*

McCULLOCH, D. S. and D. M. HOPKINS (1966): Evidence for an early recent warm interval in northwest Alaska; *Bull geol. Soc. Am.* 77, 1089—1108.

McCULLOCH, D. S., J. S. TAYLOR and M. RUBIN (1965): Stratigraphy, non-marine mollusks, and radiometric dates from Quaternary deposits in the Kotzebue Sound area, western Alaska; *J. Geol.* 73, 442—453.

MORGAN, A. V. (1969): Intraformational periglacial structures in the Nose Hill gravels and sands, Calgary, Alberta, Canada; *J. Geol.* 77, 3, 358—364.

MORGAN, A. V. (1972): Late Wisconsin ice-wedge polygons near Kitchener, Ontario, Canada; *Can. J. Earth Sci.* 9, 6, 607—617.

MULLER, S. W. (1945): Permafrost on permanently frozen ground and related engineering problems; *U. S. Engrs' office, Strategic Eng. Study Spec. Rep.* 62 (Rep. 1947, Ann Arbor, Michigan: J. W. Edwards Inc).

NEWCOMB, R. C. (1952): Origin of the Mima Mounds, Thruston County region, Washington; *J. Geol.* 60, 461—472.

PÉWÉ, T. L. (1948): Origin of the Mima Mounds; *Scient. Mon.* 66, 293—296.

PÉWÉ, T. L. (1965a): Fairbanks area. In: PÉWÉ, T. L., O. J. FERRIANS, Jr., D. R. NICHOLS, and T. N. V. KARLSTROM: *INQUA Field Conf. F*, central and south central Alaska, Guidebook. Lincoln, Nebraska Acad. Sci., 6—36.

PÉWÉ, T. L. (1965b): Middle Tanana River Valley. In: PÉWÉ, T. L., O. J. FERRIANS, Jr., D. R. NICHOLS, and T. N. V. KARLSTROM: *INQUA Field Conf. F*, central and south central Alaska Guidebook. Lincoln, Nebraska Acad. Sci. 36—54.

PÉWÉ, T. L. (1965c): Delta River area, Alaska Range. In: PÉWÉ, T. L., O. J. FERRIANS, Jr. D. R. NICHOLS, and T. N. V. KARLSTROM: *INQUA Field Conf. F*, central and south central Alaska, Guidebook. Lincoln, Nebraska Acad. Sci., 55—93.

PÉWÉ, T. L. (1966a): Paleoclimatic significance fossil ice wedges; *Bull Peryglac.* 15, 65. 72.

PÉWÉ, T. L. (1966b): *Permafrost and Its Effect on Life in the North.* Corballis: Oregon State Univ. Press.

PÉWÉ, T. L. ed. (1969): *The Periglacial Environment: Past and Present.* Montreal: McGill-Queens Univ. Press.

PÉWÉ, T. L. (in press a): *Geomorphic Processes in Polar Deserts.* Univ. of Arizona Press.

PÉWÉ, T. L. (in press b): Quaternary Geology of Alaska; *Prof. Pap. U. S. geol. Surv.*, 835.

PÉWÉ, T. L., R. E. CHURCH, and M. J. ANDRESEN (1969): Origin and paleoclimatic significance of large scale polygons in the Donnelly Dome area, Alaska; *Spec. Pap. geol. Soc. Am.* 109.

PISSART, A. (1963): Traces of pingos in the Pays de Galles (Grande Bretagne) and the plateau of the Hautes Fagnes (Beligque); *Z. Geomorph.* 7, 147—155.

PISSART, A. (1965): Pingos of the Hautes Fagnes: the problems of their origin; *Ann. Soc. Geol. Belgium* 88, 277—289.

PISSART, A. (1967a): Les pingos de l'Ile Prince Patrick (70° N—120° W); *Bull. Géogr.* 9, 189—217.

PISSART, A. (1967b): Les plygones de tundra de l'Ile Prince Patrick (Artique Canadien 76° lat. N.). In: SPORCK, ed.: *Mélanges de géographie physique, humaine, économiquée, affects à M. Omer Tulippe*, 1.

PISSART, A. (1968): Les polygones de fente de gel de l'Ile Prince Patrick (Artique Canadien —76° lat. N.); *Bull. Peryglac.* 17, 171—180.

PREST, V. K. (1969): Retreat of Wisconsin and recent ice in North America; *Can. geol. Surv.* (Map 1257 A).

RITCHIE, A. M. (1953): The erosional origin of the Mima Mounds of southwest Washington; *J. Geol.* **61**, 41–50.

RUHE, R. V. (1969): *Quaternary landscapes in Iowa*. Ames, Iowa: Iowa State Univ. Press.

SCHAFER, J. P. (1949): Some periglacial features in central Montana; *J. Geol.* **57**, 154–174.

SHEARER, J. M., R. F. MacNAB, B. R. PELLETIER, and T. B. SMITH (1971): Submarine pingos in the Beaufort Sea; *Science* **174**, 816–818.

SVENSSON, H. (1969): A type of circular lakes in northernmost Norway; *Lund Stud. Geogr., Series A Physical Geography* **45**, 1–12.

SVENSSON, H. (1971): Pingos in the outer part of Adventdalen; *Saertr. Norsk Polar Inst.*, 168–174.

TOTTEN, S. M. (1973): Glacial geology of Richland County, Ohio; *Ohio geol. Surv. Bull.*

WATSON, E. (1972): Pingos of the Cardiganshire ice limit; *Nature* **236**, 5346, 343–344.

WAYNE, W. J. (1965): Great Lakes-Ohio River Valley — day 4; *Guidebook G. 7th INQUA Congr.*, 29–36.

WAYNE, W. J. (1967): Periglacial features and climatic gradient in Illinois, Indiana, and western Ohio, east-central United States. In: CUSHING, E. J. and H. E. WRIGHT. ed.: *Quaternary Plaeoecology*. Hartford, Conn.: Yale Univ. Press.

WESTGATE, J. A. and L. A. BAYROCK (1964): Periglacial structures in the Saskatchewan gravels and sands of central Alberta, Canada; *J. Geol.* **72**, 5, 641–648.

WILSON, L. R. (1958): Polygonal structures in the soil of central Iowa; *Okla. Geol. Notes, Okla. Geol. Surv.* **18**, 1, 4–6.

WOLFE, P. E. (1953): Periglacial frost-thaw in New Jersey; *J. Geol.* **61**, 113–141.

8

Reprinted from *Bull. Geol. Soc. Amer.*, 67, 824–827, 859–861 (July 1956)

CLASSIFICATION OF PATTERNED GROUND AND REVIEW OF SUGGESTED ORIGINS

BY A. L. WASHBURN

INTRODUCTION

The study of patterned ground has been handicapped by many overlapping and synonymous terms, by the difficulty of digging in frozen ground and determining three-dimensional characteristics, and by the lack of detailed and quantitative studies. The present paper attempts to alleviate the first of these handicaps, to show the effect of the others, and to facilitate further study by systematically reviewing the many hypotheses of patterned-ground origin. Comprehensive foreign-language reviews have already been presented by Steche (1933), Troll (1944; *cf.* Antevs, 1949),[1] Jahn (1948a),[2] and Cailleux and Taylor (1954). Troll's synthesis is a classic study of the distribution of patterned ground and its geographic variations.

The present paper is based in part on Washburn (1950) but a number of corrections have been made and much new material added. Revision is appropriate because the original discussion appeared in a geographical publication with limited distribution to geologists.

Field investigations by the writer are cited from the Victoria Island and Banks Island region, arctic Canada. Most of these observations were made during the spring and summer of 1949 in the course of a joint investigation with Dr. A. E. Porsild, Director of the National Herbarium in Ottawa, and Dr. J. L. Jenness, who represented the Geographical Branch of the Canadian Department of Mines and Technical Surveys.

[1] Troll's study, translated into English by H. E. Wright, is in process of publication by the Snow, Ice, and Permafrost Research Establishment, Corps of Engineers, U. S. Army.

[2] Jahn's extensive paper in Polish was not read; it is cited on the basis of a rough translation of a brief section only. A full translation is being prepared.

ACKNOWLEDGMENTS

The writer is indebted to Dr. Gerald M. Richmond of the U. S. Geological Survey whose encouragement as Chairman of an informal committee on classification of colluvium led to the present discussion of classification and terminology of patterned ground. The writer is also grateful to the Revue Canadienne de Géographie for permission to use material which first appeared in that journal, and to all those who assisted in the preparation of that material, particularly Prof. Richard Foster Flint of Yale University, Mr. W. H. Ward, Building Research Station, Watford, England, and Tahoe Washburn. Very helpful criticism has been received from them and also from Dr. Charles S. Denny of the U. S. Geological Survey, Prof. Richard P. Goldthwait of Ohio State University, Profs. Carl F. Long and S. Russell Stearns of the Thayer School of Engineering at Dartmouth College, Mr. Robert R. Philippe of the Corps of Engineers, U. S. Army, Dr. A. E. Porsild, Director of the National Herbarium in Ottawa, and Prof. Robert P. Sharp of the California Institute of Technology, all of whom have earned the writer's deepest appreciation.

DEFINITION OF PATTERNED GROUND

Patterned ground is a group term for the more or less symmetrical forms, such as circles, polygons, nets, steps, and stripes, that are characteristic of, but not necessarily confined to, mantle subject to intensive frost action.

Patterned ground that is demonstrated to have a distinctive climatic or geographic distribution exclusive of other environments can be designated by appropriate adjectives. Thus, cold-climate patterned ground describes most, but not all, patterned ground.

The term patterned ground was suggested by Washburn (1950, p. 7–8)[3] because of the following considerations:

The terms *Rutmark, Strukturboden, Polygonboden, Polygonenboden, Zellenboden, stone circles, stone rings, stone nets, stone polygons, mud circles, soil circles, mud polygons, soil polygons, fissure polygons, tundra polygons, stone stripes, soil stripes, solifluction stripes* and others have all been used to describe features here collectively named *patterned ground* for want of a satisfactory collective term in English. The term *structure ground* (Antevs, 1932, p. 48) is awkward, and *soil structures* (Sharp, 1942[a], p. 275) is objectionable because it may imply the presence of humus and, as recognized by Sharp, a soil profile, both of which may be absent.[4] Regularity is inherent in the term *pattern*, and the writer would restrict the use of *patterned ground* to more or less symmetrical features rather than include phenomena such as stone-banked terraces, rock glaciers, etc. The term *patterned ground* thus corresponds most closely to the German term *Strukturboden* as employed by Sørensen (1935, p. 8), although *Strukturboden* was originally introduced by Meinardus (1912a, p. 256–257).

The term patterned ground has been widely adopted (Thorarinsson, 1951; Ahlmann 1952, p. 2; Arnold, 1953, p. 3; Johnson and Lovell, 1953, p. 106; Kersten, 1953; Mackay, 1953; Colton and Holmes, 1954; Blackadar, 1954, p. 20). The Highway Research Board Committee on Frost Heave and Frost Action in Soil (Hennion, 1955, p. 109) defined the term in connection with patterns resulting from frost action, but the original discussion quoted above did not specify such a limitation, and the occurrence of desiccation patterns very similar to certain patterns attributed to frost action make a narrow definition undesirable. The objection suggested by Black (1952a, p. 125) that some types of patterned ground are bedrock forms and not ground in the strict sense of the word does not seem serious in view of the broad use of the term ground (for instance, ground water is not restricted to unconsolidated deposits) and the fact that forms in bedrock (Washburn, 1950, p. 47–49) are exceptional.

The unit component of patterned ground (excepting steps and stripes)—a circle, polygon, or intermediate form—is here termed the mesh.

CLASSIFICATION OF PATTERNED GROUND

Many closely related classifications of patterned ground have been developed, such as those of Meinardus (1912a, p. 257; 1912b, p. 16), Högbom (1914, p. 308), Beskow (1930, Tabelle I, p. 629), Huxley and Odell (1924), Steche (1933, p. 195), Sørensen (1935, Tabelle 4, p. 64–65), and Troll (1944, p. 673; *cf.* Antevs, 1949, p. 232–233). The writer's classification retains the essential elements of the classifications cited above—pattern, and presence or absence of sorting—and reflects a revised terminology that is orderly and consistent.

A revised terminology is needed. Different terms have been used for similar forms and the same term for dissimilar ones (Smith, 1949, p. 1505–1506; Cailleux and Taylor, 1954, p. 9). Stone-polygons (Huxley and Odell, 1924, p. 208), stone rings (Hawkes, 1924, p. 509–510), and stone nets (Antevs, 1932, p. 48–58) are a few of the many terms used for stone-bordered polygonal features. (*Cf.* Sharpe, 1938, p. 37; Sharp, 1942a, p. 275.) The term stone-polygons has also been used (Elton, 1927, p. 188–189; Bretz, 1935, p. 176) to describe almost perfectly circular forms of possibly different origin, called stone circles by Huxley and Odell (1924, p. 224–225). Elton used the terms stone-polygons and stone-circles interchangeably in discussing the latter. The use of the name "circular . . . polygon" (Paterson, 1951, p. 18, Pl. 14) to distinguish circular from polygonal forms illustrates the problem.

Terminology difficulties also exist regarding forms of patterned ground without a marked stone border. Although the term fissure-polygons (Huxley and Odell, 1924, p. 208) is appropriate in places, it is misleading for forms lacking obvious fissures. Moreover, fissures do not necessarily indicate absence of stone borders, for some polygons with fissures have stone borders as well, coincident with the fissures (Furrer, 1954, p. 228–232; Figs. 12–14). Elton (1927, p. 165) objected to the term fissure-polygons and suggested mud-polygons, but this term is ambiguous because forms without a stone border do not invariably consist of "mud" but may consist of sand, gravel, or a

---

[3] The original pagination of Washburn (1950), followed here, is uniformly four pages higher than that of the separate.
[4] The unsuitability of the term soil in this connection is stressed by Antevs (1949, p. 232–233). The same objection applies to the term frost pattern soils (Troll, 1944, p. 547).

nonsorted mixture of fines[5] and stones, including boulders. Ice-wedge polygons, tundra polygons, and Taimyr polygons are three synonymous terms where one would suffice.

Elton (1927, p. 190) combined forms with and without a stone border under the group term soil-polygons, but it is misleading to use the term soil, as previously noted, or to refer to obviously circular forms as polygons. A variety of overlapping terms is also used for steplike and striped forms of patterned ground (Sharpe, 1938, p. 37–38, 42; Sharp, 1942a, p. 275).

Genetic terms such as Brodelboden (Gripp, 1926, p. 352), based on an unproved mode of origin, are undesirable, as emphasized by Bryan (1946, p. 634). Clearly it is not yet time for a sound genetic classification of patterned ground, as pointed out by Troll (1944, p. 620) and illustrated by the numerous hypotheses of origin.

Two commonly obvious characteristics of patterned ground that can be ascertained in the field without digging and are also usable in photo-interpretation are: (1) the pattern—whether dominantly circular, polygonal, intermediate (nets), steplike, or striped, (2) the presence or absence of obvious sorting between stones and fines. The following classification, based on Washburn (1950, p. 8–9), combines the two characteristics. The arrangement of the classification is in the direction of increasing gradient; thus most circles, nets, and polygons occur on essentially horizontal ground,[6] and steps and stripes are limited to slopes. Where steps and stripes occur together the latter are on the steeper gradient.

Circles
    Sorted (including debris islands)
    Nonsorted (including peat rings, tussock rings)[7]

Nets
    Sorted
    Nonsorted (including earth hummocks)
Polygons
    Sorted
    Nonsorted (including frost-crack polygons, ice-wedge polygons, tussock-birch-heath polygons, desiccation polygons)
Steps
    Sorted
    Nonsorted
Stripes
    Sorted
    Nonsorted

As pointed out by Black (1952a, p. 125) some patterned ground forms are gradational in pattern and sorting. With respect to pattern, some sorted circles grade into sorted polygons (Miethe, 1912, p. 242; Högbom, 1914, p. 319; Washburn, 1947, p. 99), and some sorted polygons merge into steps and stripes (Antevs, 1932, p. 53, 56–57; Sharp, 1942a, p. 277–281, 295–296). There seems to be a similar gradational series with nonsorted forms of patterned ground (Douvillé, 1917, p. 246; Washburn, 1947, p. 94, 97–99; Hopkins and Sigafoos, 1951, p. 88, 90, 96). With respect to sorting, transitional types have been reported between sorted and nonsorted circles (Washburn, 1947, p. 97) and between sorted and nonsorted polygons (Högbom, 1910, p. 56; 1914, p. 321, Huxley and Odell, 1924, p. 227–228; Elton, 1927, p. 164, 173; Salomon, 1929, Footnote, p. 10; Corbel, 1954a, p. 56; Furrer, 1954, p. 228–232, Figs. 12–14). Thus, in places it is difficult to classify a given form, but most classifications involve the same problem. A single broad heading may include similar forms of differing origin, but the only solution to this difficulty is a sound genetic classification, which requires more knowledge of formative processes than is now possessed.

The writer believes that the use of the simple descriptive terms of the present classification, supplemented by qualifying adjectives, such as miniature, large, sandy, silty, or clayey, and by detailed quantitative descriptions where necessary, will help develop a sound genetic classification. The terminology of the present classification is not intended to supplant existing terms, such as ice-wedge polygons (Elton,

---

[5] The term fines as used in this paper may include sand as well as silt and clay sizes.

[6] Some circles (debris islands), nets (earth hummocks), and nonsorted polygons occur on appreciable slopes also.

[7] Because vegetation is a common feature of nonsorted patterns, Lewis (1952) suggested it might be a useful term component in distinguishing nonsorted from sorted forms. However, vegetation also outlines some sorted patterns, and some nonsorted patterns lack it, so that vegetation cannot be used as a classification element in distinguishing between them, as recognized by Hopkins, Karlstrom, et al. (1955, p. 137–139).

1927, p. 175), peat rings, tussock rings, tussock-birch-heath polygons (Hopkins and Sigafoos, 1951, p. 52–53, 66–70, 76–87), and any other appropriate descriptive or genetic terms that can be used in a unique sense.

The following definitions are reworded and formalized from Washburn (1950, p. 9–13) except where otherwise noted. References given for synonyms clarify the sense in which a term has been used and generally, but not necessarily, include the first use.

[*Editor's Note:* Material has been omitted at this point.]

Conclusions

The descriptive classification of patterned ground adopted in this paper eliminates ambiguities and confusion resulting from the many overlapping and synonymous terms that have appeared in the literature. The classification is based on geometric pattern and presence or absence of sorting, and its main classes comprise sorted and nonsorted varieties of circles, nets, polygons, steps, and stripes. In the absence of more data on the genesis of the forms involved, any comprehensive classification but a purely descriptive one is believed to be impractical and premature.

The preceding discussion of hypotheses and processes supports the conclusion of Poser (1931, p. 226) and others concerning the polygenetic origin of patterned ground. Although this conclusion is not new, it is here fully documented. Thus, the writer believes that any over-all explanation must involve separate processes, with frost action being regarded as a complex of processes that are highly important but not the only ones represented. Not only are many different types of patterned ground involved, but it seems probable that somewhat similar forms may originate by quite different processes. Another and obvious conclusion is that the genesis of the

various kinds of patterned ground is far from established.

In polygonal patterns, it is certain that some meshes are products of drying, and very probable that others result from contraction due to low temperature. The writer believes that both types occur in cold climates and that some meshes may be combination forms reflecting the operation of both processes. In the case of circular patterns, the evidence indicates that probably local differential heaving and perhaps cryostatic movement are genetic processes of widespread significance. Possibly these processes, too, complement each other in complex fashion. For instance, cryostatic movement could bring saturated fines adjacent to coarser material with the result that the fines might be subjected to intense local heaving.

With respect to sorting, ejection of stones toward freezing surfaces by multigelation, their movement by gravity, and eluviation of fines may be key processes. They do not explain polygonal, circular, or striped patterns as the above-mentioned processes may but they are associated with these processes and may determine whether sorted or nonsorted forms are produced if other conditions remain equal.

Solifluction is probably of major significance in the origin of patterned ground on a slope. Presumably it combines with one or more of the other processes to produce the various types of patterned ground that are confined to slopes.

It is tempting to speculate that desiccation, contraction due to low temperature, local differential heaving, and cryostatic movement may combine to form a single continuous system responsible for both polygonal and circular patterns. Geometrically the end members could be represented by the corners of a tetrahedron, and the combination patterns by intermediate positions. Such intermediate positions might account for some nets, which according to the usage in this paper are neither dominantly circular nor polygonal. If to these pattern-forming processes are added ejection of stones, their movement by gravity, and eluviation as sorting processes, and solifluction as the key modifier of patterns on a slope, elements of a comprehensive explanation of patterned ground are at hand. Whether or not these processes turn out to be the key ones, the writer predicts that a complete explanation of patterned ground will involve several specific processes arranged in a similar model.

Of the remaining processes reviewed, the writer believes that in the absence of additional evidence, circulation due to ice thrusting and convection due to temperature-controlled density differences can be eliminated from further consideration in the origin of patterned ground. Frost wedging, weathering, artesian pressure, and rillwork are probably valid genetic processes in special situations but not of widespread significance. The writer regards all the other reviewed processes as speculative but nevertheless stimulating in the light of present inadequate knowledge.

Before the origin of patterned ground is fully understood and the climatic significance and interpretation of present-day and "fossil" forms correctly established, it will be necessary to have much more detailed and accurate information than is now available. Cold-room studies, field observations of subsurface phenomena at different times of the year, and cooperative work between geologists, physicists, soil engineers, pedologists, and plant ecologists are urgent desiderata in further research.

References Cited

Ahlmann, Hans W:son, 1936, Polygonal markings, *in* Scientific results of the Swedish-Norwegian Arctic Expedition in the summer of 1931, v. 2, pt. 12: Geog. Annaler, Årg. 18, p. 7–19

—— 1952, Summary of reports from the Nordic countries: 17th Internat. Geog. Cong. Proc., Comm. on Periglacial Morphology, p. 1–2 (Preprint for Geophysical Research Directorate, U. S. Air Force, 51 p.)

Antevs, Ernst, 1932, Alpine zone of Mt. Washington Range: Auburn, Maine, Merrill & Webber Co., 118 p.

—— 1949, Strukturböden, Solifluktion und Frostklimate der Erde, By Carl Troll (Review): Jour. Geology, v. 67, p. 232–235

Arnold, C. A., 1953, Searching for plant fossils in Alaska: Office of Naval Research, Research Rev., June 1953, p. 1–9

Beskow, Gunnar, 1930, Erdfliessen und Strukturböden der Hochgebirge im Licht der Frosthebung: Geol. Fören. Stockholm, Förh., band 52, p. 622–638

Black, R. F., 1951, Graphs for visual comparison of several factors in heat exchange near Barrow, Alaska (Abstract): Geol. Soc. America Bull., v. 62, p. 1546–1547

—— 1952a, Polygonal patterns and ground conditions from aerial photographs: Photogrammetric Eng., v. 17, p. 123–134

Blackadar, R. G., 1954, Geological reconnaissance north coast of Ellesmere Island, Arctic Archipelago, Northwest Territories: Canada Geol. Survey Paper 53–10, 22 p.

Bretz, J H., 1935, Physiographic studies in East Greenland, p. 159–245, *in* Boyd, L. A., The fiord region of East Greenland: Am. Geog. Soc. Special Pub. 18, 369 p.

Bryan, Kirk, 1946, Cryopedology—the study of frozen ground and intensive frost-action with suggestions on nomenclature: Am. Jour. Sci., v. 244, p. 622–642

Cailleux, André, 1948, Etudes de cryopédologie: Expéditions Polaires Françaises: Paris, Centre de Documentation Univ., 68 p.

Colton, R. B., and Holmes, C. D., 1954, Patterned ground near the Thule Air Base (Abstract): Geol. Soc. America Bull., v. 65, p. 1373

Corbel, J., 1954a, Les sols polygonaux: observations, expériences, genèse: Rev. Géomorphologie Dynamique, 1954, p. 49–68

Douvillé, Robert, 1917, Sols polygonaux ou réticulés: La Géographie, tome 31, p. 241–251

Elton, C. S., 1927, The nature and origin of soil-polygons in Spitsbergen: Geol. Soc. London Quart. Jour., v. 83, p. 163–194

Furrer, Gerhard, 1954, Solifluktionsformen im schweizerischen Nationalpark: Schweizer. naturf. Gesell. Ergebnisse der wissenschaftlichen Untersuchungen des schweizerischen Nationalparks, Band 4 (Neue Folge), no. 29, p. 201–275

Gripp, Karl, 1926, Über Frost und Strukturboden auf Spitzbergen: Gesell. Erdkunde Berlin Zeitschr., 1926, p. 351–354

Hawkes, Leonard, 1924, Frost action in superficial deposits, Iceland: Geol. Mag., v. 61, p. 509–513

Hennion, Frank (chairman), 1955, Frost and permafrost definitions: Nat. Acad. Sci.—Nat. Research Council, Highway Research Board Bull. 111, 110 p.

Högbom, Bertil, 1910, Einige Illustrationen zu den geologischen Wirkungen des Frostes auf Spitzbergen: Upsala Univ., Geol. Inst. Bull., v. 9 (1908–1909), p. 41–59

—— 1914, Über die geologische Bedeutung des Frostes: Upsala Univ., Geol. Inst. Bull., v. 12 (1913–1914), p. 257–389

Hopkins, D. M., and Sigafoos, R. S., 1951, Frost action and vegetation patterns on Seward Peninsula, Alaska: U. S. Geol. Survey Bull. 974-C, p. 51–100

Hopkins, D. M., Karlstrom, T. N. V., *et al.*, 1955, Permafrost and ground water in Alaska: U. S. Geol. Survey Prof. Paper 264-F, p. 113–146

Huxley, J. S., and Odell, N. E., 1924, Notes on surface markings in Spitsbergen: Geog. Jour., v. 63, p. 207–229

Jahn, Alfred, 1948a, Badania nad strukturą i temperaturą gleb w Zachodniej Grenlandii: Polska Akad. Umiejętności, Rozprawy Wydz., matem.-przyr., tom 72, dział A, 1946, no. 6, 121 p. (p. 63–183 of whole volume)

Johnson, A. W., and Lovell, C. W., 1953, Frost-action research needs, p. 99–124, *in* Soil temperature and ground freezing: Nat. Acad. Sci.—

Nat. Research Council, Highway Research Board Bull. 71, 124 p.

Kersten, M. S., 1953, Program for study of patterned ground: Minn. Univ. Inst. Technology, Mimeo. Rept. to U. S. Army Corps of Engineers, Snow, Ice and Permafrost Research Establishment, under Contract DA-21-018-ENG-256, 69 p.

L[ewis], W. V., 1952, Stone polygons and related phenomena: Geog. Jour., v. 118, p. 101–102

Mackay, J. R., 1953, Fissures and mud circles on Cornwallis Island, N. W. T.: Canadian Geographer, no. 3, 1953, p. 31–37

Meinardus, Wilh., 1912a, Beobachtungen über Detritussortierung und Strukturboden auf Spitzbergen: Gesell. Erdkunde Berlin Zeitschr., 1912, p. 250–259
—— 1912b, Über einige charakteristische Bodenformen auf Spitzbergen: Naturh. Ver. Preuss. Rheinlande u. Westfalens, Medizinisch-naturwiss. Gesell. Münster Sitzungsber., Sitzung 26, p. 1–42

Miethe, A., 1912, Über Karreebodenformen auf Spitzbergen: Gesell. Erdkunde Berlin Zeitschr., 1912, p. 241–244

Paterson, T. T., 1940, The effects of frost action and solifluxion around Baffin Bay and in the Cambridge district: Geol. Soc. London Quart. Jour., v. 96, p. 99–130
—— 1951, Physiographic studies in North West Greenland: Meddel. om Grønland, Bind 151, no. 4., 59 p.

Poser, Hans, 1931, Beiträge zur Kenntnis der arktischen Bodenformen: Geol. Rundschau, Band 22, p. 200–231

Salomon, Wilhelm, 1929, Arktische Bodenformen in den Alpen: Heidelberger Akad. Wiss., Math.-naturwiss. Kl., Sitzungsber., Jahrg. 1929, Abh. 5, 31 p.

Sharp, R. P., 1942a, Soil structures in the St. Elias Range, Yukon Territory: Jour. Geomorphology, v. 5, p. 274–301

Sharpe, C. F. S., 1938, Landslides and related phenomena: N. Y., Columbia Univ. Press, 137 p.

Smith, H. T. U., 1949, Physical effects of Pleistocene climatic changes in nonglaciated areas: eolian phenomena, frost action, and stream terracing: Geol. Soc. America Bull., v. 60, p. 1485–1516

Sørensen, Thorvald, 1935, Bodenformen und Pflanzendecke in Nordostgrönland: Meddel. om Grønland, Bind 93, no. 4, 69 p.

Steche, Hans, 1933, Beiträge zur Frage der Strukturböden: Sächs. Akad. Wiss., Math.-phys. Kl., Ber., Band 85, p. 193–272

Thorarinsson, Sigurdur, 1951, Notes on patterned ground in Iceland: Géog. Annaler, no. 3–4, p. 144–156

Troll, Carl, 1944, Strukturböden, Solifluktion und Frostklimate der Erde: Geol. Rundschau, Band 34, p. 545–694

9

Reprinted from *Acta Geog. Lodz.*, No. 24, 437–446 (1970)

AN APPROACH TO A GENETIC
CLASSIFICATION OF PATTERNED GROUND

A. L. Washburn *
Seattle

INTRODUCTION

There are still too many unknowns to formulate a satisfactory genetic classification of patterned ground. The genesis of many forms is problematical, not only quantitatively but even qualitatively. Nevertheless,
there is a continuing demand for an attempt at a genetic classification.
The present effort was undertaken at the request of Professor Jan D y-
l i k, Chairman of the Commission on Periglacial Morphology of the International Geographical Union.

The attempted classification is incomplete and full of assumptions
and uncertainties. However, it should help to pinpoint critical problems.
It incorporates existing terminology so that many of the designations
are self explanatory, and it is sufficiently flexible to permit major modifications.

THE CLASSIFICATION

The proposed classification is based on the following premises: 1. patterned ground is of polygenetic origin; 2. similar forms of patterned
ground may be due to different genetic processes; 3. some genetic processes may produce different-appearing forms; 4. there are more genetic processes than presently recognized terms for associated forms; 5. it
is desirable to maintain a simple and self-evident terminology.

Table 1 satisfies the premises by combining, in matrix form, existing
terms for geometric pattern with existing terms for genetic processes.
The pattern terms have been reasonably widely accepted; the process
terms are, for the most part, also widespread. The resulting combined
terms are readily understandbale in light of the matrix approach. However, some would be awkwardly long without being shortened and

---

* University of Washington.

[*Editor's Note:* Unfortunately, a certain number of the word breaks in this
article are incorrect, rendering comprehension difficult at times.]

therefore the terms *nonsorted* and *sorted* in the combined forms have been abbreviated to N and S.

Only initial processes ("first causes") with respect to origin of the basic geometric pattern (circles, nets, polygons, or stripes) are given in table I. Once a basic pattern is established, sorting processes can modify it to produce sorted forms. For instance, a cracking process in itself provides a nonsorted pattern only, but a sorting process acting on an initially nonsorted pattern can change it to a sorted pattern. These sorting processes are not specifically included in table I unless they also act as an initial patterning process. The Class I genetic processes in table 1 are cracking processes in which the cracking is the essential and immediate cause of the resulting patterned-ground forms; the Class II processes are those in which cracking is nonessential.

Involutions are omitted from the classification, since they are probably of various origins and associated with many kinds of patterned ground. The table also omits any mention of solifluction, although this process is present in many kinds of patterned ground on slopes. Through "microsolifluction" it may contribute to sorting, and through "macrosolifluction" to the development of nets, steps, and stripes. However, to what extent it is an initial process in forming patterned ground is uncertain.

No effort is made to review the large literature on the origin of patterned ground, and for a review of processes, evidence, and terms not otherwise discussed, reference is made to W a s h b u r n (1956, 1968) and the associate bibliographies.

In the following commentary, table I will be reviewed according to process to clarify certain points and problems in the classification.

CLASS I. CRACKING ESSENTIAL

GENERAL

Cracking is a very widespread and important process in the initiation of sorted as well as nonsorted patterned ground. Sorted forms initiated by cracking obviously involve the addition of a sorting process. Basic patterns are typically polygonal but solifluction, possibly ice wedging in cracks, and some volume changes associated with cracking (for instance, in gilgai formation — H a l l s w o r t h, R o b e r t s o n, and G i b b o n s 1955) can deform some polygons to nets. Moreover, given favorable slope conditions, it seems probable that small transverse cracks would be narrowed by mass-wasting, whereas cracks paralleling the slope would tend to remain open and determine the location of small

GENETIC CLASSIFICATION

			DESICCATION CRACKING	DILATION CRACKING	SALT CRACKING	SEASONAL FROST CRACKING	PERMAFROST CRACKING	FROST ACTION ALONG BEDROCK JOINTS
					CRACKING ESSENTIAL			
					THERMAL CRACKING			
						FROST CRACKING		
GEOMETRIC PATTERNS	CIRCLES	NONSORTED						
		SORTED						Joint-crack S circles (at crack intersections)
	POLYGONS	NONSORTED	Desiccation N polygons	Dilation N polygons	Salt-crack N polygons	Seasonal frost-crack N polygons	Permafrost-crack N polygons, incl. Ice-wedge polygons, sand-wedge polygons	Joint-crack N polygons ?
		SORTED	Desiccation S polygons	Dilation S polygons	Salt-crack S polygons	Seasonal frost-crack S polygons	Permafrost-crack S polygons ?	Joint-crack S polygons
	NETS	NONSORTED	Desiccation N nets, incl.? Earth hummocks	Dilation N nets		Seasonal frost-crack N nets, incl.? Earth hummocks	Permafrost-crack N nets, incl. Ice-wedge nets and sand-wedge nets ?	
		SORTED	Desiccation S nets	Dilation S nets		Seasonal frost-crack S nets	Permafrost-crack S nets ?	
	STEPS	NONSORTED						
		SORTED						
	STRIPES	NONSORTED	Desiccation N stripes ?	Dilation N stripes ?		Seasonal frost-crack N stripes ?	Permafrost-crack N stripes ?	Joint-crack N stripes ?
		SORTED	Desiccation S stripes	Dilation S stripes		Seasonal frost-crack S stripes ?	Permafrost-crack S stripes ?	Joint-crack S stripes

A. L. Washburn

Table 1

OF PATTERNED GROUND

SES

CRACKING NONESSENTIAL

PRIMARY FROST SORTING	MASS DISPLACEMENT	DIFFERENTIAL FROST HEAVING	SALT HEAVING	DIFFERENTIAL THAWING AND ELUVIATION	DIFFERENTIAL MASS-WASTING	RILLWORK
	Mass-displacement N circles	Frost-heave N circles	Salt-heave N circles			
Primary frost-sorted circles, incl.? Debris islands	Mass-displacement S circles, incl. Debris islands	Frost-heave S circles	Salt-heave S circles			
	Mass-displacement N polygons ?	Frost-heave N polygons ?	Salt-heave N polygons ?			
Primary frost-sorted polygons?	Mass-displacement S polygons ?	Frost-heave S polygons ?	Salt-heave S polygons ?	Thaw S polygons ?		
	Mass-displacement N nets, incl.? Earth hummocks	Frost-heave N nets, incl. Earth hummocks	Salt-heave N nets			
Primary frost-sorted nets	Mass-displacement S nets	Frost-heave S nets	Salt-heave S nets	Thaw S nets		
	Mass-displacement N steps	Frost-heave N steps ?	Salt-heave N steps ?		Mass-wasting N steps ?.	
Primary frost-sorted steps ?	Mass-displacement S steps	Frost-heave S steps	Salt-heave S steps	Thaw S steps ?	Mass-wasting S steps	
	Mass-displacement N stripes	Frost-heave N stripes	Salt-heave N stripes		Mass-wasting N stripes ?	Rillwork N stripes ?
Primary frost-sorted stripes ?	Mass-displacement S stripes	Frost-heave S stripes	Salt-heave S stripes	Thaw S stripes ?	Mass-wasting S stripes	Rillwork S stripes

stripes. Although small circular desiccation cracks have been observed (C h a m b e r s, 1967 p. 4, 7), there is no evidence that cracking forms
well-developed circles or steps. As mentioned under "Permafrost Cracking", the present writer has seen polygonal cracks surrounding a circular central area, but in this case the circular aspect is believed to be
secondary and the pattern still a polygonal form of patterned ground.

DESICCATION CRACKING

Desiccation cracking is fissuring due to contraction by drying. It is
probably one of the most common and important patterned-ground processes, especially for formes having diameters 1 m but not excluding
larger forms (W a s h b u r n 1956, p. 848—850; 1968). In the absence of
convincing evidence that freeze drying is a critical cracking process in
the origin of patterned ground, desiccation in the present context refers
primarily to air drying. Criteria for distinguishing patterned ground of
desiccation origin are discussed elsewhere (W a s h b u r n 1968). In general, the patterns developed are similar to those resulting from other
cracking processes. It has been suggested that desiccation cracking is
also the primary process accounting for some earth hummocks, but this
hypothesis remains to be demonstrated.

DILATION CRACKING

Dilation cracking is fissuring due to stretching of surface materials
by subsurface expansion. Probably it is not a widespread cause of patterned ground but it is clearly a genetic process in places (W a s h b u r n
1968). Although the cracking is in turn commonly consequent on unequal heaving or collapse of the ground, the terms *differential heaving*
and *differential thawing* as used in table 1 imply processes in which any
associated cracking is incidental and nonessential to the origin of patterned ground. The identification of dilation cracking as opposed to other
kinds of cracking may be obvious from associated evidence of pronounced heaving or collapse. Given the presence of dilation cracking the process might lead to the various patterned-ground forms associated with
cracking, but patterns other than nonsorted polygons are probably very
rare since even nonsorted polygons of this origin appear to be uncommon.

SALT CRACKING

Salt cracking is used here for the fissuring that initiates nonsorted
and sorted polygons in rock salt and salt crusts of warm deserts. The
exact process is not established but is probably thermal contraction (H u n t

100

and W a s h b u r n 1966, p. B120—B130). There is no evidence that it
forms other than polygonal patterns, some of which occur on slopes as
steep as 37° without becoming nets (H u n t and W a s h b u r n 1966,
p. B127—B128). Although not specifically genetic, salt cracking that any
of the other cracking processes listed can explain these forms. (However,
desiccation cracking probably initiates some associated patterns in salt-
-rich qround lacking a hard crust).

SEASONAL FROST CRACKING

Seasonal frost cracking is fissuring due to thermal contraction of frozen ground in which the fissuring is confined to a seasonally frozen layer
(W a s h b u r n 1968). In a permafrost environment this layer is the
active layer. Patterned ground due to this process is probably widespread
but may be difficult to distinguish from forms developed by other processes. Criteria are discussed elsewhere (W a s h b u r n 1968). Most of
the geometric patterns conform to those developed by other cracking
processes as modified by slope conditions and presence or absence of
superimposed sorting process. In addition, it has been suggested that
frost cracking may initiate some earth humocks. There is little evidence
as to whether or not seasonal frost cracking causes stripes but reasoning by analogy with desiccation stripes, for which the evidence is better, seasonal frost-crack stripes are plausible.

PERMAFROST CRACKING

Permafrost cracking is fissuring due to thermal contraction of perennially frozen ground; i. e. it is frost cracking that extends into the permafrost and is not confined to the active layer alone. The resulting patterned ground is very widespread in polar environments and is of considerable significance because of its temperature implications and the
fact that fossil forms are more subject than most patterned ground to
preservation in the geologic record. The literature dealing with active
and fossil forms and criteria for their recognition is prolific (W a s h-
b u r n 1956, p. 850—852; 1968). The best-known examples of the process are permafrost-crack nonsorted polygons (following the terminology of table 1) of which two varieties are commonly distinguished: ice-
-wedge polygons and sand-wedge polygons. Permafrost-crack sorted
polygons can develop by activation of a sorting process along the crack in
the active layer (P é w é 1964), and perhaps also in fossil forms by melting out of ice wedges and concurrent infilling with surface stones. Nets
initiated by permafrost cracking may exist but are probably less common

than polygons, because deformation of an originally polygonal pattern would tend to be inhibited by permafrost. It is even more questionable whether the pattern could be modified to form stripes.

FROST ACTION ALONG BEDROCK JOINTS

Frost action in the present context comprises frost wedging and frost heaving where sufficiently concentrated along bedrock joint patterns, it can initiate some forms of patterned ground but they are uncommon, judging from the few reference to them in the literature (W a s h b u r n 1956, p. 846—847; 1968). Control of the patterns by bedrock joints is the essential criterion for these forms. Sorted polygons are the best known examples; in addition, sorted circles in bedrock have been observed by W i l l i a m s (1936) and the present writer. If the latter are controlled by joint intersections as suggested, they could be termed joint-crack sorted circles. The existence of joint-crack nonsorted polygons and nonsorted stripes remains to be demonstrated, but seems probable in view of the known sorted forms and the likelihood that ice wedges would develop in pre-existing bedrock joints underlying unconsolidated material. If bedrock fissuring were due to frost cracking, however, the forms would be frost-crack polygons. The fact that joint-crack patterns are confined to bedrock joints minimizes the possibility of net patterns developing by deformation of polygons.

CLASS II. CRACKING NONESSENTIAL

GENERAL

A number of hypotheses of patterned-ground origin involve processes that do not depend on initial cracking, and the more probable of these are grouped in table I. The selection is incomplete and, in view of the uncertainties involved, some important initiating processes may have been omitted. Among the general features of the group is the question as to whether or not the processes can develop strictly polygonal, as opposed to circular, net or striped petterns. Theoretically, continued mutual impingement of circles would develop polygonal patterns simulating those of crack origin, especially if secondary bordering cracks developed. However, because proof appears to be lacking that this sequence of events does in fact occur, polygonal forms in this section of table I carry a question mark. Because steps are closely related to mass--wasting, they are questioned in relation to other processes.

PRIMARY FROST SORTING

Primary frost sorting is suggested here as a term to include the ejection of stones from fines by multigelation (repeated freezing and thawing) W a s h b u r n 1956, p. 838—840), and the gradual particle-by-particle sorting that C o r t e (1963) has termed vertical and lateral sorting. A number of frost-related sorting processes (W a s h b u r n 1968) probably modify established patterns to sorted forms. The evidence with respect to initiating basic patterns is less clear and the criteria are correspondingly vague, but primary frost sorting seems more likely that most to be also an initiating process. In listing primary frost-sorted forms in table I, the usual "S" designation is omitted as redundant. Nonsorted forms of patterned ground are excluded by definition. Primary frost sorting would develop sorted circles or, if closely spaced, sorted nets. It is also uncertain whether or not primary frost sorting would produce debris islands, since normally the coarse rubble in which they occur has a scarcity of fines near the surface and the fines of the debris islands appear to be derived from depth.

MASS DISPLACEMENT

Mass displacement as applied to patterned ground is the en-masse local transfer of thawed mineral soil from one place to another within the soil as result of frost action (W a s h b u r n 1968). Suggested processes include 1. cryostatic pressure and 2. changes in intergranular pressure (W a s h b u r n 1956, p. 842—845; 854—855). The latter process appears the more probable but both remain to be demonstrated as applied to the genesis of patterned ground. Criteria for recognizing patterned ground due to mass displacement are therefore uncertain, and in practice are mainly negative, involving elimination of other origins. Supposedly mass displacement can form both nonsorted and sorted patterns. Thus, movement of a mass of fines through other fines, or through slightly coarser but not obviously stony material, could produce nonsorted varieties would depend upon movement of fines into distinctly stony material. Here, again, most of the normal geometric patterns are expectable, but strictly polygonal ones are questioned in table I for reasons noted above. Earth hummocks and steps are included but questioned because some varieties cannot be excluded from such an origin, although there is no convincing evidence for it.

DIFFERENTIAL FROST HEAVING

Differential frost heaving, as used here, is the blisterlike expansion of the ground due to intensified freezing where there are inequalities in insulating vegetation and/or snow. In table 1, the adjective frost-heave implies this process. Such frost heaving may well start many nonsorted and sorted patterns (W a s h b u r n 1956, p. 841—842). Criteria for recognizing patterned ground of this origin are largely negative. For instance, forms consiting of nonheaving mineral soil are excluded, and forms more consistent with other origins include strictly polygonal forms (unless the development of well-developed polygons by impingement is accepted) as well as forms showing en-masse extrusionlike movement of material from depth (as opposed to surficial or very shallow features).

SALT HEAVING

Salt heaving is volume expansion due to growth of salts. The process probably initiates a variety of patterned ground in warm-arid deserts (H u n t and W a s h b u r n 1966). It has many analogies with frost heaving, and is probably capable of instituting the same general type of patterns as those associated with differential frost heaving.

DIFFERENTIAL THAWING AND ELUVIATION

Differential thawing and eluviation in the present context constitute a sorting process, as between fines and stones, that stems from variable thawing rates and from preferential development of rills in coarse material. Such sorting probably initiates some small sorted nets (W a s h b u r n 1956, p. 855—856), but criteria for recognizing patterns of this origin are unsatisfactory except for shallow forms developed on thawing ice. Nonsorted forms are excluded. Of sorted forms, polygons are again questioned; sorted stripes seem possible but are not known to occur, and other sorted forms seem unlikely.

DIFFERENTIAL MASS-WASTING

Differential mass-wasting is the faster downslope mass movement of some materials than others. Little is known about the process relative to the initiation of patterned ground, but it may account for some sorted steps and stripes (W a s h b u r n 1968), and perhaps some nonsorted ones as well.

RILLWORK

Rillwork is the process that starts small channels, and as such it may also initiate some striped forms of patterned ground. In this context the process is usually limited to small sorted stripes (W a s h b u r n 1956, p. 857—858). Although most sorted stripes nomally carry drainage, in many places this is probably a result rather than a cause of the pattern. The possibility that some nonsorted stripes are due to rill is not excluded.

REFERENCES

1. C h a m b e r s M. J. R., 1967 — Investigations of patterned ground at Signy Island, South Orkney Islands: III. Miniature patterns, frost heaving and general conclusions. *British Antarctic Survey Bull.* 12; p. 1—22.
2. C o r t e A. E., 1963 — Particle sorting by repeated freezing and thawing. *Science*, v. 142; p. 499—501.
3. H a l l s w o r t h E. G., R o b e r t s o n G. K., and G i b b o n s R. F., 1955 — Studies in pedogenesis in New South Wales: VII. The "gilgai" sols. *Jour. Soil Sci.*, v. 6; p. 1—31.
4. H u n t C. B., W a s h b u r n A. L., 1966 — Patterned ground, p. B104—B133, in: H u n t C. B., and others, Hydrologic Basin, Death Valley, California. *U. S. Geol. Survey Prof. Paper* 494—B; 138 p.
5. P é w é T. L., 1964 — New type large-scale sorted polygons near Barrow, Alaska (abs.). *Geol. Soc. America Spec. Paper* 76; p. 301.
6. W a s h b u r n A. L. 1956 — Classification of patterned ground and review of suggested origins. *Geol. Soc. America Bull.*, v. 67; p. 823—865.
7. W a s h b u r n A. L., 1969 — Weathering, frost action, and patterned ground in the Mesters Vig district, Northeast Greenland. *Medd. om Grønland*, v. 176 no. 4.
8. W i l l i a m s M. Y. 1936 — Frost circles. *Royal Soc. Canada Trans.*, Sec. 4, v. 30; p. 129—132.

ERRATA

Page 437, *line 23 from the top:* "understandbale" should read "understandable."

Page 439, *line 9 from the top:* "formes" should read "forms"; "1 m" should read "< 1 m."

Page 442, *line 4 from the top:* after "salt cracking" insert "is introduced in the classification because it seems unlikely"; *line 7 from the top:* "qround" should read "ground"; *line 17 from the top:* "humocks" should read "hummocks."

Page 443, *line 7 from the bottom:* "petterns" should read "patterns."

Page 444, *line 9 from the top:* "that" should read "than"; *line 13 from the top:* after "nets." insert "The development of clearly defined polygons by impingement is questioned for reasons discussed on page 443."; *line 17 from the top:* "pattermed" should read "patterned"; *lines 7 and 8 from the bottom:* "mon-sorted" should read "non-sorted"; after "varieties," insert "whereas sorted varieties."

Page 445, *line 7 from the top:* "consiting" should read "consisting"; *line 7 from the bottom:* after "questioned" insert "(cf. page 443)."

Page 446, *line 4 from the top:* "nomally" should read "normally"; *line 7 from the top:* "Chambers, M. J. R." should read "Chambers, M. J. E."

10

Reprinted from *Biul. Peryglac.*, No. 12, 8, 71–81, 84–86 (1963)

RELATIONSHIP BETWEEN FOUR GROUND PATTERNS, STRUCTURE OF THE ACTIVE LAYER, AND TYPE AND DISTRIBUTION OF ICE IN THE PERMAFROST

A. E. Corte

[Editor's Note: In the original, material precedes this excerpt.]

INTRODUCTION

The surface of the outwash near Thule, northwest Greenland (fig. 1) is marked by four distinctive patterns, three of which are of widespread occurrence. These patterns are shown in plate 1, labeled as follows:

Pattern type 1 is characterized by depressions in the form of kettles or valleys of about one meter relief in coarse unsorted gravel and sand, spaced from 6 to 10 meters on centers. This pattern type is not listed in the international nomenclature (Commission de Morphologie Périglaciaire 1952) for patterned ground and occurs in only a few places in the Thule area.

Pattern type 2 is the very well known ice-wedge polygon investigated by Leffingwell (1919) and Black (1951, 1952, 1954) in Alaska. In the Thule area these polygons commonly have a mesh diameter of 20 to 30 meters.

Pattern type 3 consists of sorted circles or centers of fines surrounded by coarse washed particles. There is a wide size variation in those observed about Thule, ranging from a few centimeters to several meters in diameter. This pattern and its many variations have been fully described in the literature, and many conflicting theories have been advanced on its origin. Summaries and criticisms of these theories have been published recently by Cailleux and Taylor (1954) and Washburn (1956), to which the reader is referred for background material.

Pattern type 4 is characterized by elevations and depressions of low relief, without surface sorting. The humps in this pattern are flatter and less well developed than those of type 1 and are formed in finer material. Vegetation growing in the depressions, outlines the mounds more distinctly. This pattern has been described often, but little is understood as to its origin (Washburn 1956, p. 829). It is widespread in the Thule area.

[*Editor's Note:* Certain figures and plates have been omitted owing to limitations of space.]

Pl. 1. Airphoto of study area, showing ground pattern types and outlines of excavations made

[*Editor's Note:* Material has been omitted at this point.]

SUMMARY AND CONCLUSIONS

PATTERN TYPE 1: LINEAR AND CIRCULAR DEPRESSIONS IN UNSORTED OUTWASH

This pattern, described and investigated for the first time, was studied in two variant forms in Areas 1 and 5. In Area 1 the pattern was one of closely spaced, more or less circular depressions or kettles with centers spaced 6 to 10 meters apart and a local relief of about 1 meter. The surface was unsorted coarse sandy gravel with cobbles up to 20 cm in diameter.

The pattern in Area 5 was modified to form linear valleys 15 to 20 meters long, but otherwise the same description applies to both areas. Ice wedge troughs entering the two areas pinched out abruptly upon coming in contact with the pattern of depressions.

The active layer beneath this pattern was characterized by a lack of sorting at the surface and no vertical sorting of particle sizes other than the remnants of primary depositional beds. These were relatively horizontal in some instances, but often inclined and, in some cases, markedly broken and contorted. No depositional bedding was observed in Area 5, but the larger stones were oriented with their long axes conformable to the ground and ice surface contours. Fines averaged less than 1% in both the active layer and permafrost, making them not susceptible to frost heaving.

A layer of fine particles on top of the larger stones and coarser washed material beneath occurred in both the active layer and permafrost as the result of washing. A very thin layer of fines on top of the ice in the permafrost was deposited by the same process. A thin coating of siliceous calcareous evaporite on the under surface of the stones in the active layer was the result of washing and evaporation and indicates an arid or semi-arid climate during its deposition.

Removal of the active layer exposed sinuous and interwoven bands of fairly clean and bubbly ice beneath what had been elevations at the surface, and pockets of frozen sand and gravel permafrost under the depressions. Trenching the permafrost revealed that the bands extended laterally with depth to join beneath the pockets of gravel and that, therefore, the entire area was underlain by a large body of ice whose upper surface is believed to have been sculptured by thermal and aqueous erosion. This ice, apparently not discribed previously, is termed *relict ice* based on the belief that it originated as a fragment (or fragments) of glacial ice which was insulated and preserved by deposits of outwash material as the ice cap retreated. It is therefore a relict of another environment and was not formed *in situ* as ground ice.

Ice wedges entering the areas cut into the bands of relict ice but then pinched out abruptly, as did their troughs at the surface. A fine wedge found in a sample of relict ice in Area 1 appeared to be related to a small crack in the ground surface.

Fabric studies of relict ice showed it to have irregularly shaped crystals ranging in size from 0,5 to 6 cm^2. The ice generally contained many air bubbles and Tyndall figures which gave it a milky appearance, but in some cases it appeared relatively clear. Dirt particles and stones occurred at the edges of frozen gravel pockets and, in some cases, in cracks or slight depressions in the surface. In samples far from contact with other kinds

109

of ice *c*-axis orientation tended to be nearly horizontal, with concentration maxima strong and oriented normal to the axial plane of the band of ice as exposed at the permafrost table.

Ice wedges were found to have fine cracks along their axes filled with very small horizontally oriented crystals. Outside these cracks the crystals were larger, elongated and reoriented so that *c*-axes and axes of elongation dipped steeply toward the axial plane of the wedge.

Relict ice crystals in the zone of contact with a wedge were generally smaller in size and showed elongation and reorientation to nearly vertical, as in older wedge crystals. Further from the contact, grain size increased, vertical elongation became less pronounced, and concentration maxima tended to become weaker and more horizontal until a typical relict ice fabric was found about 30 cm from the contact. In general, the strongest concentrations of *c*-axes, in both relict ice and ice masses, were found in the zone of contact with ice wedges.

A lens of clear transparent ice found in one corner of Area 1 appeared to be unrelated to any element of surface pattern, although airphotos indicate the possibility that the active layer in that spot contained a greater percentage of sand and exhibited no surface pattern. The ice was clean although a few frost-shattered rocks were suspended just above the bottom of the lens. Tyndall figures were relatively few and evenly distributed through the ice, and air bubbles were almost entirely absent. Crystal size ranged from 0,25 cm² in the vicinity of some soil particles to a maximum of 80 cm² in clean ice, with an average size of about 4 cm². *C*-axis orientation was generally inclined at about 45° and appeared to be affected locally by proximity to dirt and stone particles, although the amount of data available makes generalization of the effect liable to great inaccuracy. Although the origin of the lens is unknown, it is possible that it was formed by the freezing of a small slush pond during a later period of outwash deposition.

Many stones in the permafrost were found embedded in sockets of clean, transparent and bubble-free ice whose crystal sizes averaged about 0,5 cm².

Although much evidence points to the conclusion that relict ice is a buried segment of glacial ice eroded by heat and/or water-borne sediments, such a hypothesis does not appear to explain the horizontal orientation of *c*-axes into one or two maxima in a plane normal to the axial plane of the bands exposed at the permafrost table. If pressures within the permafrost were responsible, *c*-axes should be vertical as they are in ice wedges or in other types of ground ice affected by the lateral pressure of wedge growth.

Since the outwash in which this pattern of depressions was studied was almost free of fines (non-frost-susceptible), it would be interesting to investigate such glacial ice covered by materials containing a higher percentage of fines to see what type of pattern is formed.

PATTERN TYPE 2: POLYGONAL TROUGHS FORMED IN UNSORTED OUTWASH

Polygonal troughs, the best known form of arctic ground pattern, occurred in all areas studied, either in fully developed form or as fragments which ended upon contact with another pattern type. Easily recognized from the air, these polygons ranged in size from 3 to 30 m in diameter and appeared to be most fully developed in well-drained, somewhat elevated, areas.

The active layer associated with this pattern is ·characterized by a lack of sorting at the surface (cobbles are concentrated within the troughs, but the polygons themselves are unsorted), no vertical sorting of fines and coarse material within the active layer, more or less continuous primary depositional bedding, and an average of less than 1% fines. For descriptive and comparative purposes, an active layer with these characteristics is designated as „undisturbed". Deposition of fines on the tops and siliceous calcareous evaporite on the under surfaces of stones in the active layer, plus accumulations of coarse sand beneath them, are features of washing and evaporation.

The permafrost, composed of the same materials as the active layer, contains ice wedges beneath the surficial troughs. In the areas studied, both troughs and wedges decreased markedly in size, and in most cases pinched out entirely, upon contact with other pattern types and other kinds of ice in the permafrost. In some cases the wedge continued through the other type of ice in very decreased size until it again reached the undisturbed non-frost-susceptible active layer, where it increased to fully developed size. Where the wedge decreased to a few centimeters in width, its surface trough decreased to a small, and in some cases discontinuous, crack.

Fabric studies revealed recent thermal contraction cracks along the axial plane of the wedge filled with small horizontally oriented crystals 1 to 3 mm^2 in size. On both sides of these cracks the older wedge crystals were vertically elongated to 2 to 3 cm and had a strong vertical orientation, dipping steeply toward the wedge axial plane. A variation of about 20° was found between the directions of orientation and grain elongation in several wedge ice samples, with the direction of orientation falling on either side of the direction of elongation. The stresses involved in wedge growth

111

are most likely horizontal and would therefore account for the growth
and re-orientation of the older grains at a normal or slightly acute angle
to the horizontal.

The fact that ice wedges are wider and more fully developed in coarse,
non-frost-susceptible material than in fine material containing ice masses
belies the physical principle suggested by Black (1951) that an increase
in the coefficient of expansion is to be expected when there is an increase
in the ice content of the soil. This apparent contradiction and the disappearance of ice wedges in areas of fines may perhaps be explained in terms
of a higher tensile strength in soils containing fines and of the rate of temperature change which causes thermal contraction. Temperature changes
would be slower in an area of fines because the soil would hold more moisture than a coarser area. Unequal distribution of stresses because of the
presence of pockets of coarse material may also be a factor in a sorted active
layer.

PATTERN TYPE 3: SORTED CIRCLES OR CENTERS OF FINES

Sorted circles are most obvious and commonly found in slightly depressed
areas where the surface is otherwise composed of washed cobbles and boulders. The centers may be isolated or closely spaced, a few centimeters
up to 4 meters in diameter, and the surfaces raised, level, or depressed.

The center of fines is caused by plugs of fine material within the active
layer which break through to the surface. The active layer in such areas
is vertically sorted as a result of washing, grading from coarse at the top
to fine at the bottom. Fines are deposited on top of stones and on the permafrost table, and washed out from beneath the larger stones to leave an
accumulation of coarse sand and gravel. No evaporite deposits were found
on the underside of stones, an indication of the moisture-retaining quality
of an active layer rich in fines.

Only fragmentary sedimentary bedding was found, the layers broken,
inclined, and contorted. Plugs averaging 4% fines contained fragmentary
and contorted sand and silt beds. In light of the above characteristics,
this frost-susceptible active layer is described as ,,disturbed''.

Roundness and sphericity indices determined for soil samples of two
different grain sizes from both the undisturbed and disturbed active layer
of pattern type 2 and the disturbed active layer of pattern type 3 indicate
that the materials of both kinds of active layer were derived from a common
source and have a common history of transportation and deposition.

The permafrost beneath centers of fines is characterized by more or
less continuous amorphous ice masses containing symmetrical or con-

112

torted layers of fines. Ice wedges, if present, are considerably reduced in size.

Mass ice is transparent and large-grained, with crystals up to 10 cm² in size where unaffected by silt layers. Fine crystals of a fraction of a square millimeter occur between sand and silt grains and along dirt layers, a direct and local effect of the dirt inclusions. Grain shape is irregular in general, but tends to be slightly elongate along silt bands at the contact with an ice wedge.

C-axes tend to be oriented at angles of 25 to 45° to the vertical, with a slight concentration in a plane normal to the trace of the silt layers as seen in a diagram of vertical thin sections. One fabric diagram did show a horizontal concentration in the same plane as the trace, but this was not found in other diagrams. Vertical diagrams show a concentration minimum in the horizontal plane regardless of the orientation of the silt layers, but analysis of more thin sections is necessary before it is safe to make generalizations concerning the relationship between *c*-axis orientation and the orientation of silt layers.

At the contact between an ice mass and a wedge, mass grains are smaller in size with a strongly preferred orientation parallel to wedge crystal *c*-axes, apparently a direct result of the intrusion of the younger ice wedge. Within 10 cm of the contact, however, its influence is lost and a typical mass fabric is found.

The formation of centers of fines is possibly the result of down-washing of fines through the active layer, concentration at the permafrost table and then extrusion caused by differences in density and/or thermal properties between the coarse material at the surface and the fines beneath. Laboratory experiments are necessary, however, before any definite conclusions can be drawn. Sorted circles have been produced in experiments conducted by the author in a layer of gravel of varying thickness overlying an ice sheet (Corte 1959). In this case the sandy gravel was non-frost--susceptible and sorting was produced mechanically by differential melting of the ice and collapse of the gravel layer. Although somewhat similar patterns were formed, mechanical sorting would seem to play a minor part, if any, in the formation of centers of fines produced in a frost-susceptible active layer where vertical sorting, a difference in thermal properties and/or specific gravity, and frost-heaving apparently play the significant roles. Recently Corte (1961) have demonstrated horizontal and vertical sorting by changing the orientation of the freezing plane.

The transition zone between pattern types 2 and 3 exhibits characteristics of both disturbed and undisturbed active layers. Sedimentary beds are inclined or slightly disturbed, the surface is somewhat rough and uneven

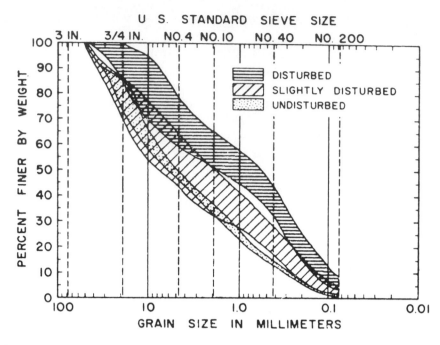

Fig. 36. Grain size range of disturbed, slightly disturbed and undisturbed active layers
of Area 3 and 4

Curves for the disturbed type are based upon analysis of plugs only

although unsorted, the percentage of fines in the active layer varies from 2 to 5%, and scattered ice masses occur in the permafrost. Vertical sorting is evident but usually incomplete, and plugs of fines begin to appear in the active layer but do not reach the surface.

Plotting average grain size curves for the three types of active layers analyzed in Areas 3 and 4 provides the range of grain sizes for each type (fig. 36). The finer limit of each range occurred in Area 3 and the coarser in Area 4. Although there is some over-lapping of ranges in the coarser fractions, particularly in the undisturbed and slightly disturbed active layers, the separation is almost complete in the fractions finer than medium sand (no. 10 sieve size). It is apparent from figure 36 that disturbance due to frost action may begin to appear in an active layer averaging only 2% fines. Recent studies by other researchers have also pointed out the advisability of lowering the frost-susceptibility limit under certain conditions to less than 2% of the fraction finer than 0,02 mm (Dücker 1958).

PATTERN TYPE 4: MOUNDS AND DEPRESSIONS OF LOW RELIEF IN FINE-GRAINED
UNSORTED OUTWASH

Although the surface was unsorted in this pattern, vertical sorting
was well-developed, both within plugs and between them, runnig as high
as 28% fines at the bottom of plugs and 5% at the top. The active layer
was highly distrubed, with traces of depositional beds present only between
plugs where soil material averaged only 2% fines. Where the percentage
of fines was higher, as in plugs, these beds were obliterated.

The permafrost beneath this pattern was composed almost entirely
of ice masses, with just a few small isolated pockets of frozen sand and
gravel. A layer of fines occurred regularly on top of the ice masses.

No relationship was apparent between features of either the active
layer or the permafrost and the surface pattern of low mounds and depressions. The one plug of fines which reached the surface did so in a depression,
but others reaching to just beneath the surface showed no relationship
to surface features. Inasmuch as the active layer was disturbed and vertically sorted to the surface by frost action and washing, the surface pattern
must also be considered the result of frost action. Failure of the surface

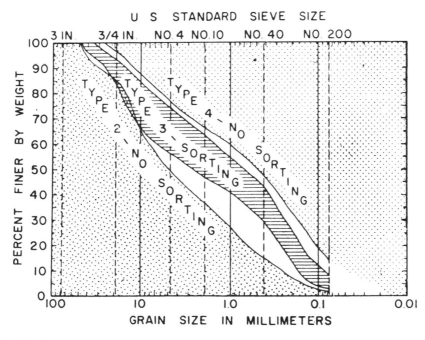

Fig. 37. Sorting in outwash as a function of grain size, based upon studies near Thule,
Greenland. Presumably further studies would result in more clearly defined borders
between sorting and no sorting in the finer grain sizes

Table VI

Comparative features of four pattern types

	Pattern Type 1	Pattern Type 2	Pattern Type 3	Pattern Type 4
Surface	Unsorted; patterned with network of circular or linear depressions up to 1 m deep	Smooth, unsorted, patterned with network of polygonal troughs	Uneven, sorted, patterned with centers of fines in coarse washed material	Largely unsorted. Few large cobbles or boulders. Pattern of low irregular mounds and depressions with relief of up to 30 cm
	Primary depositional bedding	Primary depositional bedding	No primary depositional bedding	No primary depositional bedding
	No plugs of fine material	No plugs of fine material	Plugs of fine material errupt to surface	Plugs of fine material. Do not generally reach the surface
Active layer	No vertical sorting Percent of fines {top — < 2% / mid. — < 2% / bot. — < 2%}	No vertical sorting Percent of fines {top — 2% / mid. — < 2% / bot. — < 2%}	Vertical sorting Percent of fines {top — 5% / mid. — 11% / bot. — 14%}	Vertical sorting Percent of fines {top — 5% / mid. — 12% / bot. — 28%} Plugs {top — 2% / mid. — 1% / bot. — 3%} Between plugs
	Very slight accum. of fines at permafrost table	No accum. of fines at permafrost	Accum. of fines at permafrost table	Large accum. of fines at permafrost table
	Siliceous calcareous evaporite on under surface of stones	Siliceous calcareous evaporite on under surface of stones	No siliceous calcareous evaporite on under surface of stones	No siliceous calcareous evaporite on under surface of stones
	Wedges usually pinch out upon contact with relict ice	Well-developed ice wedges up to 1 m wide	Few or no ice wedges — maximum of a few cm wide	Few or no wedges (none observed) — pinch out upon contact with masses
Ground ice	No ice masses	No ice masses	Ice masses	Dense conc. of ice masses
	Relict ice	No relict ice	No relict ice	No relict ice

A. E. Corte

Table VII

Comparative features of four types of ground ice

	Relict ice	Ice wedge	Ice mass	Ice lens
Gross shape	Large continuous body, surface sculptured into circular or linear depressions	V-shaped bands, usually connected to form polygons. Wedges up to 1 meter wide	Amorphous. May be continuous or isolated. Size variable	Lenticular in horizontal plane
Bubble content	Variable, usually fairly high	High	None	None
Crystal size	0.5 to 6 cm²	Young crystals 1 to 3 mm² Old crystals 0.5 to 3 cm²	Up to 10 cm² in clean ice, less than 1 mm² around dirt particles	0,25 cm² close to dirt particles, up to 80 cm² in clean ice
Crystal shape	Irregular	Elongate	Irregular	Irregular
Inclusions	Dust; some silt and sand or pebbles at permafrost table	None except at sides of wedge	Silt, sand and pebbles in layers or randomly distributed	Some silt streamers and frost-shattered cobbles near base of lens, but generally very clean
Fabric *	C-axes approx. 45° to the vertical in a plane normal to the axial plane of the band as exposed at the permafrost table. One or two maxima with concentrations up to 25%	Young crystals in recent thermal contraction cracks are oriented horizontally and older crystals at each side are oriented near to the vertical, dipping steeply toward the axial plane of the wedge	Relatively weak concentrations of c-axes showing random orientation. Commonly a minimum in the horizontal plane	C-axes oriented at roughly 45° to the vertical with minima in the polar and equatorial planes. Concentrations may be numerous and scattered although one is usually stronger than the others and may be up to 18%

* General fabric pattern of the individual ice type is given. In contact with another type of ground ice the fabric may be affected greatly, as may also the crystal size and shape. Dirt inclusions may also affect crystal size, shape and orientation.

to develop sorting is possibly a result of the lack of great differences in density or thermal properties between the upper and lower portions of the active layer. As the boulder and large cobble fractions were absent from this area the large density or thermal difference possible in Areas 3 and 4 could not develop and therefore could not cause extrusion of plugs of fines at the surface. Since only the material in the plugs was subject to mechanical analysis for pattern type 3, it is evident that the average percentage of fines for this type of active layer as a whole is considerably smaller than indicated by the grain size curves. Under these conditions it would seem advisable to raise the grain size limit defining fines from 0,02 to 0,074 mm and to maintain the percentage by weight limit at 2% to be relatively sure of obtaining a non-frost-susceptible soil material. It is also apparent that a large range in particle size, and particularly the presence of large quantities of cobbles and boulders is important either as a density or thermal unit in the formation of this pattern.

Comparative features of the four pattern types discussed above are given in table VI and those of the four types of ground ice in table VII. The descriptions given for each feature refer only to the archetype and do not represent the variations caused by the effect of one pattern or ice type upon another.

By plotting grain size curves from areas of pattern types 2, 3 and 4 (fig. 37), it is possible to define the limits of sorting and, roughly, of frost--susceptibility in terms of the percentage of fines. The sorted surface of the active layer is produced when the average percentage of fines in the plugs is between 3 and 8%. Although there may be a slight disturbance and vertical sorting of an active layer containing 2 to 3% fines, those containing less than 2% finer than 0,074 mm are generally not sorted or disturbed by frost action. Those containing more than 8% fines are highly susceptible to frost action although they may be unsorted at the surface. It is, however, a relatively simple matter to identify such an area from either a surface examination or from airphotos on the basis of its surface pattern.

[*Editor's Note:* Material has been omitted at this point.]

References

Black, R. F. 1951 — Structures in ice wedges of northern Alaska. *Bull. Geol. Soc. Amer.,* vol. 62; p. 1423—1424.

Black, R. F. 1952 — Growth of ice-wedge polygons in permafrost near Barrow, Alaska. *Bull. Geol. Soc. Amer.,* vol. 63; p. 1235—1236.

Black, R. F. 1954 — Ice wedges and permafrost of the arctic coastal plain of Alaska. Unpublished ms., 2 vol., 354 fig., 788 p. (Results of work under auspices of USGS).

Cailleux, A. & G. Taylor 1954 — Cryopédologie, Etude des sols gelés. 218 p., Paris. Canada, National Research Council. Division of Building Research 1957 — Frost action in soils. *Canadian Builder,* vol. 7, no. 9; p. 49.

Commission de Morphologie Périglaciaire 1952 — Rapport préliminaires pour la 8ᵉ assemblée générale et le 17 congrès international. Union Géogr. Internationale, Washington, D. C.

Corte, A. E. 1959 — Experimental formation of sorted patterns in gravel overlying a melting ice surface. U. S. Army Snow, Ice and Permafrost Research Establishment, *Research Report* 55; 15 p.

Corte, A. E. 1961 — 1962 — Laboratory and field data for a new concept on the frost behavior of soils, Part I: Vertical sorting (1961). Part II Horizontal Sorting (1962). U. S. Army Snow, Ice and Permafrost Research Establishment, Corps of Engineers, *Research Report* 85.

Dücker, A. 1956 — Gibt es eine Grenze zwischen frostsicheren und frostempfindlichen Lockergestein? *Strasse u. Autobahn,* Bd. 6; p. 78—82.

Leffingwell, E. De K. 1919 — The Canning River region, northern Alaska. *U. S. Geol. Survey, Professional Paper* 109; p. 179—243.

Washburn, A. L. 1956 — Classification of patterned ground and review of suggested origins. *Bull. Geol. Soc. Amer.* vol. 67; p. 823—865.

11

Reprinted from *Geog. Bull.*, **18**, 21–25, 52–58, 60–63 (Nov. 1962)

PINGOS OF THE PLEISTOCENE MACKENZIE DELTA AREA

J. Ross Mackay

ABSTRACT: Pingos are stable intrapermafrost ice-cored hills. Most were formed over a thousand years ago as an indirect result of the shoaling of lakes by geomorphic and climatic processes. As the lakes gradually shoaled and lake ice froze to the bottom in winter, an impermeable permafrost cover was formed, ab initio, over the unfrozen·saturated sediments beneath.

Downward aggradation of the newly formed permafrost cover was predominantly in fine to medium sands which were not susceptible to extensive ice segregation. Consequently, the pressure of expelled pore water, trapped in a closed system, was relieved by an upward doming of the thinnest part of the overlying permafrost cover, which normally coincided with the deepest part of the lake. Water in the dome froze to form the pingo ice-core. Calculations based on ground and lake temperatures and the nature of the sediments show that pingo growth was slow, probably involving tens of years for the larger pingos.

The large pingos reported in the literature as occurring in the Mackenzie delta are, in fact, not in the modern delta but in Pleistocene sediments to the east. A group of small pingos, which have not been described in detail do, however, occur in the modern delta.

MS submitted, March, 1962.

RÉSUMÉ: Les pingos sont des buttes formées de lentilles de glace qui s'étant individualisées à l'intérieur du pergélisol s'y sont maintenues par la suite. Pour la plupart, leur formation remonte à au-delà d'un siècle et sont dues indirectement à l'assèchement des lacs par les agents climatiques et morphologiques. Au cours du processus d'assèchement, les lacs gèlent complètement en hiver et, au début, amènent la formation d'une couche superficielle de pergélisol au-dessus des sédiments saturés et non gelés.

La progression en profondeur de la couche superficielle de pergélisol s'effectue dans des matériaux sableux dont la granulométrie varie de fine à moyenne,et qui ne sont pas susceptibles de former de nombreuses lentilles de glace. Par conséquent, la pression exercée par l'eau capillaire retenue dans ce circuit fermé donne naissance à un soulèvement en forme de dôme, là où la couche superficielle du pergélisol est la plus mince. Cette couche coïncide d'ordinaire avec la partie la plus profonde du lac. L'eau à l'intérieur du dôme gèle et forme alors une lentille de glace qui constitue le noyau du pingo. Les mesures de température prises sur le sol et l'eau du lac ainsi que les renseignements recueillis sur la nature des sédiments démontrent que la formation de ces pingos est très lente; dans le cas des pingos plus considérables, la période de formation serait de 10 ans.

Il est souvent fait mention dans les écrits scientifiques des gros pingos du delta du Mackenzie; leur formation ne remonte pas à l'époque du delta actuel, mais bien à l'époque pléistocène, tout comme d'ailleurs les sédiments situés plus à l'ouest. On rencontre dans le delta récent un réseau de petits pingos dont la description détaillée n'apparaît pas dans la présente étude.

Pingos are ice-cored hills, typically conical in shape. They are relatively stable intrapermafrost features, normally enduring from hundreds to thousands of years, thus differing from the smaller seasonal frost blisters or icing-mounds which form above the permafrost surface in winter. Pingos are abundant and widely distributed in the Mackenzie delta area in: (a) the distal part of the modern Mackenzie delta, and (b) the area of Pleistocene deposits to the east (Figure 1). The pingos in the two areas differ in age, size, and details of formation. This paper is concerned primarily with the older and larger pingos of the Pleistocene area; it describes their distribution and characteristics, discusses theoretical and practical aspects of their origin, and estimates their age.

TERMINOLOGY

The Eskimo word pingo, applied locally by natives to conical mounds in the Mackenzie delta area, was proposed by Porsild (1938) as a technical term for ice-cored hills. The term is now widely adopted (e.g. Muller, 1947, p. 220; American Geological Institute, Glossary of Geology and Related Sciences, 1957, p. 221). The Russian word *bulgunniakh* is synonymous with pingo (Grave, 1956). Some writers use hydrolaccolith and pingo interchangeably (Cederstrom and others, 1953, p. 8) but to do so is undesirable, because of the genetic connotation involved. As Müller (1959, p. 118) stresses, pingos are intrapermafrost mounds, lasting for many years. Therefore, annual mounds, such as frost blisters and icing-mounds, which grow in the active layer in winter, by definition are not pingos.

At the present time, two genetic types of pingos are generally recognized. Leffingwell (1919, pp. 153–155) has attributed the origin of mounds (pingos), which he studied in Northern Alaska, to hydraulic pressure that bowed up the overlying strata. Porsild (1938) has pointed out that the typical pingos of the Mackenzie delta area are different from the hydraulic type of Northern Alaska and similar mounds in East Greenland. Müller (1959), in his comprehensive study of both kinds of pingos, classifies the above types into the "open system" or East Greenland type and the "closed system" or Mackenzie type. The pingos of the Pleistocene Mackenzie delta region are of the closed system type, but some pingos do not fit neatly into either category.

DISTRIBUTION

There are over 1,400 pingos in the Mackenzie delta area (Figure 1) between the west side of the delta, at the Yukon-Northwest Territories boundary, and Nicholson Peninsula on the east. The area involved is a northeast-southwest coastal zone some 200 miles* (320 km) long and 50 miles (80 km) wide. The pingos occurring along the Yukon coast, west of

---

*Note: The cgs system has been used for computations and the presentation of data for the sake of efficiency and simplicity. Insofar as almost all measurements and graphs are concerned, data shown in metres can be read as yards without significant loss of accuracy

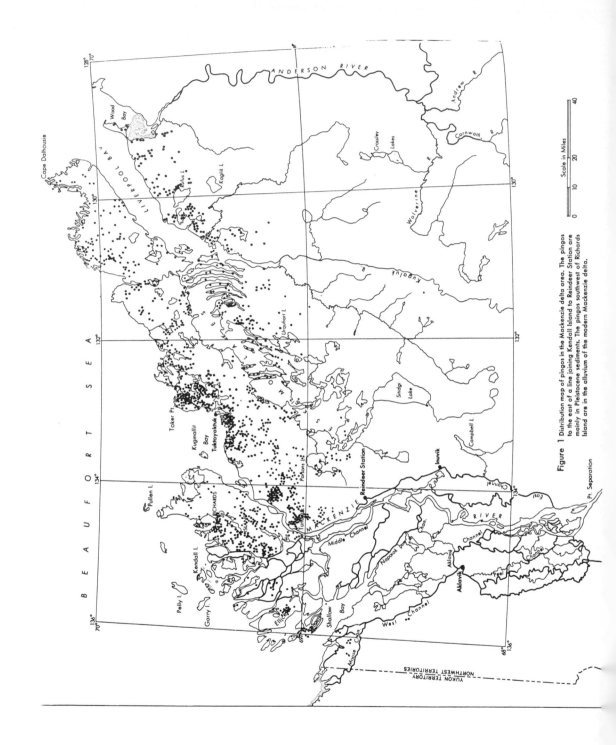

Figure 1 Distribution map of pingos in the Mackenzie delta area. The pingos to the east of a line joining Kendall Island to Reindeer Station are mainly in Pleistocene sediments. The pingos southwest of Richards Island are in the alluvium of the modern Mackenzie delta.

the Mackenzie delta, may not exceed ten. To the east of Nicholson Peninsula, there are very few pingos (Mackay, 1958a, pp. 54–56).

The largest group of pingos, numbering over 1,350, lies in a belt extending east from Richards Island through Tuktoyaktuk Peninsula to Cape
Dalhousie; this group includes those pingos on the south side of the Eskimo
Lakes. Stager (1956) has plotted the locations of about 1,380 pingos in this
area, and the writer about 1,350, the slight discrepancy arising from the
definition of size used to delimit the smallest pingo. As the figures should
be considered a minimum, rather than a maximum, the total number of
pingos is 1,350 to 1,400. and occurs in an area of Pleistocene sands and silts
with rolling relief. A second group of pingos, numbering more than 70,
occurs on the seaward part of the modern Mackenzie delta below storm
level, with their bases usually within 1.5 metres of sea level. These pingos
are within 15 miles (24 km) of the coast and lie southwest of Richards
Island. A third group of about 10 pingos is found on an older part of the
modern Mackenzie delta near the lower course of West Channel. There are
several pingo-like hills along the eastern slopes of the Richardson Mountains
at 68°29'N and near Mount Goodenough (Fraser, 1956), and one or more
pingos in the southern part of the Port Brabant map-sheet of the National
Topographic System. This paper deals with the first and largest group of
pingos, with only passing reference to other groups.

CHARACTERISTICS

Occurrence in Depressions

The most distinctive terrain characteristic of the pingos is their occurrence in flat low-lying areas. With very few exceptions, the low areas are
present or former lake basins or channels with poor drainage. Every pingo
seen in the field has been in a depression, and every mound, identifiable
from airphotos as a pingo, has also been in a depression (see Frost, 1950,
p. 34 for Alaskan pingos). Typically, the flats are broken up into tundra
polygons, normally in the depressed centre stage of development. High-
centred peaty tundra polygons are most abundant in the drier land around
the bases of the pingos. A few pingos protrude as islands, usually near the
centres of shallow lakes. No pingo has been observed rising as an island in
a deep lake, such as a lake more than 3 metres deep. About 56 per cent
(Stager, 1956, p. 16) occur as islands or are partially surrounded by water.
The remaining 44 per cent are surrounded by poorly drained tundra polygon
flats. A small percentage of the pingos, possibly less than 0.1 per cent, are

in abandoned drainage channels. Several are in modern floodplains, for example, of Holmes (Peters) Creek, but these may be of the East Greenland type. No feature, which unequivocally could be identified as a pingo, has been observed either in a large lake (e.g. 1 mile or 1.5 km in diameter) surrounded by water considerably deeper than the winter thickness of ice, nor on top of a positive relief feature, such as a mesa-like tableland. The nearly perfect correlation between pingos and present or former lake basins and channels has obvious implications on their origin.

The correlation between areas of pingo concentration with completely or almost completely drained lakes is borne out by their irregular distribution. Most drained or partially drained lakes are found in areas where a river system, even though poorly integrated, has developed in postglacial time, or else where coastal recession has initiated drainage. Areas with interior or no visible surface drainage tend to have few old lakebeds.

On Richards Island, the main concentration of pingos is on the west half where drainage is most complete (Figure 1). The eastern and northeastern part has numerous relatively deep lakes with few streams, and consequently few old lakebeds or shallow lakes. On the east side of East

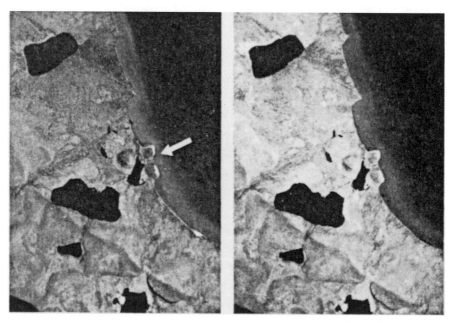

Figure 2   The stereo pair shows four pingos on the north shore of the Eskimo Lakes at 69° 02' N, 133° 12' W. The three large pingos, two of which are wave cut, exceed 50 feet in height. A tiny pingo arises on the edge of the lake in the center of the group. (RCAF photos A 12918 – 81 and 82)

Channel, between Tununuk and Holmes (Peters) Creek, the concentrations are related to stream drainage. The high density of pingos, about 10 per 10 square miles (26 square km), between Kittigazuit and Toker Point, is in a coastal region where cliff recession and stream action have partially drained large areas. Tuktoyaktuk Peninsula, east of Toker Point, has a pingo density averaging less than 1 pingo per 10 square miles. Much of the region is low and flat, with either a very poorly integrated drainage system, or no visible surface drainage. Thus, the inferred conditions suitable for pingo growth are largely absent. Pingo concentrations along the north and south coasts of the Eskimo Lakes are closely associated with shore-cliff recession and stream action.

There is a general relationship between the density of depressions "suitable" for pingo growth and the number of pingos present. The data in Figure 3 has been computed and plotted from a random sample of 25 air-photos. The area studied on each sample photo was about 30 square miles (78 square km), the total sampled area covering about 750 square miles (1950 square km). For each 30 square miles, the limits of all depressions containing pingos were outlined, the boundaries following as closely as possible the estimated extent of the depressions which generated the pingos. Then all similarly appearing areas without pingos on the air photos were marked. As shown in Figure 3, the 25 points can be represented reasonably well—for the range of data—by the line.

$$y = 10^{.071x - .283} \qquad (1)$$

where y is the pingo density per 10 square miles and x is the depression density per 10 square miles. (Note: unit areas of 10 square miles are used to avoid, as far as possible, "fractional" pingo densities of less than one pingo per unit area). Although a certain degree of subjectivity was unavoidably involved in the sampling procedure, the results suggest that the probability of finding a nearby pingo, if one pingo is selected at random, increases non-linearly with the depression density or number of "suitable" pingo depressions. For example, where the number of depressions considered suitable for pingo growth averages 5 per 10 square miles, only about one depression out of 5 (20 per cent) will usually contain a pingo. On the other hand, if there are 20 suitable depressions per 10 square miles, about 13 of the 20 depressions (65 per cent) will usually contain pingos. This implies unusually favorable conditions for the growth of pingo fields in certain areas.

[*Editor's Note:* Material has been omitted at this point.]

Hydrostatic Pressure

The generally accepted theory for pingo formation depends upon up-doming by hydrostatic pressure. The hydrostatic pressure may result from a hydraulic head associated with slopes (East Greenland type) or pressure resulting from volume expansion of water on freezing as permafrost advances into a closed unfrozen pocket (Mackenzie type).

Laboratory and field experiments carried out on the freezing of saturated sands shows that excess pore water is forced to recede before constantly growing ice crystals of an advancing freezing plane (Balduzzi, 1959; Kosmachev, 1953; Petrov, 1934). The surplus water is pressed into the unfrozen material, thus building up a hydrostatic pressure. The minimum permeability required for excess water to be squeezed out has been estimated at about 0.1 metre per day (*Foundations of Geocryology*, Vol. 11, 1959, pp. 33–35). The soils in which the pingos have grown are non-cohesive and of a grain size suitable for the rapid expulsion of pore water.

If V is the volume of unfrozen material in a closed system and η the average porosity, then the volume of expelled pore water (V_e) cannot exceed about a tenth the volume of ice.

$$V_e < 0.1 \; \eta V$$

The amount of expelled pore water could be considerably less than $0.1\eta V$, from such causes as local ice lensing in thin silty laminae and soil grains being pushed apart by some in situ freezing.

Development of a Closed System

The development of a closed system whereby expelled pore water is trapped under pressure would seem to require only the growth of a continuous permafrost seal on the lake bottom (Figure 15 B). Expelled pore water could, therefore, not escape upward through the impermeable permafrost seal; it could not go sideways because of permafrost extending out from the shore beneath the unfrozen sediments; and it could not escape downwards, because of either the presence of permafrost, saturated soils, or both. There is no need to have the unfrozen material, beneath the surface cover, completely enclosed by permafrost in order to entrap expelled pore water, because the unfrozen sediments below the lake would already be saturated so there would be no easy downward escape for expelled pore water. It would be most difficult for expelled water to escape at depth beneath the permafrost of adjacent areas and out to sea, especially since geomorphic evidence suggests relatively impermeable sediments beneath the sands.

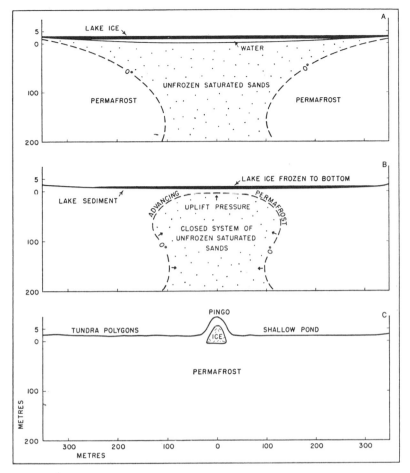

Figure 15. In diagrams A, B and C a vertical exaggeration of five has been used for the height above zero in order to show the lake ice and the open pool of water. Figure 15 A. A broad shallow lake has an open pool of water in winter with a frozen annulus around it. No permafrost lies beneath the centre of the lake. Figure 15 B. Prolonged shoaling has caused the lake ice to freeze to the bottom in winter and has induced downward aggradation of permafrost. Infilling has raised the lake bottom slightly. The deepest part of the lake, which has the thinnest permafrost, is later domed up to relieve the hydrostatic pressure. Figure 15 C. The pingo ice-core, being within permafrost, is a stable feature. The old lake bottom is occupied by tundra polygons and shallow ponds. Because of scale changes in the diagram, the volume of the ice-core should not be construed as showing a direct relationship to the initial volume of unfrozen material.

Working Hypothesis of Pingo Formation

Pingos are believed to have formed primarily in lake basins with unfrozen central portions which subsequently acquired a surface layer of permafrost to form a closed system. The initiation of permafrost along a lake bottom would seem to require either exposure of the bottom to air, as by drainage, or freezing of winter ice to the bottom for a sufficient duration each year so that mean bottom temperatures were below 0°C (Figure 15 B).

Exposure of a lake bottom to air temperatures has usually resulted from draining due to stream erosion or indirectly from coastal recession. Freezing of ice to the bottom could be initiated in many ways, besides drainage, such as from a climate change with lessened precipitation or greater evaporation or from infilling. The process of infilling in causing the formation of permafrost is believed to be relatively unimportant (*see* Porsild, 1938; Müller, 1959, p. 99). Field examination of numerous coastal sections of lake bottoms shows that the total amount of infilled lake sediments is usually of minor consequence.

For reasons previously given, a new permafrost cover beneath a lake bottom would tend to be thinnest in the centre, with a dome-shaped lower permafrost surface (Figure 15 B). With downward aggradation of the permafrost cover, and a rise of the permafrost surface beneath the unfrozen sediment, water would tend to be expelled from pores as a result of volume expansion due to freezing. As the unfrozen sediments would be in a saturated condition in a closed system, surplus water could not escape until 'room' was made available to it. Thus, hydrostatic pressure is believed to have been relieved by uplift of the permafrost layer where it was thinnest, namely at the top of the dome-shaped permafrost surface. If such were the case, then the thickness of the overburden over the ice-core would indicate the approximate thickness of permafrost when pingo growth began.

In the larger lake basins, the central portions could be completely unfrozen at all depths, although permafrost would surround the unfrozen centre (Figure 15 A). If a permafrost surface cover were to develop, the unfrozen portion would be shaped like an hourglass. Inward migration of the permafrost neck of the hourglass could produce two closed systems. Therefore, pore water could be expelled from a number of 'different' permafrost surfaces.

Lakes with irregular bottoms might have more than one deep pool, so that more than one dome-shaped lower permafrost surface might develop. Multiple pingos are believed to have grown in such lakes. Likewise, if the deeper pools were off-centre, the pingos would likewise be off-centre.

As shown in the theoretical discussion of permafrost, the rate of formation might be taken as proportional to the square root of time (equation 23). Thus the rate of downward aggradation of permafrost would be relatively rapid when perma frost was thin (e.g. .5 metres) and the temperature gradient high, but it would be relatively slow, when permafrost had extended down to greater depths, e.g. 50 metres. Therefore the expulsion of pore water, from downward freezing, would gradually slow down from this cause alone. In addition, the volume of unfrozen sediment would normally decrease with depth, so that equal increments of permafrost growth might not contribute equal amounts of expelled pore water.

As an illustration of the probable slowness of permafrost growth, consider the freezing of 15 metres of sediment with a porosity of 30 per cent and a conductivity of the frozen ground of 9×10^{-3} cgs. units. Equation 23 becomes

$$t \sim \frac{2.5 \times 10^9}{-T_T}$$

For a ground surface temperature (T_T) of $-5°C$, it is about 16 years; for $-1°C$, t is about 80 years; and for $-0.1°C$, t is about 800 years. Although the figures cannot be more than approximate ones, they do suggest that during prolonged shoaling of a lake, whether through geomorphic or climatic causes, downward penetration of permafrost will be slow, because mean lake bottom temperatures will oscillate around $0°C$ and then gradually drop below it as lake ice freezes for longer and longer periods to the lake bottom. (Note: in the derivation and use of equation 23, above, the ground beneath the advancing permafrost surface is assumed to be at $0°C$, and geothermal effects are neglected. It may be shown that the relative magnitudes of the results are not appreciably affected by these simplifying assumptions, except for temperatures near $0°C$).

With the development of permafrost on the lake bottom the permafrost surface beneath the unfrozen sediments would gradually rise in response to lowered temperatures. However, it is clear that the temperature gradient is much greater in the initial stages of permafrost aggradation, in the lake bottom layer of permafrost, and therefore its contribution to the expulsion of pore water should normally be the greater.

Pingo ice-cores seem, in general, to be of nearly pure ice, quite unlike the dirty tabular ice sheets which may occur nearby. The implication is that pingo ice forms from the freezing of a pool of water. However, there is no reason to believe that pools of water of the sizes of the ice-cores ever existed at one time, except, perhaps, in the smallest pingos. Rather, pingo ice-cores are believed to have grown slowly over many years or decades by the constant freezing of a replenished pool.

The rate of pingo growth cannot be rapid, except in very special cases. There are reports from Russia (Suslov, 1961, p. 144) of a growth of 1.5 to 2 metres for small pingos in 20 years with larger pingos reputed to develop in several dozen years, but no verifiable reports are available for the Mackenzie area. However, a consideration of the volume of pingo ice can lead to an estimate of the volume of unfrozen material required to produce the pingo, and hence to the downward penetration of permafrost. For example, if the ice-core in Ibyuk pingo is assumed to be a right cone about 40 metres high with a base 70 metres in radius, the volume would be approximately 200,000 cubic metres. If this represented a 10-per-cent-volume expansion of soil with 30-per-cent porosity, the required volume of unfrozen material would have been about 7,000,000 cubic metres. Assume that the unfrozen material was in a circular lake and that it was in the shape of a right cone with a slope of 45°. If so, the radius and depth of the cone would have been 190 metres. The freezing of such an unfrozen cone would have taken many decades. If the radius of the cone was taken as 300 metres, the depth of the cone would have been 75 metres. Again, freezing to such a depth, to which should be added the frozen layer which existed prior to the development of doming, would be long.

In Figure 16, the graph shows the volume of unfrozen sediment, assuming a 30-per-cent porosity and 10-per-cent-volume expansion, to form conical pingo ice-cores of given radii, where the radii are equal to the heights. Large pingo ice-cores involve very large volumes of unfrozen sediments. In

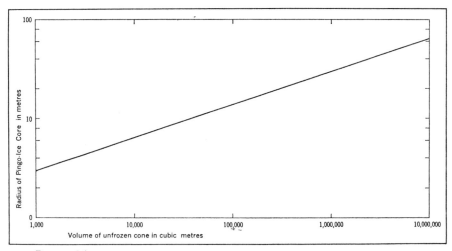

Figure 16. The graph shows the volume of sediment, with 30-per-cent porosity and 10-per-cent-volume expansion, required to form conical ice-cores with radii equal to the heights.

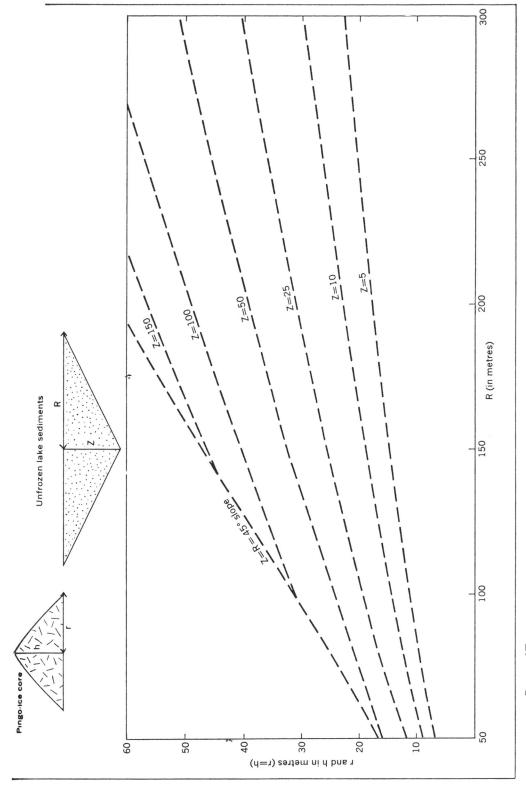

Figure 17. Nomograph for determining the dimensions of a conical volume of unfrozen sediment, given a 30-per-cent porosity and 10-per-cent-volume expansion, required to form conical ice-cores of radii equal to the heights.

Figure 17, the graph shows the relative dimensions between conical pingo ice-cores and conical unfrozen sediments, given a 30-per-cent porosity and 10-per-cent-volume expansion. Thus, a pingo ice-core 30 metres in radius and 30 metres high might have been formed from an unfrozen cone 300 metres in radius and 10 metres deep, or 200 and 25 metres, or 100 and 100 metres.

[*Editor's Note:* Material has been omitted at this point.]

SUMMARY

The pingos of the Pleistocene Mackenzie delta area number about 1,350 to 1,400. They occur typically in present or former lake basins and are most numerous in areas of drained lakes. The overburden thicknesses above the pingo ice-cores are estimated at about one-half to one-third the pingo heights. Pingo ice-cores are inferred to bottom slightly below the adjacent terrain. The pingos have grown in fine to medium sands with usually less than one or two per cent of silt and clay fractions.

Analyses of the depth of unfrozen ground beneath lakes, mean lake and ground temperatures, and three-dimensional heat conduction beneath lakes give estimates for the extent of permafrost beneath circular lakes under varying environmental conditions. As lakes shoal, either from climatic or geomorphic changes, permafrost would tend to be thinnest in the deepest parts of the lake, where open pools would remain the longest. Once a continuous permafrost cover extended over the lake bottoms, hydrostatic pressures of pore water expelled from encroaching permafrost surfaces are believed to have domed up the permafrost, in the thinnest places, to form pingos.

Pingo growth could be fast, but most of the larger pingos probably took tens of years to reach stability. Most of them are probably many hundreds to several thousands of years old.

References

American Geological Institute
    1957: Glossary of geology and related sciences. American Geological Institute, National Academy of Science—National Research Council Publ. 501.

Balduzzi, F.
    1959: Experimental investigation of soil freezing. *Mitt der Versuchsanstalt fur Wasserbau und Erdbau*, no. 44, (Trans. D. A. Sinclair NRC Trans. 912, 1960, 43 p.)

Cederstrom, D. J. and others
    1953: Occurrence and development of ground water in permafrost regions. *U.S. Geol. Surv.*, Circular 275, 30 pp.

Foundations of Geocryology
    1959: 2 vols. (Text in Russian). Permafrost Institute named in honor of V. A. Obrucheva. *Akad. Nauk CCCR*, Moskva.

Fraser, J. K.
    1956: Physiographic notes on features in the Mackenzie delta area. *Can. Geogr.*, no. 8, 18–23.

Frost, R. E.
    1950: Evaluation of soils and permafrost conditions in the Territory of Alaska by means of aerial photographs. *U.S. Army, Corps of Engineers*, v. 1.

Grave, N. A.
    1956: An archeological determination of the age of some hydrolaccoliths (pingos) in the Chuckchee Peninsula. Trans. from Russian (*Dokl. Akad. Nauk*, v. 106, 706–707) by E. R. Hope, Defence Research Board, Ottawa, June 1956, T 218 R.

Kosmachev, K. P.
    1953: Boolgoonyakhs. (Text in Russian). *Priroda*, vol. 42, 111–112.

Leffingwell, E. de K.
    1919: The Canning River region, Northern Alaska. *U.S. Geol. Surv.*, Prof. Paper 109, 1–251.

    1958a: The Anderson River map-area. Department of Mines and Technical Surveys, Geographical Branch, *Mem. 5*, 137 p.

    1958b: The valley of the Lower Anderson River, Northwest Territories. *Geog. Bull.* no. 11, 37–56.

Müller, F.
    1959: Beobachtungen über pingos. *Meddelelser om Grønland*, 153, 127 pp.

    1962: Analysis of some stratigraphic observations and C_{14} dates from two pingos in the Mackenzie delta, N.W.T. (Unpublished manuscript).

Muller, S. W.
    1947: Permafrost or permanently frozen ground and related engineering problems. J. W. Edwards, Ann Arbor, Michigan, 231 pp.

Petrov, V. G.
    1934: An attempt to determine the pressure of ground water in icing mounds. (Text in Russian). *Trudy komissii po izucheniiu vechnoi merzloty* 2, 59–72.

Porsild, A. E.
    1938: Earth mounds in unglaciated Arctic northwestern America. *Geog. Rev.*, vol. 28, 46–58.

Stager, J. K.
    1956: Progress report on the analysis of the characteristics and distribution of pingos east of the Mackenzie delta. *Can. Geogr.*, no. 7, 13–20.

Suslov, S. P.
    1961: Physical geography of Asiatic Russia. Trans. by N. D. Gershevsky, W. H. Freeman, San Francisco.

12

Reprinted from Nature, **230**(5288), 43–44 (1971)

Fossil Pingos in the South of Ireland

REMAINS of former pingos or ice-lens mounds are known in the Low Countries, Scandinavia, East Anglia[1] and Wales[2]. They are also widely distributed in the south of Ireland.

Fine examples occur near Camaross, on the road from Wexford to New Ross, 9 km west-north-west of Wexford town. A cluster of at least twenty fossil pingos north-east of Camaross Cross Roads (S891249) have been specially photographed by Dr J. K. St Joseph, Director in Aerial Photography, University of Cambridge (obliques 23/890245 and 7; verticals K17/W25–30). Part of K17/W26 is reproduced here by kind permission (Fig. 1a).

Three fossil pingos are seen. Pingo A (top right) has a basin about 60 m in diameter; the enclosing rim, which is almost complete, rises in a grassy ridge 1.5 m above the surrounding badly drained rushy ground (well seen on the south-east side). The elongated basin of pingo B does not contain open water, but the rushy vegetation inside the rim and outside it on the south-east is in contrast with the grass cover of the rim. Pingo C (bottom left) has an elongated basin which contains a wet swamp. In this vicinity, ridges of rock at about 75 m OD (ordinance datum) separate small valleys whose lower slopes (falling from north-west to south-east) are covered by soliflucted till with very poor natural drainage. The pingos formed near the base of the slopes.

Fine examples also occur at 60 m OD in Carrigeenhill Townland (W945953) on the road from Fermoy to Tallow, 14 km east-south-east of Fermoy. Here there is a rock-ridge to the north of the road, and a small valley partly filled with badly drained soliflucted till to the south. A large field abutting against the south side of the road showed at least three pingo basins and several curved ridges. When seen, the field was in the course of being drained and the lay-out of the drainage ditches had been imposed by the distribution of the pingo basins and ridges, which had disturbed the solifluction slope.

About 5 km west-north-west of Castleisland, the road from Castleisland to Tralee runs through a series of groups of fossil pingos at about 35 m OD. In Ballyegan Townland (Q962113) a small badly drained hollow is surrounded by outcrops of limestone; active quarrying is going on immediately to the west. There are at least three crescent shaped pingo ridges in the basin; one has an arc of 180° and a height of 1.5 m. In the adjoining Coolgarriv Townland (Q950117) there is a large marshy area, sloping gently to a flat valley bottom drained by the River Maine, with pingo features on both sides of the road. The road traverses one large pingo with appreciable rises where it crosses the rims which are about 100 m apart. South of the road, an almost continuous rim with an arc of 180° can be seen. In the next adjoining Maglass and Maglass East Townlands (Q945118) there are pingo features on both sides of the road.

These examples are only a few noted from main roads; many more will be revealed by survey. I will be obliged for information of other pingo features in Ireland. So far, all the examples noted lie south of the limit of the last glaciation.

G. F. MITCHELL

Department of Geology,
Trinity College,
Dublin

Received January 4, 1971.

[1] Sparks, B. W., *Bull. Geol. Assoc. Lond.,* **8** (1969).
[2] Pissart, A., *Z. Geomorphol.,* **7**, 147 (1963).

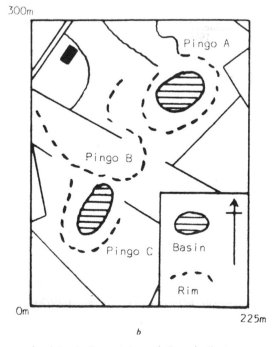

300m

Pingo A

Pingo B

Pingo C

Basin

Rim

Om

225m

Fig. 1 *a,* Vertical aerial photograph (Cambridge K17/W26, reproduced by kind permission) of three fossil pingos in Camaross Townland, Co. Wexford, Ireland. *b,* Outline map to accompany aerial photograph.

13

Reprinted from *The Periglacial Environment*, T. L. Péwé, ed., McGill-Queen's University Press, Montreal, 1969, pp. 203–215

EARTH AND ICE MOUNDS: A TERMINOLOGICAL DISCUSSION

JAN LUNDQVIST
Geological Survey of Sweden
Stockholm, Sweden

ABSTRACT. The following types of mounds related to the processes of freezing and/or thawing in the ground are defined:

(1) PINGOS. The maximum height of pingos is less than 100 m. They consist of the ice and soil of mainly mineral origin. They are formed by transfer of water and are related to freezing. They occur in the zones of continuous and discontinuous permafrost.

(2) PALSAS. The maximum height of palsas is about 10 m. They consist of ice and peat and, rarely, some mineral soil. They are formed by transfer of water and are related to freezing. Their occurrence is in the zone of sporadic permafrost, and the so-called pseudo-palsas occur outside the permafrost zones.

(3) EARTH HUMMOCKS. The maximum height of earth hummocks is about a metre. They consist of mineral and organic soil and are formed by transfer of mineral soil (with water) and related to thawing, and, therefore, only indirectly to freezing. It is possible also that they are directly related to freezing, or irrespective of the freeze-thaw processes. Their occurrence is independent of the permafrost zones.

(4) FROZEN PEAT HUMMOCKS. The maximum height of frozen peat hummocks is about a metre. They consist of ice and peat, and are formed without transfer of water or solid material; they are related to freezing. Their occurrence is within and outside the zone of sporadic permafrost.

It is emphasized that these main types must be kept strictly apart, although it is true that transitions occur, for instance, beween pingos and palsas. In the literature there has been some confusion as to the names and definitions of the types.

RÉSUMÉ. L'auteur définit les types suivants de monticules, liés aux processus de gel et de dégel dans le sol:

(1) PINGOS. La hauteur maximum des pingos est d'une centaine de mètres. Ils sont formés de glace et de sol d'origine surtout minérale. Ils sont construits par le déplacement de l'eau et sont liés au gel. Ils apparaissent dans les zones de pergélisol continu et discontinu.

(2) PALSEN. La hauteur maximum des palsen est de quelques mètres. Ils sont formés de glace et de tourbe avec, rarement, un peu de sol minéral. Ils sont construits par déplacement de l'eau et liés au gel. Ils apparaissent dans la zone de pergélisol sporadique: les soi-disant "pseudo-palsen" apparaissent en dehors des zones de pergélisol.

(3) MAMELONS DE TERRE. Leur hauteur maximum est généralement de moins d'un mètre. Ils sont formés de sol minéral et organique et construits par déplacement de sol minéral (avec de l'eau): ils sont donc liés au dégel et indirectement seulement, au gel. Il est possible aussi qu'ils soient directement liés au gel, ou indépendants des cycles gel-dégel. Leur apparition est indépendante des zones de pergélisol.

(4) MAMELONS DE TOURBE GELÉE. Leur hauteur maximum est d'environ un mètre. Ils sont formés de glace et de tourbe et construits sans déplacement d'eau ou de matériel solide; ils sont liés au gel. Ils apparaissent à l'intérieur ou à l'extérieur de la zone de pergélisol sporadique.

L'auteur insiste sur la nécessité de distinguer strictement entre ces principaux types, même si des transitions peuvent apparaître, par exemple entre les pingos et les palsen. Dans la littérature, il y a eu un peu de confusion dans la nomenclature et la définition de ces divers types.

ZUSAMMENFASSUNG. Folgende Typen von Hügel, welche mit den Prozessen von Gefrieren und Auftauen im Boden verbunden sind, werden definiert:

(1) PINGOS. Maximalhöhe in Hunderten von Metern. Aus Eis und Mineralerde bestehend. Unter Zufuhr von Wasser gebildt und mit dem Gefrieren verbunden. Vorkommen in den Zonen mit zusammenhägendem und unzusammenhängendem Permafrost.

(2) PALSEN. Maximalhöhe einige Meter. Aus Eis (Segregationseis) und Torf bestehend, ausnahmsweise auch aus Mineralerde. Unter Zufuhr von Wasser gebildet und mit dem Gefriesen verbunden. Vorkommen in der Zone mit sporadischem Permafrost. Pseudo-Palsen kommen ausserhalb den Permafrostzonen vor.

(3) ERDHÜGEL. Maximalhöhe einige Dezimeter. Aus anorganischer und organischer Erde bestehend. Unter Zufuhr von Mineralerde gebildet und mit dem Auftauen verbunden, vielleicht auch unmittelbar mit dem Gefrieren oder ganz unabhängig von diesen Prozessen. Vorkommen unabhängig von den Permafrostzonen.

(4) GEFRORENE TORFHÜGEL. Maximalhöhe einige Dezimeter. Aus Eis und Torf bestehend. Ohne Zufuhr von Wasser oder Erde gebildet. Mit dem Gefrieren verbunden. Vorkommen innerhalb und ausserhalb der Zone mit sporadischem Permafrost.

Es wird betont, dass man unbedingt zwischen diesen Haupttypen unterscheiden muss, trotzden dass Uebergangsformen wahrscheinlich vorkommen, z.B. zwischen Pingos and Palsen. In der Literatur sind früher oft die Namen und Definitionen von den oben erwähnten Hügeltypen vermischt und verwechselt worden.

CONTENTS

FIGURE

INTRODUCTION

The processes related to freeze and thaw affect the relief and structure of the ground in different ways: for instance, through the development of patterned ground. A special effect, which will be the subject of this article, is the generation of a hummocky relief. In some instances this relief should be classified as patterned ground, in others the original definition of "patterned ground" (Washburn 1956) cannot be applied. There has been much confusion in the literature, as well as in verbal discussions, as to the significance of certain terms referring to such phenomena. The relationship between the phenomena, their comparability, and possible transitions to each other have not always been made clear. The purpose of this article is to discuss some of these problems and to define a few phenomena that must be kept apart. The paper only deals with such hummocky features that can be related to freezing and thawing.

Emphasis will be laid on some different types but this does not exclude the possibility of the existence of other types. The "earth hummocks" group, especially, displays numerous other types and variations, which will not be treated here. Müller's (1959, p. 106-111) review of pingo-like phenomena suggests that there are variations within the pingo group, and transitions to palsas.

No effort is made to give the history of the investigations of the different phenomena, nor to collect all synonymous terms that have been used—not even all the instances when these terms have been used erroneously. The purpose is not to determine the origin of the processes resulting in the phenomena described.

PINGOS

The most conspicuous of the phenomena in question are pingos. Pingos are fully described and defined by Müller (1959) and need no further description here. A brief summary of Müller's account will be sufficient.

Pingos reach a maximum height of less than 100 m (Fig. 1). They are more or less cone-shaped and consist mainly of ice, often with a comparatively thin cover of mineral soil. The pingo ice is formed by injection and is a type of permafrost. Climatically, pingos occur in zones of continuous and discontinuous permafrost, especially where the continuous permafrost is thinning out.

Müller distinguished two different main types of pingos: open-system pingos and closed-system pingos. Open-system pingos, which were described from east Greenland, are related to regions with local degrada-

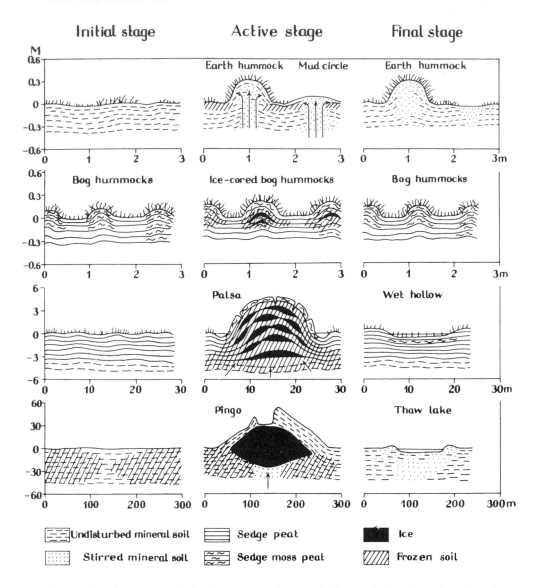

Fig. 1. Sections through the four types of mound discussed. For the sake of readability the thickness of the ice layers in the ice-cored bog hummocks and palsas is somewhat exaggerated. Notice also that there are different height scales used in the different sections.

tion of permafrost. The closed-system type, described from the Mackenzie delta, is on the contrary related to local aggradation of permafrost, normally beneath a lake.

Pingos are formed by injection when water rises through unfrozen gaps in the permafrost. The water can be forced upwards by cryostatic pressure arising when water bodies are trapped beneath a freezing surficial layer. The pressure can also be hydrostatic, that is, the ground water in surrounding, higher country can force the water upwards in the centre of depressions—an artesian phenomenon. In the present connection the origin of the pressure is not so important. The essential fact is that transfer of solid material is unnecessary for the pingo genesis except for the updoming.

Porsild (1938) distinguished two types of mounds similar to the pingos: *Sensu stricto*, which are formed "by local upheaval due to expansion following the progressive downward freezing of a body or lens of water or semi-fluid mud or silt enclosed between bedrock and the frozen surface soil." A smaller type of mound is formed when uplift takes place under the influence of hydraulic pressure. Although deviating from Müller's hypothesis, these types of mound both belong to the pingo group in the present article. The smaller of the two types looks very much like a palsa, but the explanation proposed for its origin is probably not applicable to palsas.

Pingos can be formed in different kinds of overburden. When updoming occurs, all kinds of material, even parts of the bedrock, can be heaved.

The mature pingo stage is represented by the well-known cones. A crater can arise in the pingo top by the melting of the ice in and beneath ruptures originated in the updoming.

When the ice in a pingo melts, the cone disappears. A depression will be left behind, which can be occupied by a thaw lake. After some time, when the fine structure of the superficial soil has been destroyed, the remaining lake is probably an ordinary thaw lake.

PALSAS

"Palsa" is a term that has been frequently misused. This Finnish and Lappish word refers to a very definite phenomenon. Used as a geological term it should be reserved for this phenomenon only. In this connection it may be noted that the plural form of the work in the English language ought to be "palsas." In Swedish geological literature an anglicization of the Swedish word "pals"—English plural, "palses"—is mostly used. This is not to be recommended, and very unfortunate is the German plural form "palsen," adopted even by the *Glossary of Geology and Related Sciences* (1957).

Investigations on palsas have mainly concerned their morphology and distribution, but very seldom their interior structure. This is perhaps because of the great difficulty in making borings and diggings in them. Through G. Lundqvist's (1951) investigation we know the main principles of the stratigraphy of the palsas, but little is known for certain of their genesis.

Palsas are mounds of peat and ice occurring on bogs in the subarctic region. The height of a palsa may be at least some 7 metres. The width of each mound varies from one to another, but is generally of the order of some tens of metres. Very often many palsas combine to form large complexes, which may be several hundreds of metres across. They may also be spread out over a large bog surface.

The surface of the palsas is rather dry and can be covered with a vegetation of low shrubs and lichens, or it can be a barren peat surface. Often it is traversed by open cracks in different directions. Thus the palsas differ clearly from the surrounding bog flat which is mostly very wet and overgrown with sedges, mosses, and other low fen plants. Immediately around a palsa there is often a still wetter zone with free water, the palsa lagg (G. Lundqvist 1951).

A digging in a palsa shows dry peat in the uppermost layers. At a depth of a few dm the peat is frozen with thin layers of clear ice. Still deeper the pure ice occupies a somewhat larger volume, and possibly there is in some cases a small body of clear ice. In general, however, the ice layers are not more than 2 to 3 cm thick. They are formed by segregation. The peat is the same sedge peat as forms the surrounding bog. We know from borings that the palsa body in some instances rests on the substratum of the bog and also that floating ice bodies exist. A simple calculation from the height and weight of the palsas shows that in most cases the palsas must reach down to the substratum. From recent borings (Forsgren 1964) we know that in such instances the substratum can be frozen, but unfrozen substratum has also been described (Hållen 1913).

Topographically, the palsas are generally situated in low parts of the landscape, but this is simply due to the fact that they are almost exclusively bog phenomena, and the bogs are mostly situated in depressions. Within a given depression, however, the palsas are not necessarily concentrated in the lowest part. For instance, G. Lundqvist's (1951, Fig. 3) section through a palsa bog shows them to be situated at and around a local water divide.

In exceptional cases the palsas also consist of mineral soil. G. Lundqvist (1953) mentioned some examples from northern Sweden and others are known from Finland and adjacent regions (Ruuhijärvi 1960, p. 228).

Climatically the palsas belong to the region with sporadic or discontinuous permafrost. J. Lundqvist (1962) showed that in Sweden they occur where the winter season is rather long, that is, where the temperature is below 0°C during more than 200 to 210 days a year. This implies that if the warm season is too long the ice in the palsas will melt, that is, the loss of heat in winter must exceed or counterbalance the heat supply in summer to create and preserve palsas. Where the winter precipitation is too high the snow cover will to some extent protect the ground from freezing. Thus palsas are absent where the precipitation during November–April exceeds 300 mm (J. Lundqvist 1962, p. 93).

In some respects the formation of palsas is very simple (G. Lundqvist 1951, Ruuhijärvi 1960). Their existence entirely depends on the content of ice. There are no signs of a transfer of peat from the surroundings to the palsas. The growth is caused solely by increase of the ice volume. It is not clear what causes this enrichment of ice: whether the capillarity of the peat is sufficient to account for the transfer of water, or if hydrostatic forces are required, as in the case of pingos. Such forces do contribute to the growth in some cases, as is indicated by G. Lundqvist's (1953) observations of mineral soil in the palsas. In general, the ice is separated from the peat only by segregation at the freezing point. This is a process that will create only thin ice layers. It is still not known if these layers in some way can grow thicker, or if injection is necessary for the formation of thicker ice lenses. In any case such lenses seem to occur only exceptionally.

There seems to be a mature stage of the palsas from which they decrease in size by thawing. In some instances this could be the effect of a slight climatic amelioration, but in general the causes are to be found within the palsas themselves. This is demonstrated by the fact that palsas that are apparently thawing, stable, and growing occur side by side, or at least in the same region, as in many large bogs in northernmost Sweden.

From what is known about the interior of the palsas it follows that when a palsa disappears by thawing almost nothing is left in the stratigraphy to demonstrate its former existence. If there has been peat growth on the top of the palsa, this peat will be of a different type from that beneath, and if there has been considerable wind erosion, a hiatus might be found in the stratigraphy. Both these types of "fossil palsas" are insignificant and easily overlooked if the palsa did not exist for a very long time. On the bog surface nothing will be seen of the former palsa except for a wetter part of the surface, comparable with a shallow thaw lake. The longer the palsa exists, the more the peat on the rest of the bog will grow, and the more distinct the pool will be.

It is essential that palsas can be formed without an initial stage in the form of a primary peat hummock or tussock. The bogs where the palsas occur are in general of the fen type, often with a rather smooth surface without distinct hummocks or other large unevennesses. There are, however, phenomena that could be interpreted as transition between palsas and common bog hummocks. These phenomena are more rarely observed. They are azonal features and could be named "pseudo-palsas." In these cases ice lenses exist perennially in favourable positions, such as shadowed places far outside the permafrost region (Högbom 1914, p. 307). They give rise to low mounds on the bog surface, invaded by a less hydrophilic flora than on the surrounding bog, a flora that is similar to that of ordinary bog hummocks. The pseudo-palsas are large and flat mounds with a core of perennially frozen peat and ice, which is located entirely in the peat and does not necessarily reach down to the substratum.

From the foregoing, the difference between palsas and pingos is evident: pingos are related to gaps in the permafrost, palsas are formed above a probable permafrost table. Because we know very little about the permafrost conditions beneath the palsa bogs, there is a possibility that some palsas, in fact, are related to gaps in underlying permafrost. In other cases (lower palsas) we know from direct observations (Hållen 1913) that the palsas are formed above a coherent permafrost table by freezing from above, for instance on bog roads, where there are gaps in the winter snow cover. The first palsa type could be considered a type of pingo, as hinted by Müller (1959, p. 111). The second type would be very similar to the above-mentioned pseudo-palsas.

EARTH HUMMOCKS

Earth hummocks of the type to be discussed here belong to the class "non-sorted nets" of Washburn (1956). They are very similar to ordinary peat hummocks in bogs, but they mainly consist of mineral soil. Their height is generally a few dm, but in exceptional cases it can exceed 1 metre. The diameter is of the same magnitude. The hummocks always occur in large groups and mainly on level ground. On slopes they become distorted.

The interior of the earth hummocks often displays a core of fine-grained mineral soil and a rather thick humus mantle. Involutions are typical and the mineral and organic soils are commonly distorted and pressed into each other. The whole structure clearly shows that the mineral soil has been forced upward in and through the humus cover.

Earth hummocks mainly occur on ground rich in vegetation but with low boulder content (G. Lundqvist 1964, 1949). In the Caledonides of Scandinavia they belong to a rather distinct zone of altitude (J. Lundqvist 1962, Fig. 14). The zone rises from 300 to 800 metres above sea level in northern Sweden, to 800 to 1,200 metres in central Sweden. Above this zone the phenomena are absent, but on lower levels they occur at scattered localities, such as old meadows and even down to sea level in central Sweden (Rudberg 1958). The main zone is the area of the tree-line in the mountains and upward as far as coherent vegetation reaches. The distribution indicates that the formation of earth hummocks requires a coherent vegetation which must not, however, be too thick. A cold climate favours the formation, but occurrences, even in a temperate climate, show that earth hummocks can be formed at a mean annual temperature at least as high as $+5°C$.

The vegetation cover is probably necessary to keep the silty to sandy material collected as mounds. A similar updoming of mineral soil is frequently observed at higher altitudes on barren areas. In these instances the resulting type of patterned ground will not be earth hummocks but more or less dome-shaped mud circles (for this term see Washburn 1947, p. 99). The dome shape seems to be unstable; the phenomena are only raised when newly formed. Their recent character is clearly demonstrated by extremely unstable positions of stones and lumps of earth. Older mud circles are almost flat. Most probably, rain, snow, and other erosive agents destroy the original form quickly if there is no protecting vegetation.

As to the formation of earth hummocks, Beskow (1930, p. 628) proposed the following explanation: "The origin of earth hummocks is local unevenness of the ground and vegetation. A thicker vegetation protects the soil from the frost—the surroundings thus are first frozen. The frost in the ground penetrates under the unfrozen parts, thus pressing soil material upwards. If repeated, the process will result in a hummock. Its growth will stop when it is high enough to allow the frost to penetrate from the sides" (J. Lundqvist 1962, p. 32).

Hopkins and Sigafoos (1954, p. 58) also ascribed importance to hydrostatic pressure in the formation of earth hummocks. Washburn (1956, p. 857) considered this process to have only a local effect, an opinion that the present author shared (J. Lundqvist 1962, p. 32). However, recent observations of involution structures on lake shores, as well as the occurrences of earth hummocks in temperate regions, make it probable that hydrostatic pressure can have a wider importance. The earth hummocks, as well as mud circles, can be explained as quicksand phenomena comparable with frost boils on roads: when a soil, especially if it is rich in silt and supersaturated with water, loses its excess water the discharge generally goes upward where resistance is least. If the soil is semi-liquid, it will be transferred with the water to the surface. Such a discharge commonly takes place when a frozen soil thaws from above, but may also be the effect of, for instance, a change of the water table. The water excess will be removed at random in a homogeneous soil with a uniform vegetation. If there are heterogeneities, such as gaps in the vegetation, the main excess will disappear at such weaker points. When the soil becomes frozen, openings in the frozen layer can theoretically act in the same way. Then the force is cryostatic rather than hydrostatic and the process is, in principle, similar to the formation of closed-system pingos. From this it is evident that the formation of earth hummocks is not yet completely understood and can probably be initiated in different ways. In the present connection it is of essential importance only for earth hummocks to be formed when transfer of solid material takes place by freezing or thawing, and there is no ice content in the mature hummocks. The phenomenon is not related to permafrost; it is known mainly from regions with sporadic or no permafrost.

ICE-CORED BOG HUMMOCKS

During the cold season the hummocks of bogs as well as other ground are frozen in the cold climate regions. If the summer season is warm enough, the ground thaws in the spring. If the loss of heat in the cold season exceeds the heat supply in the following warm season, the ground will persist frozen throughout the whole year. Such conditions are especially favoured in soil, such as peat, with good insulating properties. Therefore, one can often find frozen peat hummocks in bogs at the end of a cool summer, even outside the permafrost regions and when the rest of the ground has thawed. In a wide sense this is also permafrost. In Sweden such frozen hummocks are found within and along the Caledonian mountain range, even south of the zone with sporadic permafrost.

The hummocks of the type mentioned are of the same magnitude as the earth hummocks, but they consist entirely of peat (*Sphagnum* or sedge moss peat) and are located on the bogs. They are not types of patterned ground or permafrost phenomena in the true sense. The hummocks are an entirely biogenic phenomenon because of the differential growth of the vegetation and peat formation. They exist as hummocks regardless of the presence of an ice core. There is no transfer of solid material to the hummocks, nor of water, except from the transfer caused by the capillarity of the peat. The only effect of the freezing is a slight growth of the mound resulting from the volume increase on freezing. When a thaw sets in an ordinary bog hummock will be left behind.

The only reason for referring to this phenomenon, together with the aforementioned one, is the possibility of confusing it with palsas. The existence of a palsa is strictly dependent on the presence of ice: when the ice melts the palsa disappears. Peat hummocks exist whether there is ice or not, and whether there occurs any freezing at all or not.

In some cases, freezing will affect the growth of the hummocks, forming a special type called pounikkos (Ruuhijärvi 1960, p. 158, 220). Pounikkos are perennially frozen and represent a transition type between palsas and common peat hummocks. Probably, they are closely related to the phenomenon named earlier pseudo-palsas, but Ruuhijärvi's description does not fit that of pseudo-palsas. These are much more palsa-like than the pounikkos.

SUMMARY

In the foregoing, the necessity has been emphasized for distinguishing between the following four principal types of mounds related to the processes of freezing and/or thawing in the ground:

(1) PINGOS. The maximum height of pingos is less than 100 m. They consist of injection ice and soil, and are mainly of mineral origin. Formed by transfer of water and related to freezing, they occur in zones with continuous and discontinuous permafrost.

(2) PALSAS. The maximum height of palsas is about 10 m. They consist of segregated ice and peat, and rarely some mineral soil. Formed by transfer of water and related to freezing, they occur in the zone of sporadic permafrost. So-called pseudo-palsas occur azonally outside the permafrost zones.

(3) EARTH HUMMOCKS. The maximum height of the earth hummocks is about a metre. They consist of mineral and organic soil, and are formed by transfer of mineral soil (with water) and related to thawing. It is possible also that they are directly related only to the freezing. They occur independently of the permafrost zones.

(4) ICE-CORED BOG HUMMOCKS. The maximum height of ice-cored bog hummocks is about a metre. They consist of segregated ice and peat, and are formed without transfer of water or solid material, and are related to freezing. They occur within and outside the zone with

sporadic permafrost.

Pingos and palsas seem to be closely related to each other although there are differences. In fact, Müller (1959, p. 111) suggested that there is a third pingo type almost identical with palsas, according to his descriptions. Further study of the substratum of palsas, the type of their ice content, and their surroundings with regard to permafrost conditions would elucidate the similarity to proper pingos.

Earth hummocks have been confused with palsas in the literature but this is due partly to insufficient descriptions and partly to an old idea that there is an upwelling of solid material in palsas (cf. Sharp 1942, p. 420). The difference is clear enough.

Frozen peat hummocks can erroneously be interpreted as palsas and transitions do occur. A close study of the peat and vegetation and sometimes also of climatic conditions will reveal the true nature of the phenomenon.

What is said above does not exclude the existence of more complex phenomena which are formed by more than one process and thus are transitions between the main types. For instance, it is probable that peat or earth hummocks can serve as initial stages of palsas. Also a hummock can be initiated as a pingo by injection, and then develop into a palsa by further segregation of ice and peat. Even in such cases the separation of the processes involved, according to the scheme above, can help to explain the origin.

REFERENCES CITED

AMERICAN GEOLOGICAL INSTITUTE, 1960, Glossary of geology and related sciences, 2nd ed.: Washington, AGI, 325 p.

BESKOW, G., 1930, Erdfliessen und Strukturboden der Hochgebirge im Licht der Frosthebung. Preliminäre Mitteilung: Geol Fören. Förhandl., v. 52, p. 622-638.

FORSGREN, BERNT, 1964, Notes on some methods tried in the study of palsas: Geog. Ann., v. 46, p. 343-344.

HÅLLÉN, K., 1913, Undersökning af en frostknöl (pals) å Kaitajänki myr i Karesuando socken: Geol. Fören. Förhandl., v. 35, p. 81-87.

HÖGBOM, BERTIL, 1914, Über die geologische Bedeutung des Frostes: Bull. Geol. Inst. Uppsala, v. 12, p. 257-389.

HOPKINS, D. M., and SIGAFOOS, R. S., 1954, Discussion: role of frost thrusting in the formation of tussocks: Am. Jour. Sci., v. 252, p. 55-59.

LUNDQVIST, G., 1949, The orientation of the block material in certain species of flow earth: Geog. Ann., v. 31, p. 335-347.

———, 1951, En palsmyr sydost om Kebnekaise: Geol. Fören. Förhandl., v. 73, p. 209-225.

———, 1953, Tillägg till palsfrågan: Geol. Fören. Förhandl., v. 75, p. 149-154.

———, 1964, De svenska fjällens natur STF: s handböcker om det svenska fjället, Stockholm, 440 p.

LUNDQVIST, JAN, 1962, Patterned ground and related frost phenomena in Sweden: Sveriges Geol. Undersökning, ser. c, no. 583, 101 p.

MÜLLER, FRITZ, 1959, Beobachtungen über Pingos: Detailuntersuchungen in Ostgrönland und in der kanadischen Arktis: Medd. om Grönland, v. 153, no. 3, 127 p.

PORSILD, A. E., 1938, Earth mounds in unglaciated arctic Northwestern America: Geog. Rev., v. 28, p. 46-58.

RUDBERG, STEN, 1958, Some observations concerning mass movement on slopes in Sweden: Geol. Fören. Förhandl., v. 80, p. 114-125.

RUUHIJÄRVI, RAUNO, 1960, Über die regionale Einteilung der nord-finnischen Moore: Ann. Botan. Soc. "Vanamo," v. 31, no. 1, 360 p.

SHARP, ROBERT P., 1942, Ground-ice mounds in tundra: Geog. Rev., v. 32, p. 417-423.

WASHBURN, A. L., 1947, Reconnaissance geology of portions of Victoria Island and adjacent regions, arctic Canada: Geol. Soc. America Mem. 22, 142 p.

———, 1956, Classification of patterned ground and review of suggested origins: Geol. Soc. America Bull., v. 67, p. 823-865.

14

Reprinted from *Biul. Peryglac.*, No. 19, 131–142, 146–150 (1969)

THAW DEPRESSIONS AND THAW LAKES
A REVIEW

*Robert F. Black**

Madison

C o n t e n t s: Introduction — Historical perspective — Synopsis of principles — Ice content of permafrost — Thaw depressions — Thaw lakes — Oriented lakes of northern Alaska and northwest Canada — Vegetation and thaw depressions and thaw lakes — Thermal aspects of thaw lakes — Other limnological aspects of thaw lakes — Suggestions for future study — References cited.

INTRODUCTION

In permafrost regions thaw of ground ice which comprises more volume than pore space in unconsolidated sediments results in thaw depressions that may become lakes. Both thaw depressions and thaw lakes are circum-Arctic and exceedingly abundant. Many thaw lakes are large striking features characterizing the coastal plains and lowlands with continuous and discontinuous permafrost. The concept of their origin is relatively simple, but in North America quantitative data on the factors involved are meager — only a few specific case studies are available. The better known areas are in northern Alaska and northwest Canada.

This paper reviews the present status of knowledge in North America of the thaw depressions and thaw lakes of the permafrost regions, as requested by Jan D y l i k, President of the International Commission on Periglacial Geomorphology. This paper arbitrarily excludes kettles and kettle lakes, even though they are common in permafrost regions, because the mechanism of emplacement of ice in the ground specifically results from the work of glaciers. However, kettles resulting from thaw of buried glacial ice may well be identical to true thaw depressions in form, in process, and in implications regarding heat exchange.

* Department of Geology and Geophysics, University of Wisconsin, Madison, Wisconsin 53706, USA.

Details of the physical description of the thaw depressions and thaw lakes are not presented here, but can be found in the various references. Little information is new. However, a complete listing of the more important, widely scattered, more recent literature is attempted. Papers with simple reference to phenomena without new description or discussion are not listed unless needed to follow the trend of the times.

Thaw depressions are discussed after short sections provide an historical perspective, a statement of some principles, and a review of the ice content of permafrost. Thaw lakes arbitrarily are under a separate heading, and the oriented lakes of northern Alaska and northwest Canada are emphasized because they have been studied more than any others. The relationship of vegetation to thaw depressions, the thermal aspects of thaw lakes, and other limnological aspects follow. Suggestions for future study terminate this review.

HISTORICAL PERSPECTIVE

Prior to the introduction by M u l l e r (1945) of the well-established Russian concepts of thermokarst, few papers in North America even mentioned the pitting of the surface by thaw of buried n o n - g l a c i a l ice. (In contrast concepts of the origin of kettles from thaw of buried glacial ice goes back more than a century — W h i t t l e s e y, 1860; W h i t e, 1964). M u l l e r (1945, p. 84) listed the most common land forms produced: (1) surface cracks, (2) cave-ins and funnels, (3) sinks, saucers, and shallow depressions, (4) „valleys", gullies, ravines, and sag basins, and (5) cave-in lakes, windows, and sag-ponds. In the next few years other personnel of the U.S. Geological Survey working in Alaska contributed the results of various studies of Alaskan thermokarst features (C a b o t, 1947; W a l l a c e, 1948; H o p k i n s, 1949; B l a c k and B a r k s-d a l e, 1949).

In the 1950's the Geological Survey personnel continued to study thermokarst phenomena, but interest had spread to university personnel and others throughout arctic North America. The literature expanded many fold. Now most regional studies of unconsolidated sediments in permafrost areas at least mention the thaw depressions and thaw lakes present (e.g., F e r n a l d, 1964). The orientation of the thaw lakes of the Arctic Coastal Plain became the subject of considerable discussion. Case histories of the thermal regime of a few arctic lakes were compiled. Summaries of certain aspects of the thaw lakes have appeared, but no definitive monograph of the thaw depressions and thaw lakes has been attempted. Clearly some thaw lakes span considerable time, but the distribution and history of the lakes in various places are imperfectly known. Dating of events and physical, biological, and chemical limnological studies are largely in formative stages. We are still in a data-gathering period, but time is ripe to bring the various approaches together.

SYNOPSIS OF PRINCIPLES

During the freezing of unconsolidated materials to make permafrost, water is attracted to the freezing surface. Water commonly fills all pores with ice, when it freezes, and also segregates onto growing ice crystals to form all sizes of irregular masses, veinlets, and dikelets of relatively clean ice, depending on various combinations of factors like size and shape of pores, texture, amount of water, and rate of freezing. The numerous laboratory experiments and discussions of the freezing process need not be reviewed here. Suffice it to say that the upper 5 to 10 meters of permafrost, where fine-grain materials are present, commonly contain 30 to 90 percent of ice by volume. The permafrost is supersaturated, i.e., the mineral and organic matter are a suspension in ice.

After permafrost has developed, ice can be added to the upper few meters of the ground by the growth of ice wedges in polygonal networks. As the moisture comes year by year from the atmosphere, such growth ultimately can replace all mineral matter with ice. Hence, the originally supersaturated permafrost can continue theoretically to increase its ice content up to 100 percent in the upper 10 m, the maximum depth of penetration of most ice wedges. No area with that much ice is known, probably because thaw is initiated which brings on thaw depressions that destroy the ice-wedge system. When ground is once again exposed, new ice wedges must grow to start the cycle again.

Supersaturated permafrost is unstable when exposed to solar radiation and warm air and water during the summer. It melts.

The resulting liquid "sludge" has low viscosity. It flows readily or
the water can drain and evaporate away. As a consequence, the
thawed grounds has less volume, and commonly a much smaller
fraction of the volume, than the permafrost from which it was derived. Underground cavities and surface depressions result. Thaw
continues as long as the ice can gain enough heat during the summer
to raise its temperature to the melting point and overcome the
latent heat of fusion. Organic matter and the soil "waste" from
the thawed permafrost may thicken sufficiently to prevent the ice
from gaining enough heat during the summer to melt. The growth
of the cavity or depression then ceases. If water is trapped in the
depression, of course, a small pond forms which can continue to
grow into a lake. To form large thaw depressions and large thaw
lakes the permafrost requires a paucity of mineral matter. Otherwise the remaining ice in the ground is quickly covered and protected from further thaw.

This overly simplified outline underlies our concept of the
formation of thaw depressions and thaw lakes. The process can be
looked upon as a simple heat budget. Stability of the permafrost
requires that a negative temperature balance be maintained between the winter's heat loss and the summer's heat gain. The total
heat exchange is not the problem, it is the summer gain of heat
to the ground. A large winter heat loss may be compensated for
by a greater summer heat gain so thawing is initiated just as
easily as in a situation wherein a small winter heat loss is compensated for by only a slightly larger summer heat gain. The thermal
balance is, of course, more easily upset where the annual temperature approaches the melting point.

Thaw of permafrost can start a depression in almost any climate,
including the high Arctic. The nature and thickness of the active
layer and the characteristics of the vegetal cover largely determine
the heat flow and the stability of the underlying permafrost.
Without protection ice thaws during a brief interval in summer
even in extreme northern North America. Less protection is needed
than in areas further south in warmer climates. Disruption of the
vegetal mat or active layer by gravity movements, wind or water
work, or even animals is commonly sufficient to induce thaw.
The thaw continues until an insulating blanket of material is re-
established or until the supersaturated permafrost is destroyed.

ICE CONTENT OF PERMAFROST

In 1949—1950 B l a c k measured the percentage of moisture by weight in 36 samples of permafrost, excluding ice wedges, from the Barrow area of northern Alaska. It ranged from 9.6 percent in clayey-silty-sand to 91 percent in peat. Deposits of gravel and sandy gravel were supersaturated when the ice content was 16 to 18 percent by weight. Deposits of sand were supersaturated with about 10 percent or more of moisture by weight — the samples contained from 9 to 40 percent. Deposits of silt were supersaturated with 13 percent or more moisture by weight — they contained from 13 to 65 percent. Deposits of silt, clayey silt, silty peat, and peat contained the most moisture, generally between 75 and 91 percent, not counting the massive ice in ice wedges.

The volume of ice-wedge ice present in local natural exposures in various places through Alaska was estimated. In northern Alaska the oldest areas contain ice wedges averaging about 5 m wide and comprising locally 50 percent of the upper 6 m of the ground. Such places are rare. Wedges 1 to 3 m wide and comprising 10 to 30 percent of the upper 6 m are more common. Many areas have even less. Total volume of ice in the upper 10 m of ground in most parts of the coastal plain is sufficient to produce thaw depression 3 to 6 m deep and locally even deeper.

H u s s e y and M i c h e l s o n (1966) cite moisture contents of the permafrost at Barrow, both laterally and vertically, for (1) initial surface, (2) ancient drained-lake basin, (3) recent drained-lake basin, and (4) present lake bed. In the upper 6 m of permafrost the ground ice content is considerably less for the ancient drained-lake basins and recent drained-lake basins than for that of the initial surface residual. Average potential settlement as a percent of the 6 m was calculated as 55, 19, 12, and 2 respectively for the four areas listed above. Thus, it was concluded that most depressions could have formed by thaw of the permafrost.

On the assumption that the lake basins in northern Alaska are the result of thaw of ground ice, L i v i n g s t o n e, B r y a n, and L e a h y (1958) used the volume of the water and air in a basin to compute the volume of ice that must have melted. Their value of 68 percent for a total basin depth of 28 m is more than double that for any area of comparable depth in northern Alaska seen by the writer, and if true, must involve special conditions. Ice wedges normally do not grow to depths of more than 6—10 m so a special mechanism or unusual freezing conditions with abundant water must have been available during the formation of the deeper permafrost. Only the upper 5—10 m of permafrost has such high content of ice.

Ice content of permafrost has been measured in many areas for engineering purposes and is expressed as percent of ice to dry soil on a weight basis. For example, P i h l a i n e n and J o h n s t o n (1954, fig. 10) at Aklavik, N.W.T., Canada, in the delta of the Mackenzie River, plotted the moisture content of 153 samples from 16 bore holes against depth. These showed in general, but with wide variations, a decrease in ice from the surface to a depth of 3 m and a relatively constant value from 3 m to 8 m of 54 percent. In the same delta area M a c k a y (1966) describes one section about 7 m high in which ice makes up almost 80 percent of the total volume of material, but other areas had much less. At Fort Simpson, N.W.T., on an island at the junction of the Liard and Mackenzie Rivers, permafrost was encountered in 20 of 30 bore holes, at depths of 1 to 6 meters (P i h l a i n e n, 1961). Ice in 101 samples was considerably less than in permafrost at Aklavik. The number and distribution of thaw depressions in the Arctic seem to reflect the variation of ice content of the permafrost.

THAW DEPRESSIONS

Irregular hummocky topography associated with ice-wedge polygons has long been recognized, but earlier concepts considered only primary bulging upward by growth of ice rather than settling by thaw of the ground ice (L e f f i n g w e l l, 1919, p. 211—212). R o c k i e (1942) seems to have been first to discuss specifically and in some detail the pitting and settling of ground in central Alaska caused by thaw of ice wedges. P é w é (1954) followed up R o c k i e's study, adding considerable detail in description and genesis of the phenomena. P é w é cited pertinent literature (not here listed to save space) and gave discrete definitions of the terms involved. Thermokarst mounds and pits in areas cleared for agriculture were given special attention. It is clear that the mounds are the centers of former ice-wedge polygons left by the destruction of the ice wedges. The pits are in cultivated or formerly cultivated fields near the boundary between permafrost-free slopes and slopes underlain by large ice masses. Permafrost in thin isolated masses and the low water table, 5 to 30 m below the surface, localized the pits. Water from melting ice and surface water funneled down from above and circulated freely through the passageways in loess, on its way to the water table. Where the water table is close to the surface, cave-in lakes form. Cavities probably formed within 3 years, and pits appeared within 3 to 30 years after clearing.

In Seward Peninsula, Alaska, H o p k i n s (1949) described thaw depressions, small thaw lakes that occupied some of those depressions, and thaw sinks which seem to have evolved when thaw lakes pierced the permafrost and drained into subterranean openings. The depressions are started by locally deep thaw resulting **from (1) disruption of the vegetal** cover by frost heaving, (2) accelerated thaw beneath pools occupying intersections of ice-wedge polygons, and (3) accelerated thaw beneath pools in small streams.

A n d e r s o n and H u s s e y (1963) reviewed the literature of thermokarst phenomena and, among others, discussed thermokarst ravines, beaded streams, and ice-wedge intersection ponds in northern Alaska. Most of the features are related to the thaw of ice wedges. However, "badlands" topography with maximum relief of 65 m surely cannot be attributed to melting of ice wedges alone, as seems to be implied. A n d e r s o n and H u s s e y conclude that thermokarst development is the major denudation process operating on the North Slope of Alaska. This, if true, must surely be a very recent and temporary condition because the destruction of ice wedges once started proceeds rapidly to depressions. Once the supersaturated permafrost is destroyed, it must be rebuilt before thaw and collapse can reoccur. This takes several thousand years.

THAW LAKES

Thaw depressions grade into thaw ponds and thaw lakes, and the distinctions, largely arbitrary, are based on the size of the basin and the amount of water present. A basin many meters in extent is generally needed for current and wave action to become a factor in shaping the basin. Texture and ice content of the material, vegetation, climatic factors, and others determine the rate of thaw and shape of the depressions.

Of the initial group from the U.S. Geological Survey who were introduced to permafrost phenomena under Siemon W. M u l l e r, W a l l a c e (1948) deserves credit for publication of the first specific field study of cave-in or thaw lakes. In eastern Alaska he found from tree-ring counts that bank recession was 6 to 19 cm per year. A sequence of development divided into four stages was recognized, based on the size and shape of the thaw depressions — from youth to old age. H o p k i n s (1949) followed shortly thereafter with his study of thaw lakes on Seward Peninsula. The largest thaw depressions are occupied by lakes which rarely exceed 300 m in diameter and are 0.3 to 2 m deep. Gullies cut into the lake banks mark the positions of ice wedges rapidly melted. More than one cycle of thaw lakes was noted. The thaw depressions, lakes, and sinks are recorded by maps and photographs. An idealized cycle is shown.

Ice-cored mounds of various sizes are produced in the active layer and in permafrost throughout the Arctic (B i r d, 1967, p. 201—208; M a c k a y, 1966; H o l m e s, F o s t e r and H o p k i n s, 1966). On thawing many produce thaw depressions and thaw lakes. During initial destruction of the mounds, it is relatively easy to recognize the nature of the source of the ice, but once entirely melted out, this is rarely possible. Most active layer phenomena are less than a few meters in diameter and only some centimeters to a meter deep. In tundra the thaw depressions may be dry, but ice laccoliths in marshy areas commonly produce small circular ponds. They are especially common in marshy lake beds in the Arctic Coastal Plain of northern Alaska. Most survive only a few years.

The large permafrost mounds, generally called pingos, are commonly many tens of meters in diameter and several tens of meters high. They range in age from some decades to many thousands of years. On thawing they can produce lakes. Only one pingo lake has been studied in detail (L i k e n s and J o h n s o n, 1966). It is 205 m long, 165 m wide, and 8.8 m deep. As the pingo is the open system type, whereby ground water is continuously supplied to the lake, its chemistry reflects the source of the water. The lake is chemically stratified and unusually high in strontium and lithium. Its age is less than 5720 radiocarbon years, and may be as recent as a few hundred years (K r i n s l e y, 1965).

155

ORIENTED LAKES OF NORTHERN ALASKA
AND NORTHWEST CANADA

In 1946 B l a c k and B a r k s d a l e (1949) jointly studied the oriented thaw lakes of northern Alaska and completed a paper which B l a c k presented orally on January 22, 1947. That paper was published without incorporation of later literature or of data from studies by B l a c k in 1947—1950. Most of the lakes were considered to be true thaw lakes. Some resulted from the uplift and segmentation of lagoons, and the origin of others was not known. Later studies by B l a c k, by C a r s o n and H u s s e y (1962), and by H u s s e y and M i c h e l s o n (1966) confirmed that sufficient ice was in the upper part of permafrost to form the lake basins less than about 5 m deep; B l a c k doubts that sufficient ice is present to permit the lakes many tens of meters deep to form by simple thaw. In the first place, the unconsolidated sediments of the Gubic Formation (B l a c k, 1964) are too thin over most of the Arctic Coastal Plain to permit such collapse unless they are essentially solid ice. Although the units of the Gubic Formation do not all have the same ice content, none is solid ice. Exposures up to 40 m of caving banks of lakes and streams do not show such large amounts of ice in the lower parts. Moreover, the underlying bedrock is commonly saturated, but not supersaturated with ice, and does not collapse when thawed. Thaw of ice can account for only a part of the depth of some deep lake basins — particularly those near the southern margin of the Coastal Plain.

The large number of oriented lakes and their extension into Arctic Canada has been documented by other studies (B i r d, 1967, p. 211—216). These also seem to confirm the initial description by B l a c k and B a r k s d a l e (1949) of the physical features of the lakes. The major concern has been with the possible cause of the orientation of the basins. The initial thought that the orientation was by wind oriented at right angles to that of today is disputed by all who have dealt with the problem (A n d e r s o n and H u s-s e y, 1963; C a r s o n and H u s s e y, 1959, 1960a, 1960b, 1962; H u s s e y and M i c h e l s o n, 1966; L i v i n g s t o n e, 1954, 1963; M a c k a y, 1956, 1963; P r i c e, 1963, 1968; R e x, 1961, R o s e n-f e l d and H u s s e y, 1958).

L i v i n g s t o n e (1954) first called attention to a possible mechanism whereby currents account for the elongation of the lakes at right angles to the present winds. R o s e n f e l d and H u s s e y

156

(1953) point out rightly that the problem is more complicated than the simplified approach of L i v i n g s t o n e, and that his hypothesis cannot apply equally to those lakes only a few meters long or many kilometers long. They point to the possibility of fault and joint patterns controlling the orientation. It should also be mentioned that the L i v i n g s t o n e ' s hypothesis cannot be applied equally to these very shallow lakes and those with a deep central elongate basin.

C a r s o n and H u s s e y (1959) reviewed five possible hypotheses for the lake orientation and concluded that each alone was not enough, but that a composite would suffice. The five are: (1) wave action from winds parallel to elongation during earlier time, (2) present winds which produce wave-current systems which scour at right angles, (3) present winds which distribute sediment on east and west shores, insulating them from thaw, (4) orientation produced by thaw during maximum insolation at noon, and (5) lakes controlled by north-south trending ice wedges which formed in the north-south components of a right-angle fracture system. They conclude that oriented ice-wedges might develop in the fracture system; that maximum insolation would be more effective in melting the north-south trending wedges than the complementary set; that the depressions so oriented would be perpetuated and enlarged by thaw and wind (wave) oriented sediments deposited on the east-west shores. C a r l s o n, et al., (1959) also suggest that preferentially oriented ice wedges play a role in the orientation of the lakes.

C a r s o n and H u s s e y (1960a) reviewed the hypothesis of L i v i n g s t o n e (1954) and an unpublished one by R. W. R e x and then presented some current measurements from lakes near Barrow, Alaska. Their field data suggest L i v i n g s t o n e's hypothesis is not applicable, but that the approach of R e x merits further study. C a r s o n and H u s s e y (1960b) provide additional data on the hydrodynamics in three of the lakes near Barrow, including conditions when the lakes are ice free and when an ice cake is present. Their measurements show that erosion is going on at the ends of the elongated lakes by long-shore currents as predicted by the hypothesis later published by R e x (1961). In an important summary paper C a r s o n and H u s s e y (1962) bring together their field data on the lakes near Barrow in support of the circulation hypothesis of R e x (1961) and reject their earlier postulations of structural control by ice wedges. They point out that circulation in the smallest oriented ponds resembles that suggested by L i v i n g s t o n e (1954), but is too slow to erode. However, wave action on lee sides of those ponds is sufficient to erode. As lakes increase in size, a gradual change in circulation systems was observed by C a r s o n and H u s s e y (1962), approaching that of the hypothesis of R e x (1961).

R e x's hypothesis is inadequate to account for the orientation of small basins, and also according to C a r s o n and H u s s e y (1962), for the lack of erosive currents at the ends until after the basin is elongated. They develop briefly the concept of lake-basin orientation as an expression of a basin morphology controlled through interaction of slow basin subsidence by thaw and wind-oriented wave action. "These combine to produce a preferred pattern of morphology through all phases of basin development, with orientation accentuated during later stages by other processes, chiefly thermal and circulatory". Variations in natural conditions are considered to be instrumental in affecting changes in the shapes and depths of the lakes. Coalescence of lakes particularly leads to complex forms. The wind-resultant problem was raised by P r i c e (1963), but C a r s o n and H u s s e y (1963) concluded it would not change materially their earlier conclusions.

M a c k a y (1963, p. 51—55) attempted to analyze the equilibrium forms of lakes that might be produced by winds of today. "If it is assumed that winds from each of the 16 compass directions recorded in climatic records tend to develop curved bays, approximately cycloidal, than an oriented lake may be viewed as the summation or integration form of 16 cycloids, each of different size." He assumed that the diameter of the generating circle of the cycloid is equal to the resultant for a given wind direction. The computed shapes and actual shapes of lakes from the Mackenzie Delta and the Barrow, Alaska, regions agreed nicely. However, M a c k a y emphasized that the precise mechanism of lake orientation remains unexplained.

M a c k a y (1963, p. 44) also writes of a "ribbed pattern of northeast-southwest trending ridges in hundreds of large shallow lakes of the Mackenzie Delta. The sedgy ridges are minor features, 30 cm or so in height and several hundred to a thousand meters long. The ribs are perpendicular to prevailing winds and are believed to be formed primarily by wave action.

M a c k a y (1963, p. 47—50) categorized the shapes of the lakes and their central deeper basins into precise geometric forms in four categories: (1) lemniscate, (2) oval, (3) triangle, and (4) ellipse. Equations were set up for each.

B i r d (1967, p. 210—216) most recently has summarized the occurrence of thermokarst and thaw lakes in Canada, including the oriented lakes. P r i c e (1968) most recently reviewed the entire subject of oriented lakes, regardless of origin of the initial lake basin. Few originated as true thaw lakes, but they may be equally well oriented (e.g. P l a f k e r, 1964). Photo interpretation of thaw lakes is summarized by H o p k i n s, K a r l s t r o m, and others (1955).

A photographic map of the Barrow, Alaska, area on a scale of 1 : 25,000 has a 1-meter contour interval and local $1/2$-meter contours (B r o w n and J o h n s o n, 1966). The thaw lakes are portrayed beautifully. Cross-cutting age relationship of lake basins is brought out clearly by the omnipresent ice-wedge polygons. Maps showing lake basins of different ages are included in C a r l s o n, et al. (1959) and C a r s o n and H u s s e y (1960 and 1962). Other views and a summary of the lakes are given by W a h r h a f t i g (1965).

Dating of the oriented lakes has been a particularly difficult problem. L i v i n g s t o n e, B r y a n, and L e a h y (1958) by tree-ring dating showed that the gentle shore of one oriented thaw lake was exposed during the last 150 years. All observers of the lakes have witnessed shore erosion of several meters during single storms. Lateral migration can be very rapid. The truncation of existing ice wedges by lateral migration demonstrates that some lakes or parts of them must be only some decades or a few centuries in age. Radiocarbon dating of organic matter in two drained lakes suggests ages of several thousand years (B r o w n, 1965, p. 44—45). Other dates suggest most surface features near Barrow are not older than about 8,000 years, yet several older dates imply complicated events going back much farther. At 14,000 radiocarbon years ago pollen analysis shows vegetation that reflects climate somewhat colder than now (C o l i n v a u x, 1964, 1965). The Pleistocene—Recent changes of sea and of climate that affected the permafrost and thaw lakes of the Arctic Coastal Plain in western and northern Alaska are reviewed fully by H o p k i n s (1967).

[*Editor's Note:* Material has been omitted at this point.]

References cited

A n d e r s o n, Gary S., and H u s s e y, Keith M., 1963 — Preliminary investigation of thermokarst development on the North Slope, Alaska. *Iowa Acad. Sci., Proc.,* vol. 70; p. 306—320.

B i r d, J. Brian, 1967 — The physiography of arctic Canada. Johns Hopkins Press; 336 p.

B l a c k, Robert F., 1964 — Gubik Formation of Quaternary age in northern Alaska. *U. S. Geol. Survey, Prof Paper* 302-C; p. 59—91.

B l a c k, Robert F., and B a r k s d a l e, William L., 1949 — Oriented lakes of northern Alaska. *Jour. Geol.,* vol. 57; p. 105—118.

B o y d, W. L., 1959 — Limnology of selected arctic lakes in relation to. water-supply problems. *Ecology,* vol. 40; p. 49—54.

B r e w e r, Max C., 1958a — Some results of geothermal investigations of permafrost in northern Alaska. *Amer. Geophysical Union, Trans,* vol. 39; p. 19—26.

B r e w e r, Max C., 1958b — The thermal regime of an arctic lake. *Amer. Geophysical Union, Trans.,* vol. 39; p. 278—284.

B r i t t o n, Max E., 1957 — Vegetation of the arctic tundra. 18th Biology Colloquium, Oregon St. College, Corvallis, Oregon; p. 26—61.

B r o w n, Jerry, 1965 — Radiocarbon dating, Barrow, Alaska. *Arctic,* vol. 18; p. 36—48.

B r o w n, Jerry, G r a n t, C. L., U g o l i n i, F. C., and T e d r o w J. C. F., 1962 — Mineral composition of some drainage waters from arctic Alaska. *Jour. Geophysical Res.,* vol. 67; p. 2447—2453.

B r o w n, Jerry, and J o h n s o n, P. L., 1966 — U. S. Army CRREL Topographic map, Barrow, Alaska. *Cold Regions Res. & Eng. Lab., Spec. Rpt.* 101.

B r o w n, W. G., J o h n s t o n, G. H., and B r o w n, R. J. E. 1964 — Comparison of observed and calculated ground temperatures with permafrost distribution under a northern lake. *Canadian Geotechnical Jour.,* vol. 1; p. 147—154.

B r u n n s c h w e i l e r, D., 1962 — The periglacial realm in North America during the Wisconsin glaciation. *Biuletyn Peryglacjalny,* no. 11, p. 15—27.

C a b o t, E. C., 1947 — Northern Alaska coastal plain interpreted from aerial photographs. *Geog. Rev.,* vol. 37; p. 639—648.

C a r l s o n, P. R., R o y, C. J., H u s s e y, K. M., D a v i d s o n, D. T., and H a n d y, R. L., 1959 — Geology and mechanical stabilization of Cenozoic sediments near Point Barrow. in: The geology and engineering characteristics of some Alaskan Soils, *Iowa Engincering Exp. Sta. Bull.* 186; p. 101—128.

C a r s o n, Charles E., and H u s s e y, Keith M., 1959 — The multiple working hypothesis as applied to Alaska's oriented lakes. *Iowa Acad. Sci., Proc.,* vol. 66; p. 334—349.

C a r s o n, Charles E., and H u s s e y, Keith M., 1960 — Hydrodynamics in some arctic lakes. *Iowa Acad. Sci., Proc.,* vol 67; p. 336—345.

C a r s o n, Charles E., and H u s s e y, Keith M., 1960b — Hydrodynamics in three arctic lakes. *Jour. Geol.,* vol. 68; p. 585—600.

C a r s o n, Charles E., and H u s s e y, Keith M., 1962 — The oriented lakes of arctic Alaska. *Jour. Geol.*, vol. 70; p. 417—439.

C a r s o n, Charles E., and H u s s e y, K e i t h M., 1963 — The oriented lakes of arctic Alaska. A reply. *Jour. Geol.*, vol. 71; p. 532—533.

C o l i n v a u x, Paul A., 1964 — Origin of ice ages: pollen evidence from arctic Alaska. *Science*, vol. 145; p. 707—708.

C o l i n v a u x, Paul A., 1965 — Pollen from Alaska and the origin of ice ages. *Science*, vol. 147; p. 633.

D r a b b e, C. A. J. von Frijtag, 1952? — Aerial photograph and photo interpretation. Managing Director, Netherlands Topographic Service, 40 p.

D r u r y, William, H., Jr., 1956 — Bog flats and physiographic processes in the upper Kuskokwim River region, Alaska. *Gray Herbarium, Harvard Univ.*, no. 178; 130 p.

F e r n a l d, Arthur T., 1964 —Surficial geology of the central Kobuk River Valley, northwestern Alaska. *U. S. Geol. Survey, Bull.* 1181-K; 31 p.

H o l m e s, G. William, F o s t e r, Helen L., and H o p k i n s, David M., 1966 — Distribution and age of pingos of interior Alaska. *Permafrost Intern. Conf., Proc., Purdue Univ., Nat. Acad. Sci.-Nat. Res. Council, Pub.* no. 1287; p. 88—93.

H o p k i n s, David M., 1949 — Thaw lakes and thaw sinks in the Imuruk Lake area, Seward Peninsula, Alaska. *Jour Geol.*, vol. 57; p. 119—131.

H o p k i n s. David M., ed., 1967 — The Bering Land Bridge. Standford Univ. Press, 495 p.

H o p k i n s, David M., K a r l s t r o m, Thor N. V., and others, 1955 — Permafrost and ground water in Alaska. *U. S. Geol. Survey, Prof. Paper* 264-F; p. 113—146.

H u s s e y, Keith M., and M i c h e l s o n, Ronald W., 1966 — Tundra relief features near Point Barrow, Alaska. *Arctic,* vol. 19; p. 162—184.

J o h n s t o n, G. H., B r o w n, R. J. E., 1964 — Some observations on permafrost distribution at a lake in the Mackenzie Delta, N. W. T., Canada. *Arctic*, vol. 17; p. 162—175.

J o h n s t o n, G. H., and B r o w n, R. J. E., 1966 — Occurrence of permafrost at an arctic lake. *Nature,* vol. 211; p. 952—953.

K r i n s l e y, Daniel B., 1965 — Birch Creek pingo, Alaska. *U. S. Geol. Survey, Prof. Paper* 525-C; p. 133—136.

L e i f i n g w e l, Ernest de K., 1919 — The Canning River region, northern Alaska. *U S. Geol. Survey, Prof. Paper* 109; 251 p.

L i k e n s, G. E., and J o h n s o n, P. L., 1966 — A chemically stratified lake in Alaska. *Science*, vol. 153; p. 875—877.

L i v i n s t o n e, Daniel A., 1954 — On the orientation of lake basins. *Amer. Jour. Sci.*, vol. 252; p. 547—554.

L i v i n g s t o n e, Daniel A., 1963 — Alaska, Yukon, Northwest Territories, and Greenland. Limnology in North America, ed., David G. Frey, Univ. of Wis. Press; p. 559—574.

L i v i n g s t o n e, Daniel A., B r y a n, Kirk, Jr., and L e a h y, R. G., 1958 — Effects of an arctic environment on the origin and development of freshwater lakes. *Limnology and Oceanography*, vol. 3; p. 192—214.

M a c C a r t h y, Gerald R., 1952 — Geothermal investigations on the arctic slope of Alaska. *Amer. Geophysical Union, Trans.*, vol. 33; p. 589—593.

M a c k a y, J. Ross, 1956 — Notes on oriented lakes of the Liverpool Bay area, Northwest Territories. *Revue Canadienne de Géographie*, vol. 10; p. 169—173.

M a c k a y, J. Ross, 1963 — The Mackenzie Delta area, N. W. T. *Canadian Dept. Mines and Tech. Surveys, Geog. Br., Mem.* 8, 202 p.

M a c k a y, J. Ross, 1966 — Pingos in Canada. *Permafrost Intern., Conf. Proc., Purdue Univ., Nat. Acad. Sci.-Nat. Res. Council, Pub.* no. 1287; p. 71—76.

M u l l e r, S. W., 1945 — Permafrost or permanently frozen ground and related engineering problems. *U. S. Army. Special Rpt., Strategic Engineering Study*, no. 62, 2nd ed.; Reprinted 1947, Edwards Brothers, Ann Arbor, Mich., 231 p.

N i c h o l s, Harvey, 1967 — The disturbance of arctic lake sediments by 'bottom ice': a hazard for palynology. *Arctic*, vol. 20; p. 213—214.

P é w é, Troy L., 1954 — Effect of permafrost on cultivated fields, Fairbanks area, Alaska. *U. S. Geol. Survey, Bull.* 989—F; p. 315—351.

P i h l a i n e n, J. A., 1961 — Fort Simpson, N. W. T. — Engineering site information; soil and water conditions. *Nat. Res. Council, Canada, Tech. Paper* no. 126; 9 p.

P i h l a i n e n, J. A., and J o h n s t o n, G. H., 1954 — Permafrost investigations at Aklavik, 1953. *Nat. Res. Council, Canada, Tech. Paper* no. 16; 16 p.

P l a f k e r, George, 1964 — Oriented lakes and lineaments of northeastern Bolivia. *Geol. Soc. America, Bull.*, vol. 75; p. 503—522.

P r i c e, W. Armstrong, 1963 — The oriented lakes of arctic Alaska: a discussion. *Jour. Geol.*, vol. 71; p. 530—531.

P r i c e, W. Armstrong, 1968 — Oriented lakes. *Earth Sci. Encyclopedia*, vol. III, p. 784—796. Reinhold Book Corp.

R e x, R. W., 1961 — Hydrodynamic analysis of circulation and orientation of lakes in northern Alaska. Geology of the Arctic, Gilbert O. Raasch, ed. *Univ. Toronto Press*, vol. 2; p. 1021—1043.

R o c k i e, W. A., 1942 — Pitting on Alaskan farm lands: a new erosion problem. *Geog. Rev.*, vol. 32; p. 128—134.

R o s e n f e l d, G. A., and H u s s e y, K. M., 1958 — A consideration of the problem of oriented lakes. *Iowa Acad. Sci., Proc.*, vol. 65; p. 279—287.

S c h a e f e r, Vincent J., 1950 — The formation of frazil and anchor ice in cold water. *Amer. Geophysical Union, Trans.*, vol. 31; p. 885—893.

S p e t z m a n, L. A., 1959 — Vegetation of the arctic slope of Alaska. *U. S. Geol. Survey, Prof. Paper* 302—B; p. 19—58.

W a h r h a f t i g, Clyde, 1965 — Physiographic division of Alaska. *U. S. Geol. Survey, Prof. Paper* 482; 52 p.

W a l l a c e, Robert E., 1948 — Cave-in lakes in the Nabesna, Chisana, and Tanana River valleys, eastern Alaska. *Jour. Geol.*, vol. 56; p. 171—181.

W h i t e, George W , 1964 — Early Discoverers XIX: Early description and explanation of kettle holes: Charles Whittlesey (1808—1886). *Jour. Glac.,* vol. 5; p. 119—122.

W h i t t l e s e y, Charles, 1960 — On the drift cavities, or „potash kettles" of Wisconsin. *Amer. Assoc. Adv. Sci., Proc.,* 13; p. 297—301.

W i g g i n s, I. L., and T h o m a s, J. H., 1962 — A flora of the Alaskan arctic slope. Univ. Toronto Press, 425 p.

W o l f e, Peter E., 1953 — Periglacial frost-thaw basins in New Jersey. *Jour. Geol.,* vol. 61; p. 133—141.

Since January 8, 1968, when this paper was submitted in manuscript, the following paper on the oriented lakes of norther Alaska appeared:

C a r s o n, Charles E., 1968 — Radiocarbon dating of lacustrine strands in arctic Alaska. *Arctic,* vol. 21, p. 12—26.

C a r s o n concluded that transgressive expansion of the lakes reached a maximum between 4,000 and 8,000 years ago.

Part II
MASS WASTING

Editor's Comments
on Papers 15 Through 24

The patterns and smaller features typical of the periglacial environment that were considered in Part I provide distinctive phenomena that occur mainly in the cold conditions typical of the periglacial areas, and thus have valuable palaeoclimatic implications where they are found in fossil form. They do not, however, involve any very substantial movement of material or major modifications of the overall morphology of the land surface. They could be considered as ornaments on the landscape, providing intriguing geomorphological problems the solution of which has provided a great deal of valuable information concerning periglacial processes, especially in relation to the formation and characteristics of ground ice. These processes are being increasingly found to be of practical importance as the northlands are developed more and more.

The processes that are considered in Part II are capable of much greater morphological change and involve very considerable amounts of material. They are mainly processes related to the modification of slopes by the mass movement of material and the preparation of this material for movement by mass weathering. In some areas, such as southern Britain, these processes have reduced the slope gradients to values that are lower than those that present-day processes would probably produce had they acted for a long period of time. Thus the low slopes characteristic of this area are essentially stable under present-day conditions, a situation that is of great value when considering soil stability under modern agricultural methods, when the soil is left bare for parts of the year; it is thus subject to accelerated erosion unless slopes are gentler than would be expected. The transfer of these farming techniques to other areas not so favorably endowed climatically and where slopes have not been modified in this way has led in many areas to serious soil erosion.

Part II is divided into nine sections; each is devoted to a process of mass wasting that produces recognizable landforms or deposits on the slopes. These nine processes are (1) creep, (2) scree formation, (3) rock glaciers, (4) protalus ramparts, (5) block or boulder fields, (6) solifluction, (7) altiplanation terraces, (8) slope asymmetry, and (9) stratified slope deposits. Each of the nine processes is discussed in one or two papers, which give an introduction to the process. The reference list provides considerable literature covering each process.

CREEP

The slow process of creep is not confined to periglacial areas as it occurs very widely in all types of climate and on all land surfaces that have a superficial cover. The process is, however, particularly effective in the periglacial environment for several reasons. These environments tend to be damp because of the low summer temperatures and hence low evaporation, and the presence of permafrost in the colder periglacial areas renders the ground impermeable below the active layer in which creep takes place. Creep can be induced by wetting and drying of the soil, which causes alternating expansion and contraction; this process can occur in any climatic zone. In the periglacial zone, in addition to wetting and drying, the action of frost in the soil is a potent agent inducing creep of the superficial material. Creep can be defined as the downslope movement of material, the material often defining the type of creep. Thus soil creep, in fine material, can be differentiated from talus creep, which takes place in coarser material. The frost acts as a ratchet, according to Washburn, in effecting the downslope movement of material. The particles are lifted up by the frost normal to the slope surface, but they fall back vertically when the ground ice melts, thus progressing downslope at a rate that depends both on the angle of slope and on the degree of frost heave. Sometimes the full theoretical amount of creep is not achieved because cohesion of the soil results in the downward movement on thawing being at an angle between the vertical and the normal to the slope. This amount is called the "retrograde component."

This aspect of frost creep is considered in Paper 15. The detailed work done by Benedict in the periglacial area of the Front Range of the Colorado Rocky Mountains illustrated the occurrence of this type of movement. Benedict's contribution illustrates the value of detailed field observations over a fairly long time interval, which are necessary to obtain reliable results on these slow-acting periglacial processes. Work by T. Czudek (1964) exemplifies the large volume of work that has been done on periglacial processes by the East European geomorphologists, who have a wealth of relevant examples to observe and study in their cold continental climate in which periglacial processes were very active.

Periglacial creep was also active in the British Isles during the major ice advances; relics of its action are widespread in the form of the extensive and important "head" deposits that mantle so many of the gentle slopes of southern England and are particularly important on the chalklands of southeast England. The chalk breaks up readily under frost action, and the head deposits consist of angular chalk fragments that have moved downhill under the influence of frost creep. Washburn has suggested that frost creep is one of the most important processes of mass movement in a periglacial zone; he suggests that it may exceed other forms by up to 3 to 1, with absolute rates of movement measured in northern Greenland at 0.9 cm/year in areas that dried out in summer to 3.7 cm/year in areas that remained damp.

SCREE FORMATION

Head deposits tend to mantle those slopes that are formed of rocks that break up into small fragments under the action of frost processes. Screes or talus, as the coarser material is often called, occurs where the rocks break up into larger pieces. Screes are particularly common in areas that have previously been affected by active valley glaciers. These glaciers in areas of strong, hard rocks, erode deep U-shaped valleys, the sides of which are greatly oversteepened by glacial erosion. Such vertical or near vertical walls are found, for example, in the Alps, the Norwegian mountains, the Southern Alps of New Zealand and the Rocky Mountains of North America. The Yosemite valley is an excellent example in the latter area. These oversteepened rock walls are unstable when the ice retreats and the area comes under the action of periglacial processes. The process of pressure release helps to generate planes of weakness that cause cracking of the rocks as the ice load is removed. The cracks then facilitate the entry of water, and the freezing of this water helps to break off the rock, thus producing the scree deposits.

Screes vary in size according to the nature of the rocks and the way in which they break. Screes normally have an angle between 30 and 35°, but can vary between 27 and 37°. Normally the finer material occurs at the top of the scree and the coarser at the bottom, thus often resulting in a concave profile longitudinally down the scree, especially where material is not being removed from the foot of the scree. The transverse profile of the scree depends on the nature of the rock face from which it is derived. Where the rock outcrop is smooth and straight, the scree profile along the slope will tend to be fairly straight, as for example the well-known screes above Wastwater in the English Lake District.

In other areas, however, where the rock wall is irregular, with bastions and gullies, the scree tends to form fan-like masses, which have a convex profile across the slope, with runnels developing where the individual fans merge. The fans sometimes stretch up through a narrow gully and broaden out above into a broader collection area. The fan-like form is sometimes called a "talus cone."

Processes that lead to the buildup of scree deposits include rockfall, which is the fall of individual blocks of rock when they are released by frost riving or some other process. The falling rocks may break others on which they fall, thus causing comminution of the scree. Rolling, bouncing, and sliding all occur in fairly coarse material. In finer material frost creep can also be an effective process, but in the coarser material water can penetrate through too easily to render frost action very effective. Most screes are only very lightly vegetated, and many are completely bare of vegetation, so that they are exposed to processes that lead to movement.

A description of the scree slopes in the periglacial region of northern Scandinavia is provided by A. Rapp in Paper 16 from detailed firsthand measurements on scree slopes; the author has been able to produce a quantitative study of the processes that lead to scree formation and to analyze the resulting landforms. His examples are taken from the mountain walls of Tempelfjorden in Spitzbergen, and he has also studied the screes and processes associated with their formation in Kärkevagge in northern Sweden.

ROCK GLACIERS

Rock glaciers form a link between truly glacial landforms and forms developed in periglacial areas. Most rock glaciers were previously more active when their ice core was being adequately nourished by snow accumulation to be a true cirque glacier. Recently there has been a certain amount of argument concerning the relationship between ice-cored moraines and rock glaciers, a controversy that indicates the close connection between the glacial and periglacial processes in these features. A rock glacier appears superficially to resemble a very heavily debris mantled, small cirque glacier. They are characterized by a matrix of ice at depth between the boulders and they also move slowly downslope. A diagnostic superficial morphological feature is the presence of lobate wrinkles and folds on the surface near the terminus of the rock glacier. These fold structures result from the very slow movement of the rocks. Rock glaciers can reach considerable dimensions in suitable environments, and have attained lengths of 3 km, with terminal embankments up to 60 m in height; very large boulders often occur on them. The material is usually angular and coarse, with finer material tending to work down toward the bottom. In general, the upper layer moves faster than the lower layers, thus often producing a steep terminal face lying at the angle of rest of the coarse, blocky material, which produces a terminal talus slope as material rolls down the steepened front. Rates of movement of the surface have been measured in a number of areas, giving results that vary between about 0.3 m to about 1.5 m/year.

The forms of rock glaciers vary according to their surroundings. Those in Alaska have been grouped into three types: (1) lobate rock glaciers, which have a greater width than length, with dimensions often 60 to 1000 m in length and 90 to 3000 m in width; (2) tongue-shaped glaciers with greater length than width, which tend to be between 150 and 1500 m long and 60 to 800 m wide; and (3) spatulate glaciers that spread out into a broader tongue near the terminus. Rock glaciers usually occupy cirque basins high in the mountains, surrounded by rock outcrops from which they derive much of their surface material. A few rock glaciers may not be direct descendents from active cirque glaciers, but instead result from the movement of interstitial ice originally formed by the conversion of meltwater to ice or snow buried under rock falls. Some rock glaciers occur in nonglaciated mountains, such as the San Mateo Mountains of New Mexico, and these cannot be relict cirque glaciers. Rock glaciers that lie below the slopes of valleys in gulleys or along the valley side in positions not previously occupied by cirque glaciers also probably originate in this way.

Paper 17, by G. Wahrhaftig and A. Cox, is one of the most thorough and comprehensive studies of rock glaciers; it was made in the mountains of Alaska and published in 1959. Since this work was published, rock glaciers have continued to attract the attention of geomorphologists. They have been studied to provide evidence of different stages of glaciation, for example, by White (1971) in the Front Range of the Rocky Mountains in Colorado, where they are well developed. These rock glaciers formed during the Neoglacial period of more active glaciation, and are still moving slowly at about 2 to 15 cm/year. Rock glaciers are fairly widely distributed in both Alpine and Arctic environments wherever there are steep slopes, heavy snowfall, or relict glacier ice in suitable high-level hollows or gullies that lead down onto flatter surfaces where the downward-moving rock debris can accumulate in the typical wrinkled lobate form.

PROTALUS RAMPARTS

Protalus ramparts were originally described by Bryan in 1934, who named them. The features form a link between true talus slopes and rock glaciers. They are described by Bryan as ridges of rubble or rock debris that have accumulated at the foot of a perennial snowbank lying at the base of a rock slope. The material accumulates by individual rock sliding or bouncing over the snow slope to accumulate in a ridge at the base of the slope. They can be distinguished from scree or talus slopes by the hollow that exists between them and the rock wall or the rest of the talus. The french term for the features is *moraines de nevé*, a term that stresses their association with snow patches; the reference to moraine suggests that at one end they blend into true cirque glacier moraines. Protalus ramparts have been described by Watson (1971) in the mountains of north Wales, where they occur in hollows that were not quite deep enough to contain snow to form a moving cirque glacier. The thickness of snow in the hollow can be estimated from the form of the longitudinal profile, the result suggesting that it was not enough to cause flow and hence true cirque development. The protalus ramparts can form ridges along the base of the slope reminiscent of terminal moraines of cirque glaciers, or alternatively they may form ridges around the lateral margins of the snow patches, thus resembling the lateral moraines of cirque glaciers.

A good example of a feature that has all the characteristics of a lateral protalus rampart occurs in the upper Eden valley in northern England. It forms an asymmetrical ridge leading down from the outcrops of Millstone grit normal to the slope contours. The ridge is steeper on the inside, producing a slope that appears to be at the angle of rest of the angular material of which it is formed. The steep slope is probably the result of slumping as the containing snow patch eventually melted. The presence of slump material and a shattered rock wall inside the feature indicates the origin of the material that forms the lateral protalus ridge. Downslope it fans out from the hollow from which it was derived, again indicating mass movement rather than glacial formation, as lateral moraines of true cirque glaciers tend to form a rounded front as they curve round to form the terminal part of the moraine. The roundness of the particles in the protalus ridge was significantly less than that of particles from true moraine on the slopes below the feature, again indicating its formative process.

Protalus ramparts generally consist of angular and coarse blocks, because the fines do not have the necessary momentum to travel across the entire length of the snow patch to accumulate at its foot. The height to which protalus ramparts reach is often 9 m, but exceptionally high ones can reach 24 m. Protalus ramparts of this height are discussed in Paper 18, by J. W. Blagborough and W. J. Breed; it is one of the relatively few papers to be devoted to these interesting forms. The ramparts they describe are up to 14 m wide and 91 m long. They normally are arcuate and concave upslope. Their distribution is related to climate and especially to exposure, as they occur where sheltered snow patches can survive for a long time. Slight variations in the size of the snow patch from time to time can cause complexities of form, with several ridges forming where the snow patch is variable. Frost wedging is an important process in breaking off the blocks that form the rampart, so that the nature of the rock is an important attribute, as well as the pattern of thaw–freeze activity. Because of the importance of thaw–freeze processes, the features are probably more common in high-altitude, steep mountain areas rather than in the very cold permafrost areas. Permafrost is not necessary for their formation, but they have been observed in the Antartic.

Protalus ramparts provide valuable evidence of previous snow line positions, to which they are closely related, when they can be identified in fossil form. The examples from southern Utah probably formed actively in the glacial periods of the Wisconsin.

BLOCK FIELDS

Block fields are known by a variety of names, which include boulder fields, the German *Felsenmeer,* block streams, or mountaintop detritus. The last two terms indicate that block fields can occur in a variety of situations; some occur on slopes or in valley floors, and these tend to have a stream-like form or are tongue-shaped; others occur on flat summit surfaces, where they occur as boulder fields.

The block fields of the mountains of northern Scandinavia are described by R. Dahl in Paper 19; it is an example of the important contribution of Scandinavian workers to the understanding of the processes of mass weathering and movement in a periglacial environment. Another example of boulder fields from Scandinavia illustrates some of the problems associated with the interpretation of boulder-field formation. In the Åland Islands off southwest Finland two very different types of boulder field were located in close proximity. One type was formed of rounded boulders and consisted of a series of ridges running across the slope. The pattern of ridges and the arrangement of the stones, with the larger on the ridge crests and on the surface, and their high degree of roundness all suggested that the ridges could be raised shingle beach ridges formed by waves when the land was isostatically depressed. This area has risen above the sea during the postglacial period of about 9000 years in this vicinity. The pattern of ridges related to the available fetch and the sharp contact between the boulders, which included erratics, and the rocks on which they were resting or against which they were banked up, all support a marine origin.

Below the rounded boulder fields others occurred, which consisted of angular boulders. They lay at a lower angle, had no ridges, and the stones could be traced up to outcrops of Rapakivi granite, which constituted the boulders of both types of boulder field. The outcrop of granite showed newly broken off masses of rock, which had broken off along very well marked pressure release joints, giving the impression of pseudobedding planes in the rock. The angular boulder field had presumably been formed by the downslope movement of blocks rived off by frost action working along pressure-release joint planes. The blocks then moved farther downslope by the processes of frost creep, as the slopes were too gentle to allow active sliding or rolling of the blocks.

Where the slopes are rather steeper the block fields can be elongated downslope or down valley, as for example the clock streams described by Andersson in the Falkland Islands. The Hickory Run boulder field described from Pennsylvania by Smith (1953) is an example of a valley-floor boulder field. This boulder field also indicates another process that may be important—the eluviation of fines by running water in the valley bottom, which concentrates the large blocks, allows rapid drainage, and hence restricts soil formation and vegetation growth, resulting in the typical unvegetated nature of well-developed boulder fields. The block fields of the mountain summits, such as those of the Trongat Mountains in Labrador, are probably formed in situ by the mechanical breakdown of the rock material. They occur on gently sloping ground across which the material could not be readily moved. They are sometimes thought to occur only in areas that were not overrun by active glaciers, as their rate of formation is probably slow. It depends on the nature of the freeze–thaw cycles and the degree of cold. The boulders on gentle slopes show random orientation and can be distinguished from those affected by moving ice, which tend to show a preferred orientation in the direction of ice flow.

SOLIFLUCTION

Solifluction is a process that can take place in many different environments, as it occurs whenever the surface soil is rendered so wet that it can flow. This process is, however, especially likely to occur in a permafrost environment because the active layer always rests on an impermeable layer, and much water is available from the melting of snow and ground ice in the active layer. The excessive moisture in ground ice in frost-susceptible soils makes the active layer into a slurry that can move downhill on very gentle slopes.

The term "solifluction," which suggests the flow of the soil, was first introduced by J. G. Andersson in 1906. Paper 20 is a classic paper because of its historical importance and the fact that Andersson drew attention to a process that is very important in the modification of slopes by mass movement in a periglacial environment in particular. The term "gelifluction" was introduced by Baulig for the process of solifluction particularly associated with frost action. There is some advantage in using the term "gelifluction" in a consideration of periglacial processes, as it is a definitely periglacial term.

Since the pioneer work of Andersson on periglacial solifluction, the process has been studied in several areas in the field. A useful attempt to record solifluction in progress was undertaken by Williams (1963) in Norway. He used a pliable probe buried in the soil to record downslope movement when the thaw set in. The inclination of the probe was recorded, which revealed that the amount of movement was a maximum on the surface and fell to nil at the base of the active layer. Further experiments have been carried out in Rocky Mountain Front Range by Benedict,

175

using the soil pillars devised by Rudberg (1958) for similar observations. The pillars are set in the ground and their movement can be traced as the soil moves.

Surface markers have also provided useful information on the pattern of solifluction over a slope and its relation to the ground surface configuration and soil characteristics. The vertical velocity profile tends to be concave downslope. Because of the importance of moisture in solifluction, it tends to operate seasonally in areas where the soil dries out after the snow and ground ice have melted and run off. Solifluction continues for longer in the wettest areas or the soils that retain moisture longest. It is a process that can operate on very low angle slopes, with gradient of only 1 or 2°. The size of the soil is important as it affects the moisture content, and vegetation also pays an important part, due both to its effect on water and its capacity to hold the surface soil in place where it is dense. Rates of solifluction recorded in northeast Greenland by Washburn (1969) varied from 0.6 cm/year on a drier slope of 10 to 14° to 6.0 cm/year on a wetter slope. A rate of 12.4 cm per year was recorded on one wet slope of 12° gradient.

The process gives rise to specific slope forms, of which the solifluction lobe is the most conspicuous, especially in the moraine or soil-covered slopes of the permafrost zone in the sparsely vegetated tundra regions. The solifluction lobes have a steep front, and their rate of movement can sometimes be assessed by the fact that the soil moves downslope over the vegetation, which is preserved beneath the flowing material; it can be dated where it can be excavated from beneath the overlying lobe. The features associated with solifluction on slopes include the turf-banked lobe; one such terrace has overridden the ground beneath for 13 m, at a rate of about 2 mm/year.

Another feature is the stone-banked lobe. Benedict (Paper 15) described those on Niwot Ridge in the Front Range of the Colorado Rocky Mountains. Their rate of downslope movement ranged from 1.5 to 23 mm/year according to his measurements, which were made over a number of years with Rudberg pillars and marker pegs. The stones in such lobes tend to become preferentially orientated, with their long axes parallel to the slope, although those at the front of the lobes are often aligned parallel to the border of the lobe, but standing on edge where they are of platy form. Movement on all the features associated with gelifluction occurs most rapidly in the spring thaw; there may even be retrograde or upslope movement during the later summer, as the thawed soil sinks back toward the ground. There is a complex interaction between vegetation and solifluction, the former being affected by the latter as much as the other way round. Moisture control may be more important than the binding effect of vegetation, so that sometimes better-vegetated and therefore wetter slopes move more rapidly.

ALTIPLANATION TERRACES

The term "altiplanation terrace" was first introduced by Eakin in 1916 to refer to the same landform described as Goletz benches in the Russian literature. These features occur widely in the USSR in the Ural mountains and in Siberia, and have subsequently been recognized in fossil form in Brittany, Devon, and the Ungava region of Canada. They have also been widely described from eastern Europe.

Papers 21 and 22 describe these features. In Paper 21, R. S. Waters compares the fossil altiplanation terraces of Devon with currently forming and still active features in Vest-Spitzbergen. Paper 22 is an example of the work done on these features in eastern Europe by T. Czudek and J. Demek, who describe the terraces in Yakutia. They term the process that forms the terraces "cryoplanation," which implies that the action of frost processes is an essential element in their formation. Morphologically they consist of a series of benches, which are nearly horizontal, separated by fairly steep risers. The tread width varies from a few to several hundred meters. They often circle mountain summits, but may be discontinuous. The front slope may reach 20° or even 40° at times. The height of the riser in Siberia is usually 5 to 7 m, but can be higher in Alaska. The terraces are relatively independent of structure and are solid rock benches with only a thin veneer of debris, which often shows periglacial patterning. In length they can exceed 10 km.

The process of cryoplanation or flattening by frost is brought about by a combination of nivation and gelifluction. The terraces may go through three phases of development, the first of which is characterized by nivation that produces a wide bench in which snow tends to linger as it bites deeper into the hillside. The snow helps to provide moisture for the second stage, which includes solifluction across the growing tread and frost riving on the steep slope at the back of the snow patch. The debris that is moved across the slope reduces its gradient to a slope commonly of about 7°. As the step behind the terrace retreats back, it may in time remove the hilltop, at which point the third stage of development is reached and a cryoplanation surface is produced. The gradient at this stage may be reduced to 2°. The movement of the material across the slope helps to reduce its gradient with time. The process continues by positive feedback relationships, whereby the snow patch grows bigger and becomes more effective with time. Where a series of terraces occur they may well mark a sequence of nevé limits. Each terrace expands by the destruction of the scarp above it, so they gradually consume the hillside. The altiplanation terraces can form without permafrost, but they develop more efficiently in permafrost areas; those found in Devon, for example, around the Dartmoor hill summits, provide evidence of palaeoclimatic significance.

The best-developed altiplanation terraces probably take tens of thousands of years to form and occur in well-jointed, resistant rocks. Some altiplanation terraces may well have been confused with benches formed under postulated erosion subcycles. Their recognition is, therefore, of importance in a study of the development of landforms through time, and they provide good evidence of periglacial conditions where they can definitely be located.

SLOPE ASYMMETRY

Asymmetrical slopes have been reported from a number of different areas and can be due to a wide range of phenomena, not all of them related to periglacial processes. The asymmetry of slopes in periglacial areas is thought to be due to the effect of aspect on the intensity with which some of the slope processes already commented upon act. In Paper 23, H. M. French discusses the slope asymmetry in the Chiltern Hills of southeast England. Generally asymmetry is best developed in east-west trending valleys in which one side faces south and the other north. The south-facing side receives more insolation and hence has greater freeze-thaw activity and can also support a denser mat of tundra vegetation. Weathering is more active, and a finer soil develops, which can move downslope more effectively on a lower gradient. This is due partly to the greater moisture under the influence of soil size and vegetation and to the more effective freeze-thaw regime. On the north-facing slopes the debris tends to be coarser, the infiltration rate is higher, the moisture supply at the surface is less, and owing to the colder conditions vegetation is sparser. All these factors help to account for the gentler slopes that usually characterize the south-facing side of the valley.

Other variables that can affect valley-side asymmetry and which may more than counterbalance those reducing the gradient of south-facing slopes include the position of the stream at the foot of the slope. Slopes being actively undercut by a stream will tend to be steeper regardless of the slope processes operating. Slope asymmetry may also be due to exposure to the dominant snow-bearing winds. The slopes that receive the greater snowfall will tend to be of lower gradient because of the greater moisture supplied by the melting snow over a longer period. In the zone marginal to the limit of periglacial processes the reverse asymmetry may apply. In these marginal zones the north-facing slope may experience the greater effectiveness of thaw–freeze processes and hence become gentler than the south-facing slope, which may dry out more rapidly at the southern limit of permafrost in the northern hemisphere. Many different processes are involved in the development of valley asymmetry, and each area must be considered individually if the correct reason for asymmetry is to be elucidated.

STRATIFIED SLOPE DEPOSITS

Stratified slope deposits have been studied in periglacial areas. This occurrence is also known by the name *grèzes litées* in the French literature, a term introduced by Y. Guillien in 1951 (Paper 24). The term *éboulis ordonées* is also used. In a paper published by Dylik in 1960 stratified slope deposits are described; this paper represents the work of one of the best-known periglacial specialists from eastern Europe.

Individual beds in stratified slope deposits tend to thicken downslope, with dip increasing with depth. Angles between 24 and 27° have been reported; in New Zealand dips of 30 to 35° have been described. The deposits consist of frost-shattered material, and the main problem is to account for the stratification. The material ranges up to 2.5 cm in diameter, according to Guilien, but other observers have noted larger particles. The slopes tend to be smoothly concave and develop fairly rapidly; some in the upper Durance valley at Le Queyras have developed during zone III. The coarser layers may have accumulated during the colder phases. A similar type of deposit on the slopes of the Manifold Valley on the Derbyshire–Staffordshire border shows alternating coarse layers of well-cemented limestone chips with intervening, less-cemented, finer limestone debris.

It has been suggested by Prentice and Morris that the coarser layers represent cold periglacial conditions when frost action was effective, and cementation took place under the cold temperatures of these periods. The finer layers represent the warmer intervening periods. The formation of stratified slope deposits does not require permafrost, and Soons has suggested that needle ice may help to account for the sorting. Perennial snowdrifts are thought to be necessary for their origin by some workers; others stress the action of frost shattering, rill work, and sheet flow. Melt water from snow may play an important part in eluviating the coarser material of fines and depositing the coarser material. Much still remains to be learned of the details of the sorting process that leads to the occurrence of stratified screes. Some areas show preferential orientation of these slopes, but other areas do not. It is generally agreed, however, that this distinctive slope deposit is characteristic of the periglacial zone, but not necessarily of the permafrost zone.

15

Reprinted from *Arctic Alpine Res.*, **2**(3), 165–183, 219–226 (1970)

DOWNSLOPE SOIL MOVEMENT IN A COLORADO ALPINE REGION: RATES, PROCESSES, AND CLIMATIC SIGNIFICANCE

JAMES B. BENEDICT

Institute of Arctic and Alpine Research
University of Colorado
Boulder, Colorado 80302

CONTENTS

ABSTRACT

Soil-movement rates, processes, and landforms were studied above timberline on the east slope of the Colorado Front Range. Maximum rates of downslope movement measured in sorted stripes, turf-banked lobes and terraces, and stone-banked lobes and terraces at nine experimental sites on Niwot Range ranged from 0.4 to 4.3 cm/yr. Rates of displacement were strongly influenced by differences in moisture availability and gradient, but were relatively unaffected by differences in soil texture and temperature. Movement is currently confined to the upper 50 cm of soil; columns of small cement rods placed in vertical drillholes showed no downslope displacement below this depth after four years of burial.

Periodic measurements of downslope movement and frost heaving during the 1965-66 annual freeze-thaw cycle suggest that solifluction is a more effective process than frost creep in the saturated axial areas of turf-banked lobes in wet sites, but is less effective than frost creep at their edges. Potential frost creep at one experimental site exceeded theoretical values calculated from heave and slope measurements. Retrograde movement was larger than anticipated. At a site where frost creep is the dominant movement process, stone-banked lobes moved three times as rapidly as the finer-textured soil between them; at a site where shallow solifluction is important, stripes of coarse debris moved only half as rapidly as stripes of fine material.

Turf-banked lobes and terraces are the result of intense solifluction beneath a cover of vegetation; they form where downslope movement is impeded, and are normally associated with a decrease in gradient such as occurs on concave lower slopes. The shape of the front (linear or lobate) is determined by the uniformity of moisture distribution parallel to the contour of the slope. At least two generations of turf-banked lobes and terraces occur in the Niwot Ridge area. Terraces with gentle, subdued fronts and patterned treads date from the late Pleistocene, and were used as camping and butchering areas by prehistoric man as long ago as 7,650 ± 190 radiocarbon years. Most bear an Altithermal soil: five dates for the soil range from 5,800 ± 125 to 5,300 ± 130 BP. Lobes and terraces with overhanging fronts and unpatterned treads postdate the Altithermal interval: radiocarbon and stratigraphic evidence suggest that they formed late in the Temple Lake Stade of Neoglaciation. The front of one small Neoglacial turf-banked lobe has advanced

at an average rate of 0.19 cm/yr during the past 2,340 ± 130 radiocarbon years.

Stone-banked lobes and terraces are caused by frost creep and are favored by an absence of vegetation; they commonly develop where moving sorted stripes or blockfields encounter a decrease in gradient. Sorting is partially inherited, but is accentuated as the lobe or terrace moves downslope. Stone-banked lobes and terraces in the Niwot Ridge area developed late in the Temple Lake Stade of Neoglaciation. A series of radiocarbon dates from the buried A horizon in one stone-banked terrace suggests that its front has advanced at an average rate of 0.34 cm/yr during the past 2,470 ± 110 radiocarbon years. Movement was slow during the Temple Lake-Arikaree* interstade (2,650 to 1,850 BP), a time of soil formation and intense cavernous weathering, and during the Arikaree glacial maximum, when lichen measurements show that the slope was covered with an insulating blanket of perennial snow. Disappearance of the snowbank led to an eight-fold increase in the rate of terrace advance between about 1,150 and 1,050 BP.

The widespread occurrence of stone-banked lobes and terraces, sorted polygons, and sorted stripes on the treads of turf-banked lobes and terraces suggests a general decline in the availability of moisture and the effectiveness of solifluction since the end of the Middle Stade of Pinedale (Late Wisconsin) Glaciation. During the latest Pinedale and earliest Neoglacial ice advances frost creep replaced solifluction as the dominant movement process on many slopes. Neither process is particularly effective today, except in specialized microenvironments that are saturated with meltwater in the autumn.

---

*The term Arikaree, as used in this context, has been preempted. No alternative name has been established (Editor).

INTRODUCTION

An eight-year study of soil-movement processes above timberline in the Colorado Front Range has recently been completed. The objectives of the study were threefold:

(1) To describe, classify, and determine the origins of landforms produced by moving soil on slopes in the Front Range alpine region.

(2) To measure modern rates of downslope

soil movement, evaluate the processes involved, and relate modern rates of movement to present environmental conditions.

(3) To determine former rates of movement, date previous intervals of intense frost action, and infer past changes in the Front Range environment.

FIGURE 1. Oblique aerial view of Niwot Ridge, looking west toward the continental divide. The photograph was taken on May 9, 1962, and shows the ridge under conditions of maximum snow accumulation. Variations in snow cover result from the redistribution of snow by strong northwesterly winds. The striated texture of the ridge at tree line reflects the influence of wind upon patterns of tree growth and snow drifting. U.S. Air Force photograph, courtesy of Ronald Foreman and William S. Osburn.

CHARACTERISTICS OF THE STUDY AREA

Field studies were centered on the crest and sides of Niwot Ridge, a windswept tundra upland in the Indian Peaks region of the Colorado Front Range (Figures 1 and 2). The ridge begins on the Continental Divide, at the summit of Navajo Peak (4,087 m), and extends eastward as a narrow arête for about 2 km before its crest abruptly widens and its flanks become more gentle. Low, rounded knolls and shallow saddles reflect the influence of prolonged mass wasting upon the structurally complex, igneous and metamorphic basement. Bedrock is largely hidden by a thick mantle of soil. Subdued topography persists to timberline at the eastern end of the ridge, approximately 9 km east of the Continental Divide. Except for its length, the ridge is representative of other alpine ridge systems in the Front Range, and the conclusions reached in this report are believed to be applicable to the Front Range as a whole.

Timberline lies at an average elevation of about 3,350 m in the Indian Peaks region. Stunted Engelmann spruce (*Picea engelmanni*), subalpine fir (*Abies lasiocarpa*), and limber pine

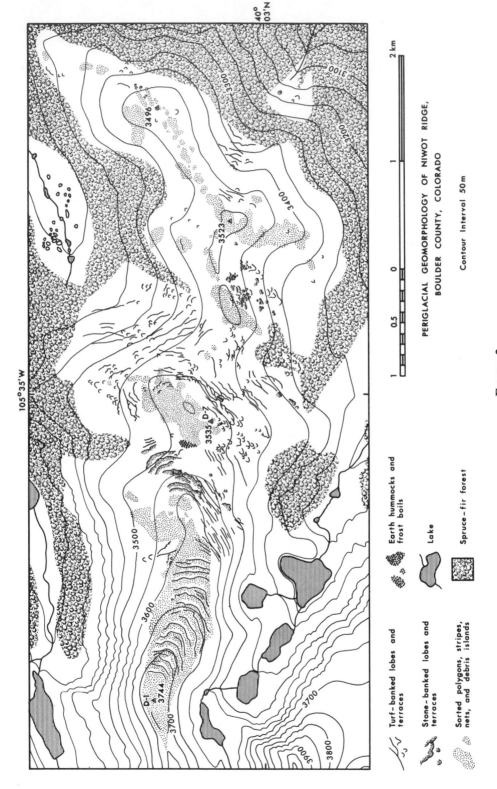

Turf-banked lobes and terraces

Stone-banked lobes and terraces

Sorted polygons, stripes, nets, and debris islands

Earth hummocks and frost boils

Lake

Spruce—fir forest

PERIGLACIAL GEOMORPHOLOGY OF NIWOT RIDGE, BOULDER COUNTY, COLORADO

Contour Interval 50 m

FIGURE 2.

(*Pinus flexilis*) persist to higher elevations, but the dominant vegetation above timberline consists of grasses, sedges, and low perennial herbs. Marr (1961) refers to the local tundra as "a mosaic of stand-type ecosystems," in which individual units are characteristically small, sometimes less than 1 m² in area. Differences in the depth and duration of snow, the degree of disturbance by frost processes, wind erosion, and small mammals, the availability of fine-textured soil, and the amount and distribution of growing-season moisture all contribute to the patchiness of the alpine vegetation. Differences in rock type and exposure to solar radiation are not generally reflected by the plant cover.

For the period 1953 to 1964, Marr *et al.* (1968) report a mean air temperature of −3.3°C at a weather station near the western end of Niwot Ridge (Station D-1, elevation 3,744 m). Absolute maximum and minimum temperatures recorded during this interval were 19.4 and −36.7°C. Observations of perennially frozen ground are inconclusive, but temperatures appear to be low enough to produce permafrost, or at least to maintain it, in favorable sites. Direct precipitation at the D-1 station is relatively low (487 to 825 mm/yr), but *effective* precipitation is many times greater in protected sites where winter snowfall is concentrated by wind. Prevailing winds are northwesterly, and blow with an average velocity of 8 m/sec; velocities are highest during the winter months, when gusts commonly exceed 50 m/sec.

The Front Range glacial sequence (Table 1) is a convenient chronological framework for the study of geomorphic processes. During the Wisconsin stage of Pleistocene glaciation, two major ice advances, the Bull Lake and Pinedale glaciations (Richmond, 1965; Madole, 1969), occurred in valleys to the north and south of Niwot Ridge. Neither covered the crest or upper flanks of the ridge itself. Pinedale glaciers withdrew from Front Range valleys about 7,500 years ago. Conditions during the subsequent Altithermal interval were generally warm, with wide fluctuations in precipitation. During the past 4,500 years, the region has experienced repeated small-scale cirque glaciation; three Neoglacial ice advances—the Temple Lake, Arikaree, and Gannett Peak Stades—are recorded in deposits on the floors of Front Range cirques (Benedict, 1967, 1968).

The combination of low temperatures, locally high *effective* precipitation, and large snow-free expanses of upland tundra has produced an environment in which frost processes are highly effective. Active sorted polygons, nets, and stripes occur on the floors of ephemeral ponds and in areas that receive meltwater drainage from persistent snowbanks; inactive polygons occur on windswept knolls and in dry saddles (Figure 2), and reflect at least one former period during which conditions were moister than they are at present. Earth hummocks and frost boils blanket gentle slopes in wet areas where winter snow cover is light to negligible. On steeper slopes, frost creep and solifluction have produced the terraces and lobes of slowly moving soil that are the subject of this paper.

TABLE 1
Front Range glacial chronology[a]

	Period	Approx. age (years BP)
		0
		100
	Gannett Peak Stade	
		350
	Interstade	
		950
Neoglaciation	Arikaree Stade	
		1,850
	Interstade	
		2,650
	Temple Lake Stade	
		4,500
	Altithermal Interval	
		7,500
	Late Stage	
		10,000
	Interstade	
		12,000
Pinedale Glaciation	Middle Stade	
	Interstade	
	Early Stade	
		25,000
	Interglaciation	
		32,000
	Late Stade	
Bull Lake Glaciation	Interstade	
	Early Stade	

[a]Modified from Richmond, 1965, Table 2, and Benedict, 1968, Figure 7.

MOVEMENT PROCESSES

Frost creep and solifluction are the two main processes responsible for downslope soil movement in the Colorado Front Range. Because of a lack of consistency in the definition of these terms by different authors, their use in this paper is explained below.

FROST CREEP

Frost creep is defined as *the net downslope displacement that occurs when the soil, during a freeze-thaw cycle, expands normal to its surface and settles in a more nearly vertical direction.* Expansion, or heaving, is proportional to the total thickness of segregated ice layers that form in the freezing soil, and is generally directed at right angles to the ground surface. Heaving is favored by saturated conditions and by slow, deep freezing, and is important only in soils that contain sufficient fine-textured material to permit water to move upward to the base of the frozen layer.

In analyzing movement data from Niwot Ridge, frost creep was broken down into two components: (1) *potential frost creep,* the downslope displacement caused by frost heaving during the fall and winter freeze, and (2) *retrograde movement,* the apparent upslope displacement caused by non-vertical settling during the spring and summer thaw (*see* Washburn, 1967, Figure 5). Potential frost creep can be calculated from slope angles and heave measurements.

SOLIFLUCTION

Following Andersson (1906), solifluction is defined as *"the slow flowing from higher to lower ground of masses of waste, saturated by water. . . ."* This definition does not restrict the process to a periglacial environment (Butrym *et al.,* 1964), and special terms such as "gelisolifluxion" (Baulig, 1956), "gelifluction" (Washburn, 1967), and "congelifluxion" (Dylik, 1967) are sometimes used for solifluction operating over a frozen subsoil.

Solifluction is favored by an impermeable substratum, such as frozen ground, which limits the downward movement of water through the soil. The process is most effective where melting ice layers weaken the soil and provide a source of excess water that reduces *internal friction and cohesion* (Williams, 1957, 1959). The importance of ice-lensing is illustrated by the restricted occurrence of solifluction on Niwot Ridge. Although much of the alpine region is saturated during the spring thaw, significant solifluction occurs only in areas where the water table remains high enough during the fall freeze to permit thick ice-lens development. A. L. Washburn (pers. comm., 1969) has cautioned that autumn moisture may *not* be required for solifluction in other periglacial regions.

In the analysis of movement data from Niwot Ridge, all downslope displacement occurring during the spring thaw has been attributed to solifluction. Frost creep also occurs during this interval, but is too shallow to affect the movement of wooden stakes used in measuring displacement.

CLASSIFICATION AND DESCRIPTION OF LANDFORMS

INTRODUCTION

Frost creep and solifluction commonly operate together. The importance of each process varies from site to site and from year to year, and has varied, on a much larger scale, during the climatic oscillations of late Pleistocene and Recent time. As a result, most landforms produced by downslope soil movement are polygenetic. In classifying landforms, we should avoid the use of genetic names that stress the importance of one process at the expense of another (solifluction terrace, gelifluction lobe, etc.), unless the processes involved are clearly separable. This is rarely possible. In the present paper I have adopted a descriptive terminology based on the parameters used by Washburn (1956) in his classification of patterned

ground. These are surface expression and the presence or absence of sorting (Table 2). Moving soils that lack distinct topographic expression are included in the classification, but are not discussed in the paper.

TURF-BANKED TERRACES

Turf-banked terraces (Lundqvist, 1949; Galloway, 1961; Embleton and King, 1968) are defined as *bench-like accumulations of moving soil that lack conspicuous sorting.* Synonymous terms include "soil terraces" (Sigafoos and Hopkins, 1952), "solifluction terraces" (Billings and Mark, 1961; Rapp and Rudberg, 1960), and "nonsorted terraces" (Osburn *et al.,* 1965). Representative turf-banked terraces are illustrated in Figure 3.

FIGURE 3. Turf-banked terraces at tree limit on Niwot Ridge. A lush cover of sedges gives the moist treads of the terraces a dark appearance. Winter snow accumulation is moderately heavy on this 9 to 11°, northeast-facing slope. The mountains on the skyline are (left to right) unnamed, Pawnee Peak, Mount Toll, Paiute Peak, and Mount Audubon. July 27, 1967.

FIGURE 4. Miniature *Dryas*-banked terraces on Albion Ridge. The fronts of the terraces are covered by *Dryas octopetala,* but their treads are relatively free of vegetation. *Dryas* terraces are aligned either parallel, or at right angles, to the local prevailing winter wind direction, whichever orientation comes closest to paralleling the contour of the slope. The terraces form as a result of surficial frost creep, modified by wind and by the restraining influence of the vegetation cover. August 1, 1967.

TABLE 2
Classification of landforms produced by downslope soil movement in the Colorado Front Range

	No surface expression	Lobate	Terrace-like
Nonsorted	Nonsorted sheet	Turf-banked lobe	Turf-banked terrace
Sorted	Blockfield	Stone-banked lobe	Stone-banked terrace

"*Dryas*-banked terraces," such as those shown in Figure 4, are miniature turf-banked forms produced by the interaction of vegetation, wind, and surficial frost creep.

Turf-banked terraces in the Niwot Ridge area occupy slopes of 2 to 19°, with an average gradient of 10°. The terraces are largely restricted to south- and east-facing snow-accumulation slopes (Figures 5 and 6), where moisture is abundant and evenly distributed along the contour, and where the deposition of windblown soil eroded from exposed west-facing slopes replenishes the

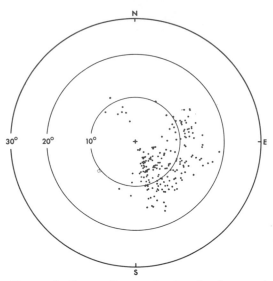

FIGURE 5. Scatter diagram showing the slopes and exposures of 195 turf-banked terraces on Niwot Ridge. The terraces show a preference for gentle southeast-facing snow-accumulation slopes.

supply of fine-textured material. Concave lower slopes are favored, but terraces also occur in convex slope positions.

For about 1.5 km east of the D-1 weather station, turf-banked terraces form giant stairsteps, whose arcuate fronts trend parallel to the contour of the ridgecrest (Figure 2). The largest terrace is 580 m long, 95 m wide, and 4 m high. Farther to the east, on steeper slopes, the terraces are almost as long and high, but are considerably narrower; Wilson (1952) has noted a similar relationship between gradient and terrace width on Jan Mayen Island. The fronts of the Niwot Ridge terraces slope at angles of 10 to 50° and rarely bulge or overhang; in plan they may be either straight or irregularly lobate, and commonly trend obliquely across the contour of the slope.

Where winter snow accumulation is deep, sorting processes are ineffective, and turf-banked terraces are covered with dense vegetation. Where winter snow is shallow, the moist treads of the terraces show the effects of intense frost activity. Inactive sorted polygons, 3 to 10 m in diameter, occur near the fronts of terraces at the western end of Niwot Ridge. Elsewhere in the study area, small active polygons, 1 to 4 m in diameter, occupy the floors of shallow ponds on terrace treads. Most of the terrace ponds are irregularly shaped, but some are elongated parallel to the prevailing wind direction. Earth hummocks occur where winter snow is deep enough to provide protection from

wind erosion, but shallow enough (generally 10 to 50 cm) to permit deep frost penetration; in less protected sites the hummocks are replaced by frost boils, which form in narrow, relatively snow-free zones atop the risers of the terraces. Paralleling the fronts of the terraces, and associated with the frost boils, are tension cracks caused by differential heaving. Many of the cracks are filled with stones; rock-filled depressions mark the junctions of branching cracks.

Representative transects across terraces in areas of moderate and light snow accumulation (Figure 7) illustrate the effects of topography upon snow cover, vegetation, and patterned ground.

TURF-BANKED LOBES

Turf-banked lobes (Galloway, 1961; Embleton and King, 1968) are defined as *lobate accumulations of moving soil that lack conspicuous sorting.* Other terms commonly used for these features include "soil lobes" (Sigafoos and Hopkins, 1952), "soil tongues" (Williams, 1959), "solifluction lobes" (Washburn, 1947; Dahl, 1956; Holmes and Colton, 1960; Rapp, 1960; Jahn, 1961; Rudberg, 1964), "nonsorted lobes" (Osburn *et al.*, 1965), "nonsorted congelifluction lobes" (Dutkiewicz, 1961), and "gelifluction lobes" (Washburn, 1967). A representative turf-banked lobe is illustrated in Figure 8.

Turf-banked lobes on Niwot Ridge occur on slopes of 4 to 23° (Figure 9). Like turf-banked terraces, their average slope angle is 10°. The lobes are best developed in winter snow-free areas, where moisture, instead of being uniformly distributed across the slope, is confined to linear drainageways. Lobes form one below the other wherever moisture is channeled along definite drainage routes (Figure 6). North- and south-facing slopes that receive runoff from nearby snow-accumulation areas provide the ideal combination of snow-free winter conditions and abundant moisture.

On Niwot Ridge turf-banked lobes are 3 to 50 m wide and extend downslope for distances of 3 to 100 m. Spoon-shaped recesses lie at the rear of most lobes. Frontal banks are 0.5 to 3.5 m high, and generally slope at angles of 10 to 35°; the fronts of a few particularly active lobes bulge outward over the soil that they are overriding. Treads are gently inclined downslope at angles of 2 to 14°.

Because of their restriction to winter snow-free areas, turf-banked lobes are particularly susceptible to erosion by wind. Wind-erosion

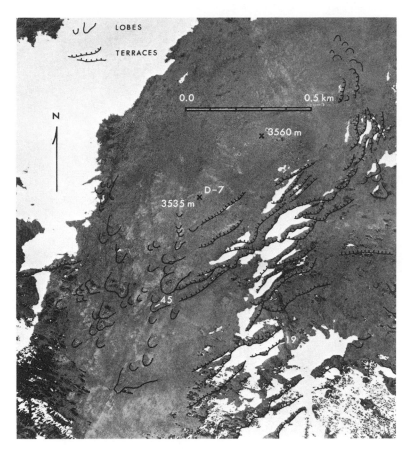

FIGURE 6. Vertical air photograph showing the distribution of turf-banked lobes and terraces on the flanks of the D-7 knoll, Niwot Ridge. The winter snow cover was still in the process of melting off turf-banked terraces when this photograph was taken on June 19, 1962. Other terraces remain buried beneath the large snowbank (upper left). Lobe 45 and terrace 19 are movement-study sites. U.S. Air Force photograph courtesy of Ronald Foreman and William S. Osburn.

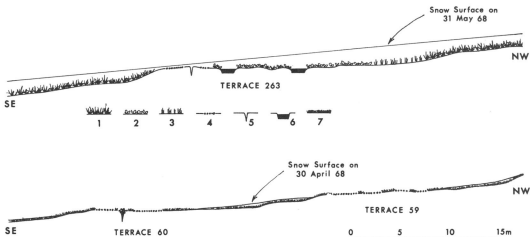

FIGURE 7. Topographic profiles across the fronts of turf-banked terraces in areas of moderate (top) and light (bottom) winter snow accumulation. Frost boils and frost cracks occur only where winter snow is shallow or absent. Plant distribution is controlled by differences in snow depth and growing-season moisture; these factors are related to terrace topography and orientation. 1—*Deschampsia caespitosa-Sibbaldia procumbens* snowbank community. 2—Wet *Carex scopulorum-Caltha leptosepala-Salix anglorum* meadow. 3—*Juncus drummondii-Carex nigricans* meadow. 4—Frost boil area, sparsely vegetated. 5—Frost crack. 6—Pond. 7—Dry *Kobresia myosuroides-Carex rupestris* meadow.

FIGURE 8. View looking downvalley onto the surface of a turf-banked lobe. This is lobe 26, a movement-study site on Niwot Ridge. September 23, 1967.

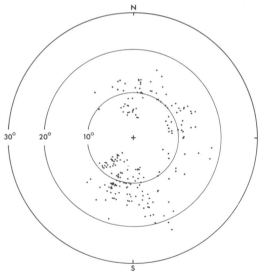

FIGURE 9. Scatter diagram showing the slopes and exposures of 208 turf-banked lobes on Niwot Ridge. Winter snow-free north- and south-facing slopes with gradients of 5 to 15° are favored.

scarps occur on steep, west-facing frontal banks; oriented ponds, aligned in the direction of the prevailing winter winds, occupy the gently sloping treads. Patterned ground is active at the surfaces of all but the driest and most steeply sloping lobes. Frost boils occur in exposed sites with abundant moisture, and earth hummocks are prominent in sheltered locations. Frost cracks are also common; unlike

the cracks on turf-banked terraces, they tend to be aligned in the direction of slope. The cracks are caused by tensional forces resulting from differential frost heaving (Benedict, 1970). Stones move into the frost cracks during periods of daily freezing and thawing, producing sorted stripes or crude polygonal patterns. Except for areas of patterned-ground activity, turf-banked lobes on Niwot Ridge are completely covered with vegetation.

STONE-BANKED TERRACES

Stone-banked terraces (Lundqvist, 1949; Galloway, 1956; Benedict, 1966; Embleton and King, 1968) are defined as *terrace- or garland-like accumulations of stones and boulders overlying a relatively stone-free moving subsoil* (Figure 10). Synonymous terms include "stone garlands" (Antevs, 1932), "block-banked terraces" (Thompson, 1961), "boulder steps" (Lundqvist, 1962), and "sorted terraces" (Osburn *et al.,* 1965).

Stone-banked terraces on Niwot Ridge occupy south- and east-facing snow-accumulation slopes with an average gradient of 16° and a range of 9 to 23° (Figure 11). There is partial overlap between the slope and exposure requirements of turf-banked and stone-banked terraces; where both occur on the same slope, the latter commonly override the former, suggesting a difference in age or in rate of movement. In Figure 12, profiles b and c show the positions of stone-banked terraces on representative snow-accumulation slopes, and

FIGURE 10. Stone-banked terrace on Niwot Ridge. The terrace is composed of coalescing lobes of blocky debris, and occupies a 14°, southeast-facing snow-accumulation slope. This is terrace 328, a movement-study site. September 13, 1968.

illustrate a common slope catena in the Front Range alpine region. The catena begins with sorted polygons at the crests of windswept knolls. With increasing slope, the polygons become elongated and their borders spread to form a blockfield with scattered debris islands. Irregularly branching sorted stripes extend from the lower edge of the blockfield; where movement is retarded, normally by a decrease in gradient, the stripes spread and merge to form stone-banked lobes and terraces. Gentle lower slopes tend to be smoothed by sheetwash, and are broad and featureless except where turf-banked terraces are present.

The fronts of stone-banked terraces are lobate, steep, and rocky. On Niwot Ridge, they slope at angles of 20 to 50°, and have a maximum height of 2.7 m. The terrace treads may be as wide as 60 m, and extend along the contour of the slope for distances of 200 to 400 m. Material at the rear of the tread is a mixture of soil and rock, covered with vegetation except where sorted stripes are present or where crudely sorted polygons are currently active. Closer to the front, stones and boulders form an arcuate strip of openwork rubble, 3 to 20 m wide, in which fines are lacking. Most boulders near the front range from 20 to 50 cm in maximum dimension, but some are as large as 180 cm. Tabular gneisses and blocky granites, syenites, and monzonites are rock types that occur in stone-banked terraces on Niwot Ridge. As noted by Lundqvist (1949), the long axes of stones tend to be oriented at right angles to

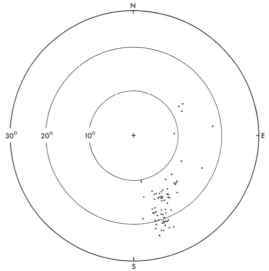

FIGURE 11. Scåtter diagram showing the slopes and exposures of 66 stone-banked lobes and terraces on Niwot Ridge. Relatively steep, south-southeast-facing slopes are favored.

the terrace front except immediately behind the riser, where they are rotated to an orientation paralleling the front. Rocks on the surfaces of the terraces commonly are cavernously weathered, with pitted surfaces and grotesquely irregular shapes. Maximum-diameter measurements suggest that lichens became established on most stone-banked terraces at the close of the Arikaree Stade of Neoglaciation, 900 to 1,000 years ago.

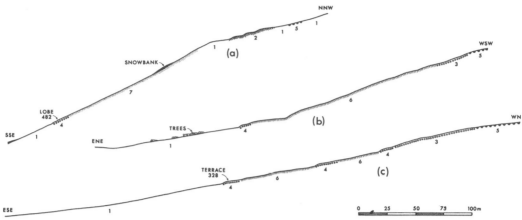

FIGURE 12. Topographic profiles of snow-accumulation slopes in the Indian Peaks region. Profiles a and c are from Niwot Ridge. Profile b is from the crest of a ridge east of the summit of Pawnee Peak. 1— Nonsorted sheet. 2—Turf-banked terraces. 3—Block-field with debris islands. 4—Stone-banked lobe or terrace. 5—Sorted polygons. 6—Sorted stripes. 7— Unstable scree in late snowbank area.

FIGURE 13. Lobes of debris with steeply-sloping fronts and pavement-like surfaces occur in stream channels at high elevations in the Indian Peaks region. These lobes were photographed in the valley north of Niwot Ridge. The terminus of the Navajo Glacier is visible beneath low clouds in the background. September 26, 1966.

STONE-BANKED LOBES

Stone-banked lobes (Galloway, 1961; Embleton and King, 1968) are defined as *lobate masses of rocky debris underlain by relatively stone-free, fine-textured, moving soil.* Other names for these features include "stone-banked flow earth cones" (Lundqvist, 1949), and "stone-banked solifluction lobes" (Rudberg, 1964).

Stone-banked lobes occur on south- and east-facing snow-accumulation slopes with gradients of 12 to 24° (Figure 11). The average slope angle is 17°. Stone-banked lobes and terraces frequently occur together in the same areas, and many of the terraces have developed from the merger of closely spaced stone-banked lobes (Figure 10). Stone-banked lobes also occur at the fronts of isolated "stone streams," and along the lower margins of scree slopes below persistent snowbanks (Figure 12a). Lobate accumulations of boulders found in stream channels near and above timberline (Figure 13) may be related in origin, although they have formed under conditions of total saturation.

The treads of stone-banked lobes are composed of stones and boulders without visible fine material. Cavernous weathering is conspicuous where the lobes are composed of medium-textured crystalline rocks. Maximum-diameter lichen measurements suggest that the lobes became available for lichen colonization at the close of the Arikaree Stade of Neoglaciation.

The fronts of stone-banked lobes are rarely more than 1 m high. In other respects, the lobes are comparable in size to the individual lobate components of stone-banked terraces.

EXPERIMENTAL SITES

INTRODUCTION

Annual rates of soil movement were measured in three turf-banked terraces, chosen for their locations in widely different moisture environments. Additional studies were conducted on two of the lobes in order to evaluate the processes involved in downslope movement, to determine flow patterns, and to measure differences in movement rates with depth. One of the lobes was subsequently trenched for profile description and radiocarbon sampling.

Annual movement surveys were also conducted in a stone-banked lobe and a stone-banked terrace, and in sorted stripes at two localities. The stone-banked lobe and terrace were trenched for soil profile studies. Lichen measurements provided information about former variations in snow cover.

LOBE 499

Description

Lobe 499 is located on the riser of a large and relatively stable turf-banked terrace east of the D-1 weather station on Niwot Ridge. Its elevation is 3,640 m, and its slope is to the northeast at 13°. Both the location and the rapid movement of this small turf-banked lobe are the result of a small stream that originates in the interconnected borders of sorted polygons on the terrace surface, leaves the terrace through a rocky sorted stripe,

FIGURE 14. Lobe 499. The bulging, overhanging front of this small turf-banked l o b e contrasts sharply with the subdued terrace riser on which it has developed. Flowing water is present at the surface of the lobe throughout the ice-free season. July 10, 1962.

FIGURE 15. Winter view of lobe 499. The surface of the lobe remains free of snow despite its location in the lee of a large turf-banked terrace. Movement stakes have been heaved and tilted downslope. January 20, 1966.

and flows across the surface of lobe 499. The stream derives its water from summer rainfall, melting snow, and ground ice. It flows throughout the warmer months of the year, keeping the central portion of the lobe saturated during the critical spring thaw and fall freeze periods.

The surface of the lobe slopes northeastward at an angle of about 9°, steepening as it approaches the front, where it bulges outward and overhangs the ground surface by as much as 20 cm (Figure 14). Filamentous green algae grow in flowing water along the axis of the lobe. In better-drained peripheral areas, *Carex scopulorum* and *Caltha leptosepala* are the dominant species. Winter snow accumulation is heavy in the lee of the terrace

riser, but the surface of the lobe stands high enough to remain snow-free throughout the year (Figure 15). The surface of the lobe is featureless. Its lichen cover appears to have become established at the beginning of the Arikaree-Gannett Peak interstadial interval.

Annual Movement Surveys

Annual movement surveys were begun in 1962, when 10 wooden stakes were installed along a theodolite line crossing the axis of the lobe (Figure 16). Benchmarks for the study consisted of 1-m-long steel rods, cemented in concrete below a depth of 60 cm, and surrounded by coarse sand and gravel. Stakes were inserted to depths of 25

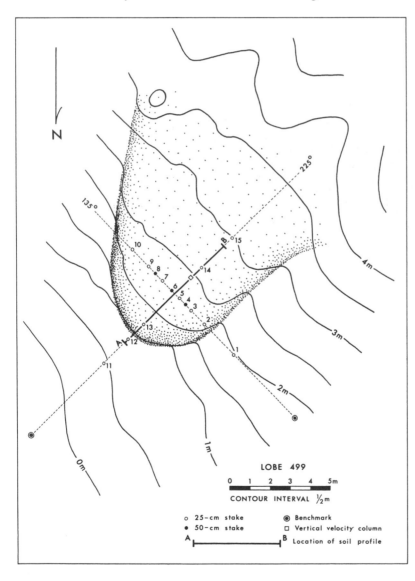

FIGURE 16. Map showing the positions of movement markers in lobe 499. After movement studies were completed, the lobe was trenched along line A-B.

and 50 cm. Displacement was measured parallel to the ground surface between a reference line drawn on each stake and a vertical straight edge positioned on the theodolite range line. Stakes were resurveyed annually between 1962 and 1967; frost-heaved stakes were reinserted to their original depths in 1963, but not in subsequent years.

The results of annual soil-movement measurements on lobe 499 are shown in Figure 17. During the five-year period beginning in August 1962, average rates of movement varied from 3 mm/yr, at the edge of the lobe, to 43 mm/yr along its axis. A stake on the slope northwest of the lobe moved at an average rate of about 1 mm/yr. Although the depth of burial of stakes became progressively shallower and more uniform after 1963 as a result of frost heaving, data from the first two years of the study indicate greater displacement of 25-cm than of 50-cm stakes. This suggests that the surface soil is moving more rapidly than the soil at depth.

Heaving of stakes relative to the ground surface is proportional to the total thickness of segregated ice lenses in the layer of soil surrounding the stakes. As a result, stakes inserted to depths of 50 cm were thrust farther out of the ground than stakes driven to depths of only 25 cm. The effects of winter heaving do not become apparent until the spring thaw, when the subsiding ground surface slips downward around stakes whose bases are still firmly cemented in frozen ground at depth.

Evaluation of Movement Processes

In order to evaluate the relative contributions of frost creep and solifluction to total movement in lobe 499, additional theodolite surveys were made during the fall of 1965 and spring of 1966. Results are graphed in Figure 18 and are discussed below.

July 30, 1965 to November 6, 1965. This period included much of the summer thaw, together with the first few weeks of the fall freeze. Downslope displacement was the result of potential frost creep in the surface soil, partially counteracted by retrograde movement due to late summer thawing and settling of the soil at depth. The results of a study on nearby lobe 45 suggest that solifluction is not a factor in this environ-

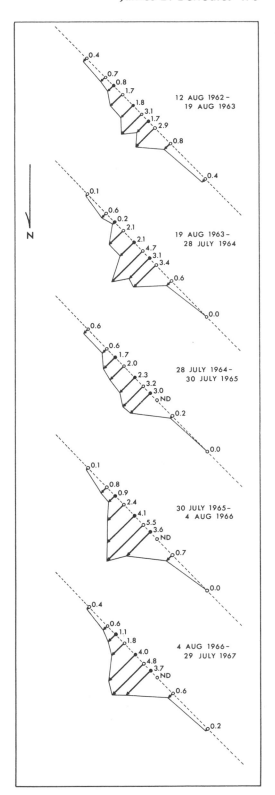

FIGURE 17. Annual displacement (cm) of wooden stakes along a line crossing the axis of lobe 499. Black dots represent stakes inserted to depths of 50 cm, and open circles represent stakes inserted to depths of 25 cm. The slope is to the northeast.

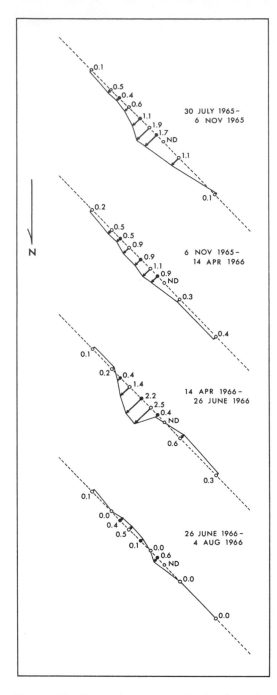

FIGURE 18. Seasonal movement (cm) of wooden stakes in lobe 499, 1965-66. Movement during the fall and winter was the result of potential frost creep. Displacement during the spring thaw was the result of solifluction, partially counteracted by retrograde movement.

ment after late July. Net movement was predominantly downslope, reaching a maximum value of 19 mm near the axis of the lobe, where the surface soil was wettest and frost heaving most pronounced.

November 6, 1965 to April 14, 1966. Downslope movement during this period of continued heaving can be attributed entirely to potential frost creep caused by freezing of the soil at depth. Maximum displacement amounted to 11 mm. Movement was greater at the axis of the lobe than at its edges, but differences were not so pronounced as during the earlier stages of freezing, suggesting that moisture distribution becomes increasingly uniform with depth.

April 14, 1966 to June 26, 1966. This interval was characterized by rapid thawing of the surface soil. Although retrograde movement due to settling of the thawing soil has minimized the apparent effects of flow, a maximum displacement of at least 25 mm along the axis of the lobe indicates that solifluction is important in saturated areas. In drier, peripheral parts of the lobe, solifluction either did not occur, or was so feeble that its effects were masked by retrograde movement.

June 26, 1966 to August 4, 1966. This was a period of continued thawing at depth. Movement was minor, and for the most part was retrograde.

Conclusions. During the winter of 1965-66, maximum potential frost creep in lobe 499 amounted to approximately 3.0 cm. Due to an unknown amount of retrograde movement during the spring and summer thaw, actual frost creep must have been considerably less. The maximum measured displacement caused by solifluction was 2.5 cm. An undetermined amount of retrograde movement occurring during the spring thaw makes this value a minimum estimate for total movement due to solifluction.

Both processes contribute significantly to net downslope movement in lobe 499. Solifluction is apparently the more important process in saturated areas along the lobe axis, but only frost creep operates where the soil is drier.

Movement Along the Lobe Axis

In 1963, additional wooden stakes were installed in a line along the axis of lobe 499 (Figure 16). The stakes were resurveyed in 1964; measurements were made by taping the distance between each stake and a benchmark located 7 m downslope from the front of the lobe. Corrections were made for temperature effects in the steel tape, and measurements were taken at a constant tension of 15 pounds (6.8 kg).

Data for 1963-64 (Figure 19a) show that

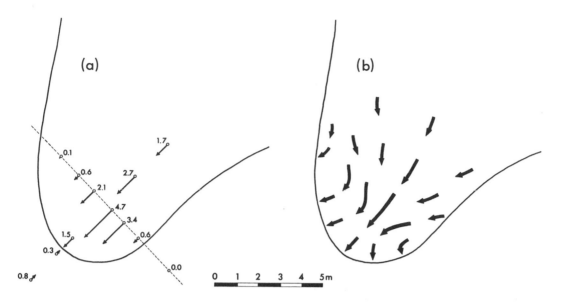

FIGURE 19. Rates and inferred directions of soil movement in lobe 499, August 19, 1963 to July 28, 1964. (a) Measured rates of movement (cm) are greatest along the axis of the lobe, reaching maximum values a few meters behind its front. The front rates of soil movement increase downslope along the axis of the lobe, reaching maximum values a few meters behind the front. Closer to the front, the lower water table and restraining turf layer cause movement to be retarded.

itself advances relatively slowly. Only 25-cm-long stakes are plotted. (b) Directions of soil movement inferred from the tilting of stakes and the orientations of plant roots. The lengths of the arrows are roughly proportional to movement velocities.

Directions of Movement

Directions of soil movement have been inferred from the orientations of plant roots that penetrate deeply into the moving soil. The only deeply-rooted species growing on lobe 499 is the marsh marigold, *Caltha leptosepala*, whose roots are able to withstand short-term strains of 25 to 30% before breaking. This elasticity enables the plants to be carried downslope without injury, even though the tips of their roots may be anchored in stationary or slowly moving soil at depths of 1 to 1.5 m. Directions of soil movement in lobe 499 were estimated from the positions of plants with respect to their root tips. Additional information was obtained from annual measurements of the tilting of wooden movement stakes.

Both lines of evidence suggest that movement at the rear of lobe 499 is directed obliquely downslope and *toward* the axis of the lobe (Figure 19b). Nearer the front, directions of movement shift so that the soil is flowing *away* from the lobe axis. Directions of flow are directly related

to variations in velocity. Where movement along the axis is accelerating, the ground surface subsides, and a local depression develops. Soil moves inward from the edges of the lobe to replace soil that has moved out of the axial area. Near the front of the lobe, where movement is retarded, the ground surface bulges upward, creating a local slope away from the axis of the lobe; the soil moves outward in response to this gradient, and an equilibrium surface profile is maintained.

Subsurface Movement

To determine rates of soil movement beneath the ground surface, a "vertical velocity column" (Rudberg, 1958, 1962, 1964; Iveronova, 1964; Dutkiewicz, 1967) was installed along the axis of the lobe in 1963. Its location is shown in Figure 16. The column consisted of eight cement rods, each of which was 2.2 cm in diameter and 7.6 cm long. The rods were placed in a section of thin-walled copper pipe, closed at the bottom by a piece of tin that was fitted so that it would snap away with slight pressure. An auger hole 2.5 cm in diameter was drilled vertically into the surface of the lobe. The pipe containing the cement rods was lowered into the vertical hole, plumbed, and then withdrawn, while the rods were held in place with a wooden dowel stick. The position of the upper cement marker was measured with respect

FIGURE 20. Vertical velocity profile at the axis of lobe 499. The column of cement rods was installed in a vertical auger hole on October 9, 1963, and recovered on August 23, 1967. Maximum displacement during the 4-year period averaged 2.4 cm/yr. Movement was restricted to the upper 50 cm of soil. Humates from the upslope end of the buried organic layer visible at lower right were dated at 2,340 ± 130 BP (I-4045).

to the 135° theodolite line (Figure 16). The column was reexcavated in 1967 so that the new position of each rod could be determined.

Displacement of the cement rods (Figure 20) supports the conclusion that solifluction is a more effective process than frost creep in the axial portion of lobe 499. Differential heaving was insignificant, and was limited to the upper 35 cm of soil. Instead of being disrupted by frost creep, the column gives the impression of having been bent gently downslope by flow, with the cohesive turf layer carried as a unit by the moving soil beneath it. The restraining effect of a surface root zone has also been noted by Kirkby (1967) and Rudberg (1964). Maximum displacement occurred just above the base of the turf layer, and was 9.6 cm (2.4 cm/yr). Movement decreased with depth, and no appreciable displacement was noted deeper than 50 cm.

The orientations of *Caltha* roots near the vertical velocity column suggest that movement has extended to depths greater than 50 cm during the lives of the plants. The roots are vertical as they pass through the surface turf layer, but become inclined as they enter the moving soil beneath it. At a depth of 70 to 80 cm the roots regain their vertical orientations. The lower limit of effective soil movement undoubtedly varies from year to year; during the lives of the plants it has ranged between about 50 and 75 cm.

Profile Characteristics

In 1967, after movement measurements had been completed on lobe 499, the lobe was drained by digging a trench across its upper end. With the stream diverted, the soil became dry enough that it was possible to dig a trench along the axis of the lobe and across its front (Figure 16). Soil profile description included textural analyses and studies of the microfabric of sand grains in oriented soil samples (Benedict, 1969). A sample of buried organic material was collected for radiocarbon dating.

A sketch of the soil profile is shown in Figure 21. The modern A horizon is a very dark grayish brown (10 YR 3/2) gravelly sandy loam,[1] friable and structureless, but held together by a cohesive network of *Carex* roots.

The moving soil beneath the modern turf layer is a dark grayish brown (2.5 Y 4/3) to dark yellowish brown (10 YR 4/4) sandy loam, locally mottled gray (5 Y 5/1) to greenish gray (5 GY 5/1). Mottling is especially pronounced adjacent to *Caltha* roots. The texture of the soil is slightly finer at the front than at the rear of the lobe. The entire unit is very cohesive. Its structure is weakly platy and locally vesicular.

Beneath the moving subsoil, a buried A1 horizon extends upslope from the front of the lobe for a distance of 4.4 m. The buried humus layer is complex and difficult to interpret. It is darker (10 YR 2/2), less gravelly, and more compact than the modern A horizon. Immediately upslope from the front of the lobe, the buried A horizon consists of both a normal and an inverted soil profile. The zone of weakness separating these two superimposed organic layers can be traced upslope for 0.5 m before the two layers become indistinguishable. Infolding of the humus layer shows that the lobe advances by a "rolling over" process rather than by simple overriding of the ground surface in front of the lobe.

Material underlying the buried A horizon is

[1]Soil texture was determined with a Bouyoucos hydrometer. Textural classes are used as defined in the USDA *Soil Survey Manual* (Soil Survey Staff, 1951). Colors were measured under field-moisture conditions using *Munsell Soil Color Charts* (Munsell Color Company, Inc., 1954).

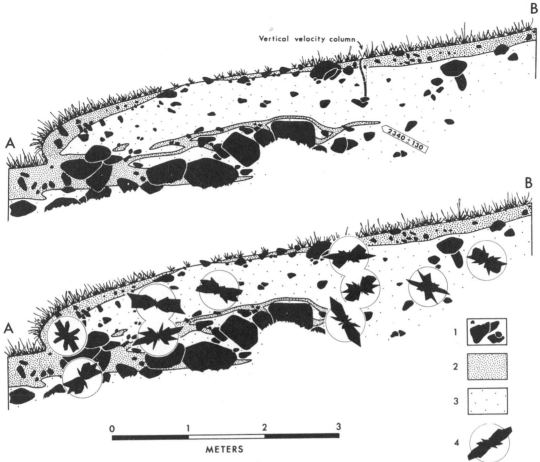

Vertical velocity column

2340 ± 130

0   1   2   3
METERS

1
2
3
4

FIGURE 21. Soil profile, lobe 499. 1—Stones, drawn to scale. 2—Very dark grayish brown to very dark brown gravelly sandy loam. Mucky and less gravelly where buried. 3—Dark grayish brown to dark yellowish brown sandy loam, becoming finer textured near the front of the lobe. Locally mottled. Cohesive, with weakly platy structure. 4—Rose diagrams showing the long-axis orientations of elongate sand grains, measured in thin section. Each diagram is a summary of 200 individual measurements, with circles representing frequencies of 10%.

similar in color to the moving soil above it, but is stonier and more gravelly.

The orientations of elongate sand grains (Figure 21) generally parallel the long-axis orientations of cobbles and boulders. Within the layer of moving soil, the fabric becomes increasingly imbricated as the front of the lobe is approached.

[*Editor's Note:* Material has been omitted at this point.]

CONCLUSIONS

CLASSIFICATION AND DISTRIBUTION

(1) Moving soils in the Colorado Front Range are classified on the basis of their surface expression and sorting. The deposits may be sorted or nonsorted; they may be uniform and sheetlike, lobate, or terrace-like. Intergradations in surface expression are common, but intergradations in sorting do not occur.

(2) Turf-banked lobes and terraces in the Niwot Ridge study area occupy gentle slopes with an average gradient of 10°. Terraces are confined almost entirely to snow-accumulation slopes, whereas lobes are best developed along meltwater drainage routes that pass through winter snow-free areas.

(3) Stone-banked lobes and terraces occur on snow-accumulation slopes with average gradients of 16 to 17°; they commonly form at the lower edges of blockfields, sorted stripes, stone streams, and unstable scree slopes.

INTERNAL STRUCTURE

(1) Beneath their surface humus layers, turf-banked lobes and terraces in the Front Range consist of stones and soil, mixed together without conspicuous sorting. The soil is commonly mottled or gleyed and has a pronounced platy structure. Its texture becomes gradually finer toward the front of the lobe or terrace. One or several buried organic layers are commonly present. The long axes of elongated sand grains, cobbles, and boulders tend to lie parallel to the ground surface or to be slightly imbricated.

(2) In profile, stone-banked lobes and terraces in the Front Range consist of (a) a surface layer of cobbles and boulders, thickening toward the front of the lobe or terrace; (b) an underlying unit with distinct platy structure and very few stones, becoming thinner and finer-textured toward the front; (c) a buried A horizon, locally disrupted or absent because of deep freezing and thawing; and (d) the buried subsoil. Rocks in the rubble layer tend to be orientated with their long axes pointing in the direction of slope and their flat faces parallel to the lobe or terrace surface. Near the fronts of individual lobes, the long axes of stones are rotated into orientations paralleling the fronts, and tabular stones are moderately to strongly imbricated.

MODERN RATES OF MOVEMENT

(1) Average maximum rates of downslope soil movement at nine experimental sites on Niwot Ridge ranged from 0.4 to 4.3 cm/yr. Comparison with other areas is difficult because of the inconsistency with which results are often reported and the variety of measurement techniques that have been used. Rates of movement are generally comparable to rates measured in other alpine and arctic regions, are higher than rates measured on grassy and forested slopes in temperate and semiarid regions, and are lower than rates of surface displacement on stony, unvegetated slopes in temperate and semiarid climates.

(2) Movement rates vary erratically along the treads of stone-banked and turf-banked terraces; movement is most rapid where moisture is most abundant.

(3) Movement rates vary systematically in turf-banked and stone-banked lobes. Movement is greatest along the axis of the lobe, with maximum values occurring just below its midpoint. Variations in movement rates are related to groundwater levels, which are typically highest along the axis and near the midpoint of the lobe. Directions of movement tend to preserve an equilibrium surface profile: flow is directed toward the lobe axis where movement is accelerating, and away from the axis where movement is retarded.

(4) Differential movement within the surface rubble layer of stone-banked lobes and terraces on Niwot Ridge is minor under present conditions, and occurs mainly in parts of the lobe or terrace that adjoin areas of finer-textured soil.

(5) Important frost heaving is currently restricted to microenvironments that are saturated at the beginning of the fall and winter freeze. Heaving at the axis of a turf-banked lobe on Niwot Ridge exceeded 36 cm during the winter of 1965-

66. Although small stones on the ground surface are affected by short-term freeze-thaw cycles during the spring and fall, only the annual cycle influences the movement of larger stones and wooden stakes.

(6) Differential heaving and downslope movement at experimental sites on Niwot Ridge are currently confined to a layer of soil 50 to 75 cm thick. The shape of the vertical velocity profile reflects differences in the relative importance of frost creep and solifluction, as well as the presence or absence of a restraining turf layer.

EVALUATION OF PROCESSES

(1) Downslope soil movement is the resultant of potential frost creep, solifluction, and retrograde displacement. Retrograde movement and solifluction both occur during the spring thaw. Each minimizes the effect of the other, making it impossible to accurately determine their individual contributions to net downslope movement.

(2) Under present conditions, frost creep is the dominant movement process on Front Range slopes. Only in specialized microenvironments, such as the saturated axial areas of turf-banked lobes in wet sites, do the effects of solifluction exceed the effects of frost creep. Along the axes of turf-banked lobes on slopes of 6 and 13°, displacement due to solifluction exceeded 2.5 and 2.2 cm/yr respectively. Potential frost creep amounted to 3.0 and 2.2 cm/yr, but much of this displacement was only temporary, and was cancelled by retrograde movement during the spring thaw. At the edges of both lobes, movement was entirely the result of frost creep.

(3) Saturation during the spring thaw is not, in the Front Range, sufficient to cause solifluction. In order for solifluction to occur, the soil must have previously been "conditioned" by ice-lens formation, which requires saturation at the beginning of freeze.

(4) Measured rates of potential frost creep in a turf-banked lobe on Niwot Ridge were as much as seven times greater than values calculated from slope and frost-heave measurements. Soil movement during the fall freeze is a more complicated process than previously suspected.

(5) In wet sites, retrograde movement may significantly exceed potential frost creep. It is uncertain whether this is a result of annual differences in the depth of summer thaw, of desiccation, or of other factors.

(6) Fine-textured soil sometimes moves faster and sometimes moves more slowly than nearby coarse material, depending upon the relative importance of different movement processes. On a saturated 12 to 13° slope subject to intense solifluction, fine-textured soil moved twice as rapidly as stripes of coarse debris (Figure 43). On a relatively dry 16 to 18° slope subject only to frost creep, and perhaps to wetting and drying, stone-banked lobes moved three times as rapidly as adjacent nonsorted soil (Figure 49).

RELATIONSHIP OF MOVEMENT TO ENVIRONMENT

(1) Rates of downslope soil movement in the Front Range are largely determined by the depth to groundwater at the beginning of the autumn freeze. At experimental sites on Niwot Ridge, the displacement of 25-cm-long wooden stakes exceeded 2 cm/yr only where the water table lay within 0.2 m of the ground surface at the beginning of freezing, and exceeded 1 cm/yr only where it lay within 0.5 m of the ground surface.

(2) Gradient exerts a secondary influence upon rates of downslope movement. Under comparable moisture conditions, movement increases as a direct function of slope angle.

(3) Differences in soil texture and temperature affect soil movement indirectly, by influencing the availability of moisture, but otherwise have little control upon rates of downslope movement in the Front Range alpine region.

ORIGIN

(1) Turf-banked lobes and terraces develop where rapid downslope movement is impeded, normally by a decrease in gradient. Terraces with linear fronts develop in snow-accumulation areas, where moisture is more-or-less uniformly distributed along the contour of the slope. Where snow accumulation is patchy, the fronts of terraces are irregularly lobate. On winter snow-free slopes, where moisture is confined to narrow drainageways, lobate forms dominate the landscape.

(2) The absence of sorting in turf-banked lobes and terraces together with their restriction to relatively moist lower slopes reflect the importance of solifluction in their formation. Solifluction largely obliterates the results of vertical frost sorting and produces deposits in which stones and fines are mixed together without obvious layering. Because solifluction is no longer an important process in most lobes and terraces in the Front Range, sorted polygons have developed and are preserved at their surfaces. The formation of turf-banked lobes and terraces requires a coherent plant cover, and is favored by perennially frozen ground at shallow depth.

(3) Stone-banked lobes and terraces develop

where a decrease in gradient causes rocky debris to accumulate and spread laterally. Topographic form is less an expression of moisture distribution than of the source of the stones that supply the lobe or terrace. Lobes develop where isolated stone streams and widely-spaced sorted stripes encounter a decrease in gradient. Terraces form at the lower edges of sloping blockfields and scree slopes, and where sorted stripes are so closely spaced that the lobes at their fronts coalesce and merge. Stone-banked lobes and terraces can develop only where coarse debris is moving more rapidly than adjacent finer-textured soil; thus they are characteristic of a frost-creep, rather than a solifluction, environment.

(4) The pronounced vertical sorting of stone-banked lobes and terraces, and their occurrence on steep, upper slopes, reflect the importance of frost creep in their formation. Most of the sorting currently observed in stone-banked lobes and terraces is inherited from the blockfields or sorted stripes that formerly supplied debris to their surfaces. Sorting is accentuated as stones that fall from the terrace fronts are recycled upward to the base of the rubble layer by frost heaving.

Ages of the Deposits

(1) Stratigraphic and archaeological evidence from Niwot Ridge and the surrounding area suggest that many turf-banked lobes and terraces with gentle fronts and patterned surfaces formed during the interval between the Middle Pinedale glacial maximum and the beginning of the Altithermal. The closing phases of the Middle Stade of Pinedale Glaciation were a likely time. Lobes and terraces undoubtedly also formed during earlier periods of favorable climate. Sorted polygons and stripes developed on terrace treads during the final stade of Pinedale Glaciation, prior to 7,650 radiocarbon years ago.

(2) Turf-banked lobes with overhanging fronts and fresh topography developed in the Front Range during Neoglaciation. A radiocarbon date of 2,340 years on organic material from the upslope end of the buried A horizon in a small turf-banked lobe on Niwot Ridge suggests formation during the Temple Lake Stade. Costin *et al.* (1967) have shown that comparable features were forming in the Australian mountains at approximately the same time.

(3) Stratigraphic, lichenometric, and radiocarbon evidence indicate that stone-banked lobes and terraces in the Front Range originated in late Temple Lake time. A minimum age for their development is provided by a date of 2,470 years for organic material overrun by a large stone-banked terrace on Niwot Ridge. Conditions would have been favorable for frost creep and intense sorting when lee slopes were emerging from beneath perennial snowbanks and when vegetation was absent.

Past Movement Rates

(1) A radiocarbon date from the upslope end of the buried humus layer in a turf-banked lobe on Niwot Ridge shows that the front of the lobe has advanced at an average rate of 1.9 mm/yr during the past 2,340 years. This value is comparable to the estimated modern rate of frontal advance.

(2) Radiocarbon dates for samples of buried humus collected from beneath a stone-banked terrace on Niwot Ridge suggest an average rate of terrace advance of 3.4 mm/yr during the past 2,470 years. Rates of movement have varied with changing climatic conditions. An inferred initial period of rapid movement came to a close at the beginning of the Temple Lake-Arikaree interstadial, a time of general slope stability, vegetation growth, and soil formation. Movement was slow during the Arikaree glacial maximum, when all but the upper sections of the slope were covered with perennial snow. Movement rates increased by a factor of almost eight at the close of the Arikaree Stade, when the slope again became free of snow in the fall. During the past 1,050 radiocarbon years, average rates of movement have been relatively low. Wide spacing of dated samples has made it impossible to determine whether expanded snowbanks during the Gannett Peak Stade caused increased rates of downslope soil movement.

Climatic Significance

(1) Modern rates of downslope movement in the Colorado Front Range reflect differences in gradient and the availability of autumn moisture. Because gradient has remained virtually constant through time, former periods of rapid downslope movement are inferred to have been times when moisture was plentiful at the beginning of freeze. Such intervals occurred at the close of the Middle Stade of Pinedale Glaciation, and during the waning phases of the Temple Lake and Arikaree Stades of Neoglaciation (Figure 57).

(2) Soil-moisture distribution in the Front Range is intimately related to winter snowdrift patterns. The locations of terraces on Niwot Ridge suggest that prevailing winter winds were from the northwest during Middle Pinedale time,

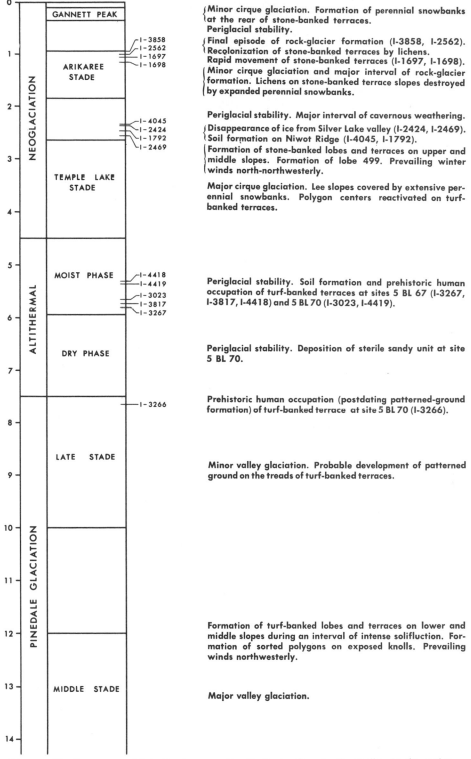

FIGURE 57. Summary of late Quaternary climatic fluctuations in the Indian Peaks region. Dates for the close of the Middle Stage and beginning of the Late Stage of Pinedale glaciation are from Richmond (1965, Table 2). The time scale is in thousands of years BP.

and from the north-northwest at the close of the Temple Lake Stade of Neoglaciation.

(3) The relative importance of frost creep and solifluction varies with the moisture content of the soil. Frost creep is generally associated with upper slope positions, and solifluction with lower slope positions. The line separating the zones dominated by each of these two processes has shifted upward and downward with changes in climate. During the late Pleistocene, solifluction was the dominant movement process on all but the driest ridgecrests and the upper flanks of knolls. Neither process seems to have been particularly important during the Altithermal. During Neoglacial intervals of rapid downslope movement, solifluction was restricted to lower slopes, and frost creep became the most important movement process in middle slope positions formerly dominated by solifluction. Under present climatic conditions, solifluction occurs only in specialized, saturated microenvironments; frost creep is the dominant movement process on upper, middle, *and* lower slopes.

(4) Sorted landforms characteristic of frost creep are often superimposed upon landforms characteristic of solifluction. The occurrence of sorted polygons and stripes, stone-banked lobes, and stone-banked terraces on the treads of turf-banked lobes and terraces reflects a general decline in the availability of moisture and the effectiveness of solifluction since the close of the Middle Stade of Pinedale Glaciation.

(5) A tentative reconstruction of climatic fluctuations in the Front Range alpine region is given in Figure 57.

ACKNOWLEDGMENTS

Many people have contributed to this project. My special thanks go to William C. Rense, Kenneth L. Petersen, Dennis Rasmussen, and Bryon Olson, whose help in the field was made possible by INSTAAR summer research participation programs sponsored by the National Science Foundation. Charles Schweger, Ronald Foreman, Mrs. Katharine T. Benedict, Mrs. Diane G. Benedict, and Mrs. Barbara Madole provided voluntary assistance on a number of occasions. The map of periglacial landforms on Niwot Ridge (Figure 2) is the result of fieldwork by Mrs. Jody Fitzgerald during the summer of 1965. David F. Murray and Donald L. Pattie identified plants collected at several of the movement sites. Transportation problems, particularly in winter, were minimized thanks to the talents of Skip Greene. Surveying equipment was made available by the University of Wisconsin Geology Department and by the Geological Sciences and Geography Departments at the University of Colorado. Discussions with Robert F. Black and William C. Bradley led to the adoption of several techniques that were used in the study. Much of the fieldwork was accomplished while I received financial aid in the form of National Science Foundation Graduate Fellowships and a University of Colorado INSTAAR Postdoctoral Fellowship; the Society of the Sigma Xi and the Geological Society of America contributed additional support. INSTAAR provided living accommodations at its Mountain Research Station, as well as other forms of assistance during the period of study. Robert F. Black, Nel Caine, Jack D. Ives, Cuchlaine A. M. King, A. Lincoln Washburn, and Peter J. Williams made helpful comments on the manuscript. My thanks go to these individuals and institutions, and others not mentioned.

REFERENCES

Andersson, J. G.
  1906 : Solifluction, a component of subaërial denudation. *J. Geol.*, 14: 91-112.
Antevs, Ernst
  1932 : *Alpine zone of Mt. Washington Range.* Merrill & Webber, Auburn, Maine, 118 pp.
Baulig, Henri
  1956 : Pénéplaines et pédiplaines. *Bull. Soc. Belge d'Études Géog.*, 25(1): 25-58.
Benedict, J. B.
  1966 : Radiocarbon dates from a stone-banked terrace in the Colorado Rocky Mountains, U.S.A. *Georg. Annaler*, 48A (1): 24-31.
  1967 : Recent glacial history of an alpine area in the Colorado Front Range, U.S.A., I. Establishing a lichen-growth curve. *J. Glaciol.*, 6(48): 817-832.
  1968 : Recent glacial history of an alpine area in the Colorado Front Range, U.S.A., II. Dating the glacial deposits. *J. Glaciol.*, 7(49): 77-87.
  1969 : Microfabric of patterned ground. *Arctic and Alpine Res.*, 1(1): 45-48.
  1970 : Frost cracking in the Colorado Rocky

Mountains. *Geog. Annaler* (in press). In preparation: Prehistoric man and environment in the Colorado Rocky Mountains.

Beschel, R. E.
1961 : Botany: and some remarks on the history of vegetation and glacierization. *In* Müller, B.S.(ed.), *Jacobsen-McGill Arctic Research Expedition to Axel Heiberg Island. Preliminary Report 1959-1960*, 179-199.

Billings, W. D., and Mark, A. F.
1961 : Interactions between alpine tundra vegetation and patterned ground in the mountains of southern New Zealand. *Ecology*, 42: 18-31.

Büdel, Julius
1961 : Die Abtragungsvorgänge auf Spitzbergen im Umkreis der Barentsinsel. *In: Deutscher Geographentag Köln: Tagungsbericht und wissenschaftliche Abhandlungen*. Franz Steiner Verlag, Weisbaden, 337-375.

Butrym, J., Cegła, J., Dzułynski, S., and Nakonieczny, S.
1964 : New interpretation of "periglacial structures." *Folia Quaternaria*, 17: 1-34.

Caine, T. N.
1963 : Movement of low angle scree slopes in the Lake District, northern England. *Rev. Géomorph. Dyn.*, 4: 171-177.
1968 : The log-normal distribution and rates of soil movement: an example. *Rev. Géomorph. Dyn.*, 18(1): 1-7.

Costin, A. B., Thom, B. G., Wimbush, D. J., and Stuiver, Minze
1967 : Nonsorted steps in the Mt. Kosciusko area, Australia. *Geol. Soc. Amer. Bull.*, 78: 979-992.

Curry, R. R.
1969 : Holocene climatic and glacial history of the central Sierra Nevada, California. *In* Schumm, S. A. and Bradley, W. C. (eds.), *United States Contributions to Quaternary Research*, Geol. Soc. Amer. Spec. Paper, 123: 1-47.

Czeppe, Zdzisław
1960 : Thermic differentiation of the active layer and its influence upon the frost heave in periglacial regions (Spitsbergen). *Bull. Acad. Polonaise Sci.*, Série des sci., géol., et géog., 8(2): 149-152.
1966 : Przebieg głownych procesow morfogenetycznych w poludniowo-zachodnim Spitsbergenie. *Zeszyty Naukowe Uniwersytetu Jagiellonskiego*, 127: 1-129 (Polish with English summary).

Dahl, Eilif
1956 : Rondane: Mountain vegetation in south Norway and its relation to the environment. *Skrifter utgitt Det Norske Vid.-Akad. i Oslo, Mat.-Nat. Klasse*. No. 3: 273-282.

Dege, Wilhelm
1943 : Über Aussmass und Art der Bewegung arktischer Fliesserde: *Z. Geomorph.*, 11: 318-329.

Dutkiewicz, Leopold
1961 : Congelifluction lobes on the southern Hornsund coast in Spitsbergen. *Biul. Peryglac.*, 10: 285-289.
1967 : The distribution of periglacial phenomena in NW-Sörkapp, Spitsbergen. *Biul. Peryglac.*, 16: 37-83.

Dylik, Jan
1967 : Solifluxion, congelifluxion and related slope processes. *Geog. Annaler*, 49A (2-4): 167-177.

Embleton, C. and King, C. A. M.
1968 : *Glacial and periglacial geomorphology*. St. Martin's Press, New York, 608 pp.

Emmett, W. W.
1965 : The Vigil Network: Methods of measurement and a sampling of data collected. *In:* Symposium of Budapest, *Representative and experimental areas*, 1. Int. Assoc. Sci. Hydrol., Publ. No. 66: 89-106.

Everett, K. R.
1963 : Slope movement, Neotoma Valley, southern Ohio. Ohio State University, Inst. Polar Studies, Rep. No. 6: 1-62.
1966 : Slope movement and related phenomena. *In* Wilimovsky, N. J. and Wolfe, J. N. (eds.), *Environment of the Cape Thompson region, Alaska*, U.S. Atomic Energy Comm., Div. Tech. Info., 175-220.

Galloway, R. W.
1961 : Solifluction in Scotland. *Scot. Geog. Mag.*, 77(2): 75-87.

Goldthwait, R. P.
1960 : Development of an ice cliff in northwest Greenland. U.S. Army CRREL, Tech. Rep. 39. 106 pp.

Gradwell, M. W.
1957 : Patterned ground at a high-country station. *N.Z. J. Sci. and Tech.*, 38(sec. B, no. 8): 793-806.

Hamilton, T. D.
1963 : Quantitative study of mass movements, southeastern Wisconsin. Unpublished M.S. thesis, University of Wisconsin, Madison. 103 pp.

Holmes, C. D. and Colton, R. B.
1960 : Patterned ground near Dundas (Thule Air Force Base), Greenland. *Medd. om Grønland*, 158(6): 1-15.

Iveronova, M. I.
1964 : Stationary studies of the recent denudation processes on the slopes of the R. Tchon-Kizilsu Basin, Tersky Alatau

ridge, Tien-Shan. *Z. Geomorph.*, N.F. Suppl. Bd. 5: 206-212.

Jahn, Alfred
1960 : Some remarks on evolution of slopes on Spitsbergen. *Z. Geomorph.*, N.F. Suppl. Bd. 1: 49-58.
1961 : Quantitative analysis of some periglacial processes in Spitsbergen. *Uniwersytet Wrocławski im Bolesława Bieruta Zeszyty Naukowe Nauki Przyrodnicze*, Ser. B, Nr. 5: 1-54.

Kirkby, M. J.
1967 : Measurement and theory of soil creep. *J. Geol.*, 75(4): 359-378.

Leopold, L. B., Emmett, W. W., and Myrick, R. M.
1966 : Channel and hillslope processes in a semiarid area, New Mexico. U.S. Geol. Sur. Prof. Paper, 352-G: 193-253.

Lundqvist, G.
1949 : The orientation of the block material in certain species of flow earth. *Geog. Annaler*, 31(1-4): 335-347.

Lundqvist, Jan
1962 : Patterned ground and related frost phenomena in Sweden. *Sveriges Geologiska Undersökning*, 55(7): 1-101.

Macar, P. and Pissart, A.
1964 : Études récentes sur l'évolution des versants effectuées à l'Université de Liège. *Z. Geomorph.*, N.F. Suppl. Bd. 5: 74-81.

Madole, R. F.
1969 : Pinedale and Bull Lake Glaciation in Upper St. Vrain drainage basin, Boulder County, Colorado. *Arctic and Alpine Res.*, 1(4): 279-287.

Marr, J. W.
1961 : Ecosystems of the east slope of the Front Range in Colorado. *Univ. Colorado Stud.*, Ser. Biol., No. 8: 1-134.

Marr, J. W., Clark, J. M., Osburn, W. S., and Paddock, M. W.
1968 : Data on mountain environments III. Front Range, Colorado, four climax regions, 1959-1964. *Univ. Colorado Stud.*, Ser. Biol., No. 29: 1-181.

Munsell Color Company, Inc.
1954 : *Munsell soil color charts*. Baltimore.

Olson, B. L.
1969 : An Archaic site in the Colorado Front Range (abstract). *J. Colorado-Wyoming Acad. Sci.*, 6(2): 43.

Osburn, W. S., Benedict, J. B., and Corte, A. E.
1965 : Frost phenomena, patterned ground, and ecology on Niwot Ridge. *In* Schultz, C. B. and Smith, H. T. U. (eds.), *Guidebook for one-day field conferences, Boulder area, Colorado*, VIIth INQUA Congress, 21-26.

Østrem, Gunnar
1965 : Problems of dating ice-cored moraines. *Geog. Annaler*, 47A(1): 1-38.

Owens, I. F.
1967 : Mass movement in the Chilton Valley. Unpublished M.A. thesis, University of Canterbury, N.Z. 92 pp.
1969 : Causes and rates of soil creep in the Chilton Valley, Cass, New Zealand. *Arctic and Alpine Res.*, 1(3): 213-220.

Pissart, A.
1964 : Vitesses des mouvements du sol au Chambeyron (Basses Alpes). *Biul. Peryglac.*, 14: 303-309.

Rapp, Anders
1960 : Recent development of mountain slopes in Karkevagge and surroundings, northern Scandinavia. *Geog. Annaler*, 42: 71-200.
1963 : Solifluction and avalanches in the Scandinavian mountains. Proc. Permafrost Int. Conf., 11-15 November 1963, Lafayette, Indiana, 150-154.
1967 : Pleistocene activity and Holocene stability of hillslopes, with examples from Scandinavia and Pennsylvania. *In: L'Evolution des versants*, Les Congrès et Colloques de l'Université de Liège, 40: 229-244.

Rapp, Anders and Rudberg, Sten
1960 : Recent periglacial phenomena in Sweden. *Biul. Peryglac.*, 8: 143-154.

Richmond, G. M.
1965 : Glaciation of the Rocky Mountains. *In* Wright, H. E., Jr. and Frey, D. G. (eds.), *The Quaternary of the United States*, Princeton University Press, Princeton, 217-230.

Rudberg, Sten
1958 : Some observations concerning mass movements on slopes in Sweden. *Geologiska Föreningens i Stockholm Förhandlingar*, 80(1): 114-125.
1962 : A report on some field observations concerning periglacial geomorphology and mass movement on slopes in Sweden. *Biul. Peryglac.*, 11: 311-323.
1964 : Slow mass movement processes and slope development in the Norra Storfjäll area, southern Swedish Lappland. *Z. Geomorph.*, N.F. Suppl. Bd. 5: 192-203.

Sandberg, Gustaf
1938 : Redogörelser för undersökningar utförda med understöd av sällskapets stipendier. *Ymer*, 58: 333-337.

Schumm, S. A.
1964 : Seasonal variations of erosion rates and processes on hillslopes in western Colorado. *Z. Geomorph.*, N.F. Suppl. Bd. 5: 214-238.
1967 : Rates of surficial rock creep on hillslopes in western Colorado. *Science*, 155(3762): 560-562.

Sigafoos, R. S. and Hopkins, D. M.
    1952 : Soil instability on slopes in regions of perennially-frozen ground. *In: Frost action in soils,* Highway Research Board, Spec. Rep. No. 2: 176-192.
Smith, Jeremy
    1960 : Cryoturbation data from South Georgia. *Biul. Peryglac.,* 8: 73-79.
Soil Survey Staff
    1951 : *Soil Survey Manual.* U.S. Department of Agriculture Handbook No. 18, U.S. Government Printing Office, Washington, 503 pp.
Thompson, W. F.
    1961 : The shape of New England mountains, Part II. *Appalachia,* June: 316-335.
Washburn, A. L.
    1947 : Reconnaissance geology of portions of Victoria Island and adjacent regions, Arctic Canada. *Geol. Soc. Amer. Memoir,* 22: 1-142.
    1956 : Classification of patterned ground and review of suggested origins. *Geol. Soc. Amer. Bull.,* 67: 823-866.
    1967 : Instrumental observations of mass-wasting in the Mesters Vig District, Northeast Greenland. *Medd. om Grønland,* 166(4): 1-297.
Williams, P. J.
    1957 : The direct recording of solifluction movements. *Amer. J. Sci.,* 255: 705-715.
    1959 : An investigation into processes occurring in solifluction. *Amer. J. Sci.,* 257: 481-490.
    1966 : Downslope soil movement at a subarctic location with regard to variations with depth. *Can. Geotech. J.,* 3(4): 191-203.
Wilson, J. W.
    1952 : Vegetation patterns associated with soil movement on Jan Mayen Island. *J. Ecol.,* 40: 249-264.
Young, Anthony
    1960 : Soil movement by denudational processes on slopes. *Nature,* 188(4745): 120-122.
    1963 : Soil movement on slopes. *Nature,* 200: 129-130.
Zhigarev, L. A.
    1960 : Eksperimental'nye isseldovaniiā skorostei dvizheniiā gruntovykh mass na solifliuktsionnykh sklonakh. *Trudy Inst. Merzlotovedeniiā im V.A. Obrucheva,* 16: 183-190.

207

16

Reprinted from *Norsk Polarinstitutt Skr.*, No. 119, 4–7, 90–96 (1960)

TALUS SLOPES AND MOUNTAIN WALLS AT TEMPELFJORDEN, SPITSBERGEN A GEOMORPHOLOGICAL STUDY OF THE DENUDATION OF SLOPES IN AN ARCTIC LOCALITY

Anders Rapp

[*Editor's Note:* In the original, material precedes this excerpt.]

1.2 TALUS SLOPES, THEIR FORM AND DEVELOPMENT A SHORT GENERAL PRESENTATION

A talus slope (or scree slope) consists of rock débris which has fallen down more or less continuously[1] from a weathering mountain wall[2] (scarp) and formed an accumulation whose surface slopes about 30-40° and has a straight or slightly curved lateral profile.[3] The material is usually angular stones and boulders, which often lie assorted according to size with the smaller particles at the top and the larger boulders at the base of the talus slope.[4] This so-called fall-sorting is caused by the fact that the larger boulders, owing to their greater kinetic energy (according

[1] In very large, instantaneous rockslides there are formed not talus but rock-slide tongues (Heim, 1932, p. 106; Kolderup, 1955, p. 211; Rapp, 1957, p. 182).

[2] Mountain wall = steep, rocky slope, from which the weathered particles move by a free or a bounding fall (after Lehmann, 1933, p. 83). The inclination is about 40—90°.

[3] Many authors have measured and discussed the so-called maximum talus angle (angle of repose) and its dependence on the kind of rock, the particle size and the shape of the particles (see Piwowar, 1903; Stiny, 1926; Burkhalow, 1945; Malaurie, 1949).

[4] This sorting has been described by a very large number of observers. Behre (1933, p. 624 f.) describes a sorting in reverse, with coarse at the top and fine at the base, and interprets this as "normal" for talus. Bryan (1934, p. 655) makes a conclusive criticism of Behre, but despite this the latter author's erroneous generalization has been made the basis of the description of talus formations in handbooks by Sharpe (1938, p. 32) and Lobeck (1939, pp. 81 and 92).

[*Editor's Note:* Certain figures and plates have been omitted owing to limitations of space.]

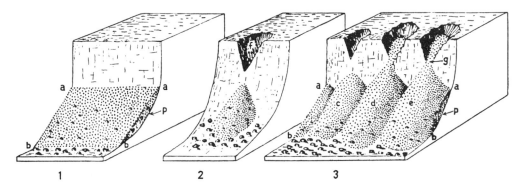

Fig. 1. The basic shapes of the talus formations. 1, Simple talus slope alongside a simple mountain wall. 2, A talus cone. 3, Compound talus slope alongside a dissected wall with rock-fall funnels.

a-a, talus crest. b-b, talus base with a base fringe of boulders in front of it. c, talus cone with a sheer top contour, as distinguished from d and e, which are proximally extended cones (hour-glass shaped). f, rock-fall funnel. g, funnel gorge (mouth of the funnel).

The talus accumulation is called by the comprehensive name of "talus mantle" irrespective of the forms. In a proximally extended talus cone the part which lies below the funnel gorge is called the "cone mantle".

to the formula $e = mv^2/2$, where $e =$ kinetic energy, $m =$ mass and $v =$ velocity) and their greater radius, can roll further down than the small particles in a rock-fall.

Certain mountain walls weather and retreat more or less uniformly over the whole of the free wall surface. They can suitably be called *simple walls.* They undergo *uniform retreat.* It would, however, seem to be more usual for the wall to retreat unequally through the formation of rock-fall chutes[5] or funnels, which dissect the wall all the more. This type can most simply be designated *dissected walls* and the corresponding retreat *dissection retreat.* Mount Templet's walls (Plate I) afford fine examples of regular dissection retreat.

Alongside a simple mountain wall is formed a *simple talus slope* (Fig. 1:1), alongside a wall with rock-fall chutes are formed *talus cones,* one in front of each chute. If the talus cones grow together laterally, they

---

[5] Rock-fall chute (shortened to "chute" (Blackwelder, 1942, p. 328, and Baulig, 1956, p. 70)) = a gully or wide cleft in a steep, rocky slope, through which débris is moving by free or bounding fall (German "Steinschlagrinne"). Rock-fall funnel (shortened to "funnel") = a half-funnel-shaped excavation in a rock wall, wide above, narrow below, through which débris is moving by rock-falls (German "Steinschlagtrichter"). Large "funnels" are classified as cirques.

form a *compound talus slope* (cf. Bargmann, 1895, p. 70, and Poser, 1954, p. 143).

Many observations show that the talus material usually lies as a relatively thin layer of some few metres' thickness on top of a substratum of solid rock with approximately the same inclination as the talus surface. Morawetz (1933, p. 36) reports a thickness of 1-3 m, Malaurie (1950, p. 15) 2-5 m, and Young (1956, p. 125) 2-8 m in descriptions of talus formations from different environments. Even greater thicknesses have been reported, for example, by Fromme (1955, p. 39) up to 25 m and by Rapp (1957, p. 197) up to 35 m, but not even these maximum values amount to more than about $1/_{10}$ of the height of the talus slope in question.

From a dynamic point of view the talus slopes can be designated as a first re-loading place for the weathered material on its way from the mountain wall to the sea or the lake. The movement of material in and from a talus slope can suitably be divided into the following stages:

Supply	Fall of débris from the wall down to the surface of the talus. It is terminated by the primary deposition of the particles.
Shifting	The movement of the material down the talus slope after the primary deposition.
Removal	Movement of the material away from the talus slope.

Supply usually occurs through smaller rock-falls from the wall and is terminated by the boulders and stones bouncing and rolling a short distance on their edges, before they remain lying on or below the talus surface.

Shifting and removal take place through processes which can work at a lower inclination than about 30—35°, for example, talus-creep, slides, snow avalanches, running water, etc. (see Rapp, 1957, p. 179 f.). Through the shifting and removal of the material the talus formation is levelled out, takes on a more and more concave profile (see Fig. 2) and may be transformed, for example, into an alluvial cone.[6]

In principle the development of a talus slope is conceived of in the following way (cf. Lehmann, 1933; Bakker, 1952; and others). A steep mountain wall weathers. The debris forms a talus slope at the foot of the mountain. According as the talus slope is supplied with more material, it increases in height with inclination preserved and an almost straight

---

[6] In the question of talus in an Arctic environment solifluction is considered to be a removal process of the greatest significance (Högbom, 1914; Jahn, 1947; Büdel, 1948; Dege, 1949; Malaurie, 1950) but practically no measurements of its rate and capacity have been reported.

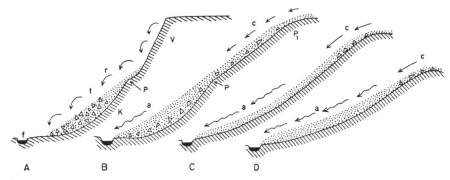

Fig. 2. Sketch showing the principles of slope development, modified by the author after Derruau (1956, p. 59); the later phases are according to the theories of Baulig and Birot.

A. The mountain wall (V) retreats on account of smaller falls (r). The talus slope (t) grows. Beneath it lies the rock core (K), which here is assumed to have been formed by earlier glacial erosion and has therefore been drawn concave. In the upper part of the rock core a "Richter's slope (P) has begun to be formed on the retreat of the wall.

B. The talus phase is finished. The slope gets a convex-concave equilibrium profile on account of weathering and creep (c) at the top and wash (a) (Baulig, 1940) or creep + congestion at the bottom (Birot, 1949, p. 20). The material is removed by a river (f) at the base. P-P$_1$, the "Richter's slope".

C, D. The slope is lowered. Wash (a) predominates more and more over creep (Baulig, 1940).

profile. The wall retreats and finally comes to be entirely covered with débris. On the retreat of the wall there is formed a more level slope of solid rock, which is covered over by the talus slope and which has about the same declivity. The new mountain slope, which forms the substratum of the talus slope is called by Lehmann (1933) a "rock nucleus" (German "Felskern"), by W. Penck (1924) "Haldenhang", by Howard (1942, p. 27) "subtalus buttress" and by Twidale (1959, p. 65) "sub-debris slope". Another term for it is "Richter's denudation slope" (Bakker 1952), here shortened to "Richter's slope" (Fig. 2).

The talus phase outlined above is an important element in the development of the slope from mountain wall to level surface. Our knowledge is still small as concerns, for example, which transport processes direct the slope's retreat in the talus phase and later (see Fig. 2). Above all we know very little about how rapidly the development goes on. It is these problems, *inter alia,* which are to be elucidated in this thesis.

[*Editor's Note:* Material has been omitted at this point.]

CHAPTER 7

Conclusions

7.1 CONCLUSIONS AS TO THE DEVELOPMENT OF THE WALL RELIEF

The author's studies of mountain walls and talus slopes in the Tempelfjorden area have led to the following conclusions:

(1) The rock-fall funnels in the walls were essentially formed before post-glacial times. They were probably formed chiefly by nivation and small rock-falls.

(2) They were possibly established inter-glacially (pre-glacially?) and in the main widened in glacial (late glacial?) times. Their widening continued during post-glacial times.

Consequently their genesis is possibly as follows: (A) inter-glacial establishment; (B) glacial widening, alternatively (1) "covering ice" for a longer or shorter glaciation optimum but not total destruction of the funnels, and afterwards widening of them during a late-glacial nunatak stage, or (2) not "covering" ice, only valley glaciation and widening of the funnels during a nunatak stage of long duration; (C) post-glacial widening probably corresponding to an average wall retreat of about 3-5 m (Bjonahamna).

These conclusions are in keeping with the suppositions of De Geer and Wråk, but imply a considerably slower development than that which Högbom and other authors pre-suppose (cf. Chapter 6.1).

7.2 CONCLUSIONS AS TO THE DEVELOPMENT OF THE TALUS CONES

The main interest of the thesis was directed towards the talus slope formed by about 10 large cones at Bjonahamna. The development of these cones has probably run its course according to the following scheme:

A. *The deglaciation phase.* The upper part of the mountain wall free from ice, the lower part covered with moving glacier ice (corresponding to the conditions at the present day at the Langtuna nunatak). Almost all debris was carried away by the glacier, on which was formed a large lateral moraine on the surface of the ice.

212

B. *The abrasion phase.* When the ice had disappeared from the valley side, talus cones began to be accumulated on the top of bedrock and moraine material. At certain places these cones were formed on top of remaining lateral terraces of moraine (and glacifluvium?). The sea reached up to around the present 90-metre level and consequently abraded the lower part of the cones, yet not T9, which stood on a terrace. The removal of material from the cones by abrasion was probably not of significant proportions, partly because most of the cones have a shallow and sheltered situation (except T1), partly because the higher lying cones came relatively soon to be elevated above the reach of the waves. The assortment on the talus (accumulation of the coarsest material at the base) also counteracted the removing effect of the abrasion.

C. *The "dry phase".* The cones are dry in so far as they are beyond reach of the effect of the waves. The new C^{14} datings (cf. p. 26) indicate that the dry phase has lasted about 10.000 years for the cone T1 etc.

Recent development

The supply of debris in modern times (1882-1954) to the cones probably corresponds to a horizontal, "smoothed" wall retreat of about 0.02-0.2 mm per year. Pebble falls and small boulder falls are quantitatively most important for the supply, large boulder falls less important.

The shifting corresponds to the supply and takes place essentially in the form of momentary, dry slides. Next in the matter of quantitative importance come mudflows and then avalanches. As regards the importance of talus creep, it appears from picture examination etc. that this form of transport has played no appreciable role on the lower part of the cones. How its effect operates on their upper parts could not be ascertained.

Removal seems at present to occur chiefly by solution and is considerably less than supply and shifting, for which reason the cone mantles are slowly growing in area at the present day, to a large extent with inclination preserved and a straight profile. The exceptions are certain cones affected by mudflows and avalanches, which are becoming slightly concave at the base.

A comparison with the rate of breaking down on nunatak walls alongside a present-day glacier (Tunabreen) indicates that one ought not to reckon with a more rapid breaking down than at most 0.5 mm of wall retreat per year at Bjonahamna during the close of the deglaciation phase and possibly the beginning of the abrasion phase.

The moderate order of magnitude of the present-day disintegration at Bjonahamna and at Tunabreen is probably representative of the Permo-Carboniferous mountains on a large scale on Spitsbergen.

The disintegration at the present day is consequently relatively slow, but the talus cones are large in spite of this, if one makes a comparison with many other climatic environments. This lack of agreement between present-day supply and the size of the cones cannot be explained by the talus slopes themselves having choked the supply of debris by growing up and covering their former collection areas, because the free wall surfaces are still large. It is possibly conceivable that the explanation of the disproportion may lie in the fact that the supply of debris has been much heavier during certain past time-phases (colder climate? climatic reversal? earthquakes?) of a 10,000-year post-glacial period. But the explanation may also lie in a more continuous development during a post-glacial time which has lasted considerably longer than 10,000 year.[1]

As regards the *methods* used by the author the comparative photograph examination especially has proved to be very useful for the study of the rate and type of slope development, even if this development was more slow than the author at first had expected. No doubt this method offers one of the best ways of getting more exact knowledge of the development of steep slopes in various rocks and climates.

---

[1] *A short comment on the new C¹⁴ datings* (p. 26) may be added (April, 1960). The datings refer to raised terraces at Billefjorden but are probably valid also for Templet at Tempelfjorden, as the distance is only 20-30 km between these localities and as the upper marine limit runs at about 90 m in both fjords (cf. Feyling-Hanssen, 1955, p. 47).

The datings give the main trend of the land-rise curve from the altitude of 56 m down to the present sea-level. They indicate a rapid uplift at first and a slow one later on. The curve from Billefjorden fits very well with a land-rise curve from Nordaustlandet, based on numerous C¹⁴ datings on samples collected by W. Blake. (Personal communication by I. Olsson.)

The new datings from Billefjorden in the author's opinion indicate the following as regards the post-glacial development at Templet.

(1) The outer parts of Billefjorden and Tempelfjorden were already deglaciated 10,000 years ago. A rapid land uplift was occurring at that time.

(2) We do not yet know if the land-rise curve can be extrapolated from 56 m to 90 m with the same gradient or if there exists one or several steps in the higher part of the curve.

(3) The unknown land-rise interval from about 90 m to about 56 m corresponds at Bjonahamna to theoretical shorelines running over the middle and lower parts of the terrace slopes (T9, S2, S6) discussed on p. 56 f. above. If the "unknown" period (90-56 m) was very long, the terraces ought to have been completely obliterated by wave abrasion. As this is not the case at all the author's belief is that a *rapid land uplift also occurred in the 90-56 m interval*. If this supposition is correct, the post-glacial period at Bjonahamna started about 10,000 years ago or a little earlier.

References

Ahlmann, H. W. 1919: Geomorphological studies in Norway. *Geograf. Annaler,* Årg. I. Hft. 1—2. Stockholm.˙

— 1948: Glaciological research on the North Atlantic coasts. *Royal Geograph. Soc. Research Series 1.* London.

— 1953: Glacier variations and climatic fluctuations. *Amer. Geograph. Soc.* New York.

Ahlmann, H. W. and Tryselius, O. 1929: Der Kårsa-Gletscher in Schwed. Lappland. *Geograf. Annaler,* Hft. 1. Stockholm.

Bakker, J. P. 1952: A remarkable new geomorphological law. *Proceed. Koninkl. Nederl. Akad. Wet.,* Vol. LV, ser. B. Amsterdam.

Balk, R. 1939: Disintegration of glaciated cliffs. *Journal of Geomorphology.,* Vol. II. No. 4.

Bargmann, A. 1895: Der jüngste Schutt der nördlichen Kalkalpen. Wissenschaftl. Veröffentl. d. Vereins f. Erdkunde. Leipzig.

Baulig, H. 1940: Le profil d'équilibre des versants. *Annales de Géograph.*

— 1956: Vocabulaire de géomorphologie. Paris.

Behre, C. H. Jr. 1933: Talus behaviour above timber in the Rocky Mountains. *Journal of Geology,* 41.

Berset, O. 1953. Hilmar Nøis, storjegeren fra Svalbard. Eides forlag. Oslo.

Birot, P. 1949: Essai sur quelques problèmes de morphologie générale. Centro de estud. geográf. Lisbonne.

Blackwelder, E. 1942: The process of mountain sculpture by rolling debris. *Journal of Geomorphology,* Vol. V, No. 4.

Blanck, E. und Rieser, A. 1928: Die wissenschaftlichen Ergebnisse einer bodenkundlichen Forschungsreise nach Spitzbergen — — —. *Chemie der Erde,* Bd. 3. Jena.

Bryan, K. 1934: Geomorphic processes at high altitudes. *Geogr. Review,* Vol. 24, p. 655.

Bryan, K. and La Rue, E. L. 1927: Persistence of features in an arid landscape. The Navajo Twins, Utah. *Geogr. Review,* Vol. 17, No. 2.

Burkhalow, A. van, 1945: Angle of repose and angle of sliding friction. *Bull. Geol. Soc. Am.,* Vol. 56.

Büdel[1], J. 1948: Die klima-morphologischen Zonen der Polarländer. *Erdkunde,* Bd. III, Lfg 1—3. Bonn.

— 1953: Die periglazial-morphologischen Wirkungen des Eiszeitklimas auf der ganzen Erde. *Erdkunde,* Bd VII, Hft. 4. Bonn.

Bögli, A. 1956: Der Chemismus der Lösungsprozesse und der Einfluss der Gesteinsbeschaffenheit auf — — —. Rep. Comm. Karst Phenom. *18th Intern. Geogr. Congr.,* Frankfurt am Main.

---

[1] See also p. 96.

Cailleux, A. and Taylor, G. 1954: Cryopedologie. Exp. Polaires Franç. IV. Paris.

Corbel, J. 1957: Les karsts du nord-ouest de l'Europe. Lyon.

Dege, W. 1941: Die Schwankungen des Von Post Gletschers auf Spitzbergen. *Zeitschr. f. Gletsch.kunde*, Bd. 27.

— 1949: Welche Kräfte wirken heute umgestaltend auf die Landoberfläche der Arktis ein? *Polarforschung*, Bd II/1949.

De Geer, G. 1896: Rapport om den svenska geologiska expeditionen till Spetsbergen — — —. *Ymer*, hft. 4. Stockholm.

— 1910: Guide de l'excursion au Spitzberg. *XIe Congr. Géol. Int.*, Stockholm.

Derruau, M. 1956: Précis de géomorphologie. Masson et Cie. Paris.

Eriksson, J. V. 1929: Den kemiska denudationen i Sverige. *Medd fr. Stat. Meteorol. Hydrol. Anst.*, Bd 5, Nr 3. Stockholm.

Feyling-Hanssen, R. W. 1955: Stratigraphy of the marine late-pleistocene of Billefjorden, Vestspitsbergen. *Norsk Polar Inst. Skrifter*, Nr 107. Oslo.

Flint, R. F. 1957: Glacial and pleistocene geology. Wiley and Sons. New York.

Freise, F. W. 1933: Beobachtungen über Erosion an Urwaldgebirgsflüsssen des brasilianischen Staates Rio de Janeiro. *Zeitschr. f. Geomorphol.*, VII. Leipzig.

Friedel, H. 1935: Beobachtungen an den Schutthalden der Karawanken *Carinthia*, II. Klagenfurt.

Fromme, G. 1955: Kalkalpine Schuttablagerungen als Elemente nacheiszeitlicher Landschaftsformung im Karwendelgebirge. *Mus. Ferdinandeum*, Bd. 35. Innsbruck.

Frödin, J. 1914: Geografiska studier i St. Lule älvs källområde. Stockholm.

Gee, E. R., Harland, W. B. and McWhae, J. R. 1953: Geology of Central Vestspitsbergen. *Trans. Roy. Soc. Edinburgh.*, Vol. LXII, Part II.

GHM, see Gee etc. 1953.

Groom, G. E. 1959: Niche glaciers in Bünsow Land, Vestspitsbergen. *Journal of Glaciol.*, Vol. 3, No. 25.

Groom, G. E. and Sweeting, M. M. 1958: Valleys and raised beaches in Bünsow Land, Central Vestspitsbergen. *Norsk Polar Inst. Skrifter* Nr. 115, Oslo

Heim, A. 1932: Bergsturz und Menschenleben. *Vierteljahrsschr. d. Nat. forsch. Gesellschaft Zürich*, Nr 20, Jahrgang 77.

Hesselberg, T. and Birkeland, J. 1940: Säkulare Schwankungen des Klimas von Norwegen. *Geofys. Publ.*, Vol. XIV No 4. Oslo.

Hesselberg, T. and Johannessen, T. W. 1957: The recent variations of the climate at the Norwegian Arctic Stations. *Journal of Atmosph. and Terrestr. Physics*, Polar Atm. Symposium. London.

Holmes, A. 1945: Principles of physical geology. Nelson and Sons. London.

Howard, A. D. 1942: Pediment passes and the pediment problem. *Journal of Geomorphol.*, Vol. V, No. 1.

Högbom, B. 1911: Bidrag till Isfjordsområdets kvartärgeologi. *Geolog. Fören. Förhandl.*, Bd 33. Stockholm.

— 1914: Über die geologische Bedeutung des Frostes. *Bull. Geol. Inst.*, XII. Uppsala.

Jahn, A. 1947: Note of the talus in the Polar regions. *Przeglad Geograf.*, XXI, 1—2. Warszawa.

— 1958: Geomorphological and periglacial researches carried out north of Hornsund fiord in summer 1957. *Przcglad Geofizyczny, d. Rozn. III.* Warszawa.

King, L. C. 1956: Research on slopes in South Africa. Prem. Rapp. de la Comm. pour l'Etude des Versants. U.G.I. Amsterdam.

Kolderup, N.-H. 1955: Raset i Modalen 14. august 1953. *Norsk Geolog. Tidsskr.*, Bd 34, hft. 2—4. Oslo.

Lehmann, H. 1956: Der Einfluss des Klimas auf die morphologische Entwicklung des Karstes. Rep. Comm. Karst Phenom. *18th Intern. Geogr. Congr.*, Frankfurt am Main.

Lehmann, O. 1933: Morphologische Theorie der Verwitterung von Steinschlagwänden. *Vierteljahrsschr. d. Nat.forsch. Gesellschaft Zürich*, Jahrgang 78.

Liljequist, G. 1956: Meteorologiska synpunkter på istidsproblemet. *Ymer*, hft. 1. Stockholm.

Lobeck, A. K. 1939: Geomorphology. McGraw-Hill. New York.

MacCabe, L. H. 1939: Nivation and corrie erosion in West Spitsbergen. *Geograph. Journal*, Vol. 94.

Macar, P. and Birot, P. 1955: Commission on the evolution of slopes. *I.G.U. Newsletter*, Vol. VI.

Malaurie, J. 1949: Évolution actuelle des pentes sur la côte Ouest de Groenland. *Bull. de l'Ass. de Géograph Franç.*, No. 198—199.

— 1950: La baie de Disco, côte Ouest du Groenland. *Bull. de l'Ass. de Géograph. Franç.* No. 206—207.

Mannerfelt, C. M. 1945: Några glacialmorfologiska formelement. *Geograf. Annaler*, Årgång XXVII. Stockholm.

Matznetter, J. 1956: Der Vorgang der Massenbewegungen an Beispielen des Klostertales in Vorarlberg. *Geograph. Jahresber. aus Österreich*, Bd XXVI. Wien.

Michaud, J. 1950: Emploi des marques dans l'étude des mouvements du sol. *Rev. Géomorph. Dynamique.*

Morawetz, S. O. 1933: Beobachtungen an Schutthalden, Schuttkegeln und Schuttflecken. *Zeitschr. f. Geomorphol.*, Bd. VII. Leipzig.

Mortensen, H. 1928: Über die klimatischen Verhältnisse des Eisfjordgebietes. *Chemie der Erde*, Bd III, Hft 3—4. Jena.

— 1956: Über Wanderwitterung und Hangabtragung in semiariden und vollariden Gebieten. Prem. Rapp. de la Comm. pour l'Etude des Versants. U.G.I. Amsterdam.

Nathorst, A. G. 1883: Kartläggningen af Tempelbay, ett bidrag till Spetsbergens geografi. *Ymer*. Stockholm.

— 1884: Redogörelse för den tills. med G. de Geer år 1882 företagna expeditionen till Spetsbergen. *Bihang till Kungl. Svenska Vet.Akad. Handl.*, Bd 9, No 2. Stockholm.

— 1910: Beiträge zur Geologie der Bäreninsel, Spitzbergens — — —. *Bull. Geol. Inst.*, X. Uppsala.

Nussbaum, F. 1957: Über rezente Erdrutsche und Felsstürze in der Schweiz. *Geographica Helvetica*, XII, Nr 4. Bern.

Olsson, I. 1959: Uppsala Natural Radiocarbon Measurements I. *Am. Journ. of Sci. Radiocarb. Suppl.* Vol. 1.

Orvin, A. K. 1940: Outline of the geological history of Spitsbergen. *Skrifter om Svalbard o. Ishavet*, Nr. 78. Oslo.

Penck, W. 1924: Die morphologische Analyse. Stuttgart.

Philipp, H. 1914: Ergebnisse der W. Filchnerschen Vorexpedition nach Spitzbergen 1910. *Pet. Mitteil.*, Erg. hft. 179. Gotha.

Pillewizer, W. 1939: Die kartographischen und gletscherkundlichen Ergebnisse der deutschen Spitzbergen-Expedition 1938. *Pet. Mitteil.*, Erg. hft. 238. Gotha.

Piwowar, A. 1903: Über Maximalböschungen trockener Schuttkegel und Schutthalden. *Vierteljahrsschr. d. Nat.forsch. Gesellschaft Zürich*, Vol. 48.

Poser, H. 1954: Die Periglazial-Erscheinungen in der Umgebung der Gletscher des Zemmgrundes. *Göttinger Geograph. Abhandl.*, Hft 15. Teil II.
—   1957: Klimamorphologische Probleme auf Kreta. *Zeitschr. f. Geomorphol.*, Bd I, Hft 2. Berlin.
Rapp, A. 1957: Studien über Schutthalden in Lappland und auf Spitzbergen. *Zeitschr. f. Geomorphol.*, Bd I, Hft 2. Berlin.
Sharp, R. P. 1942: Mudflow levées. *Journal of Geomorphol.* Vol. No. 3.
—   1948: The constitution of valley glaciers. *Journal of Glaciol.* Vol. 1.
Sharpe, J. F. S. 1938: Landslides and related phenomena. New York.
Stiny, J. 1926: Neigungswinkel von Schutthalden. *Zeitschr. f. Geomorphol.* Bd I. Leipzig.
Sweeting, M. M. and Groom, G. E. 1956: Notes on the glacier fluctuations in Bünsow Land, Central Vestspitsbergen. *Journal of Glaciol.* Vol. 2, No. 19.
Tricart, J. 1953: Resultats préliminaires d'expériences sur la desagregation des roches — — —. *Comptes Rend. d. Séances de l'Acad. Sci.*, Paris.
—   1956: Étude expérimentale du problème de la gelivation. *Biuletyn Peryglacjalny.* Nr. 4. Lodz.
Troll, C. 1944: Strukturboden, Solifluktion und Frostklimate der Erde. *Geol. Rundschau*, XXXIV.
Twidale, C. R. 1959: Evolution des versants dans la partie centrale du Labrador-Nouneau-Quebec. *Annales de Géographie*, No 365. Paris.
Williams, P. J. 1959: The development and significance of stony earth circles. *Norsk Vidensk. Akad. i Oslo. I. Mat.-Naturv. Klasse.* No. 3.
Wråk, W. 1916: Sur quelques "Rasskars" dans les escarpements des vallées glaciaires en Norvège. *Bull. Geol. Inst. Uppsala*, XIII.
Winkler-Hermaden, A. 1957: Geologisches Kräftespiel und Landformung. Springer. Wien.
Young, A. 1956: Scree profiles in West Norway. Prem. Rapp. de la Comm. pour l'Etude des Versants U.G.I. Amsterdam.

The following two works have been published during the printing of this thesis.

Büdel, J. 1960: Die Frostschutt-Zone Südost-Spitzbergens. *Colloquium Geograph.*, Bd. 6. Bonn.
—   1960: Die Gliederung der Würmkaltzeit. *Würzburger Geograph. Arbeiten*, Hft. 8. Würzburg.

17

Reprinted from *Bull. Geol. Soc. America,* **70,** 383–387, 389–390, 392, 430–436 (Apr. 1959)

ROCK GLACIERS IN THE ALASKA RANGE

By Clyde Wahrhaftig and Allan Cox

Abstract

Field studies and examination of aerial photographs of approximately 200 rock glaciers in the Healy (1:250,000) quadrangle in the central Alaska Range showed that there are three types of rock glacier in plan: lobate, in which the length is less than the width (200–3500 feet long and 300–10,000 feet wide); tongue-shaped, in which the length is greater than the width (500–5000 feet long and 200–2500 feet wide); and spatulate, tongue-shaped but with an enlargement at the front. Lobate rock glaciers line cliffs and cirque walls and probably represent an initial stage; the other two move down valley axes and represent more mature stages.

The rock glaciers are composed of coarse, blocky debris that is cemented by ice a few feet below the surface. The top quarter of the thickness is coarse rubble, below which is coarse rubble mixed with silt, sand, and fine gravel. Fronts of active (moving) rock glaciers are bare of vegetation, are generally at the angle of repose, and make a sharp angle with the upper surface. Fronts of inactive (stationary) rock glaciers are covered with lichens or other vegetation, have gentle slopes, and are rounded at the top. Active rock glaciers average 150 feet in thickness, inactive rock glaciers, 70 feet.

The upper surface of most rock glaciers is clothed with turf or lichens. Sets of parallel rounded ridges and V-shaped furrows—longitudinal near the heads of some rock glaciers and transverse, bowed downstream, on the lower parts of others—and conical pits, crevasses, and lobes mark the upper surfaces of many rock glaciers.

The upper surface of a rock glacier at the head of Clear Creek moved 2.4 feet per year between 1949 and 1957, and the front advanced 1.6 feet per year.

Heights of the upper edges of the talus aprons along the fronts of rock glaciers average 45 per cent of the heights of the fronts. Each of these observations implies that motion is not confined to thin surface layers but is distributed throughout the interiors of the rock glaciers, which in this permafrost region are probably frozen. "Viscosity" has been calculated for rock glaciers at between 10^{14} and 10^{15} poises; for glacial ice it has been estimated at between 10^{12} and 10^{14} poises. Maximum average shear stresses within active rock glaciers range from 1 to 2 bars; these values are much larger than those calculated for solifluction and creep features.

Rock glaciers occur on blocky fracturing rocks which form talus that has large interconnected voids in which ice can accumulate. They are rare on platy or schistose rocks whose talus moves rapidly by solifluction.

The rock glaciers lie in an altitudinal zone about 2000 feet thick, centered on the lower limit of existing glaciers[1]. Although the firn lines on glaciers rise 1200 feet in a distance of 25 miles northward across the Alaska Range, the lower limit of active rock glaciers rises only 800 feet. The firn line on southward-facing glaciers is 2000 feet higher than that on northward-facing glaciers, yet in any given area southward-facing rock glaciers average only 200 feet higher than northward-facing rock glaciers. Insulation by the debris cover is believed responsible for the difference in altitudinal ranges between rock glaciers and glaciers.

It is concluded that rock glaciers move as a result of the flow of interstitial ice and that they require for their formation steep cliffs, a near-glacial climate cold enough for the ground to be perennially frozen, and bedrock that is broken by frost action into coarse blocky debris with large interconnected voids. The longitudinal furrows are thought to result from the accumulation of ice-rich bands in the swales between talus cones at the head of the rock glaciers and the subsequent melting of this ice as the rock glacier moves down-valley. The transverse ridges are thought to result from shearing within the rock glacier that would occur where the thickness increases or the velocity decreases downstream.

An average of 30 feet of bedrock was removed from source areas to form the present rock glaciers, indicating an average rate of erosion of 1–3 feet per year when they are active.

---

[1] In this paper *glacier*, used alone, always refers to a body of ice, relatively free from debris, derived largely from compaction of snow.

The climatic history of the Alaska Range that can be deduced from the presence of active and inactive rock glaciers is: (1) deglaciation following the ice advance of 9000 B. C., leading to the thermal maximum; (2) a cold period, following the thermal maximum, during which the now-inactive rock glaciers formed; (3) a warm period; (4) a cold period, indicated by active rock glaciers, continuing to the present, with climatic amelioration during the last 50 years. The first cold period was somewhat colder and longer than the second.

CONTENTS

[*Editor's Note:* Certain figures and plates have been omitted owing to limitations of space.]

CONTENTS   385

TABLES

INTRODUCTION

In the half century since rock glaciers were recognized as distinct geomorphic forms they have been found in high mountains throughout the world. They were first discussed by Spencer (1900), who briefly described rock glaciers in the San Juan Mountains. Cross and Howe (1905) described the rock glaciers of the San Juan Mountains as rock streams. Rock glaciers in the Alps were described by Chaix (1919) and Cailleux (1947, p. 323–325) as *coulées de blocs*, by De Martonne (1920, p. 261–264) as *glaciers rocheux*, by Salomon (1929, p. 1–31) as *Blockglätscher*, by Domaradzki (1951) as *Blockströme*, and by Nangeroni (1954, p. 35) as *colate di pietre*. The term rock glacier was first used by Capps (1910) in his classic study of the rock glaciers in the Wrangell Mountains of Alaska.

Rock glaciers have also been reported from the Rocky Mountains by Patton (1910, p. 666); Hole (1912, p. 722); Emmons, Irving, and Loughlin (1927, p. 20); Atwood and Mather (1932, p. 162); Parsons (1939); R. L. Ives (1940); Richmond (1952); and Griffiths (1958) and from the Sierra Nevada of California (Kesseli, 1941), from the Carpathians (De Martonne, 1920), from the Altai (Matveev, 1938), and from the Andes (Lliboutry, 1953). The descriptions and photographs of these high-montane morphologic forms leave no doubt that similar phenomena have been described in widely separated places. However, theories of their origin are almost as numerous as the reports of their occurrence—evidence of the challenge that these interesting forms have presented to the imaginations of geomorphologists.[2]

Cross and Howe (1905, p. 25) advanced two theories for the two types of rock glaciers they described, considering the common streamlike ones to be unusual types of moraines. Howe (1909, p. 52) later advanced the theory that rock glaciers are the debris of violent landslides and cited the Elm landslide as a modern example.

Capps (1910, p. 364) suggested an entirely different mechanism, one anticipated by Spencer (1900). Capps thought that water from melting ice and snow and from rains sinks into the detritus at the lower edge of glaciers and freezes, gradually filling the interstices with ice up to the level where melting equals freezing. Incipient glacial movement is started by melting of the ice followed by refreezing and expansion of the water. Tyrrell

[2] After this paper went to press, we learned of two important recent discussions of rock glaciers, one by Roots (1954, p. 24–29), and the other by Flint and Denny (1958, p. 131–133).

(1910, p. 552–553) elaborated on Capps' theory, suggesting that the water which is later frozen comes from springs, and that the characteristic concentric ridges develop by the downward sliding of the rocks after melting of the ice in the spring.

Chaix (1923, p. 32) regarded rock glaciers as terminal moraines composed of blocky debris, the interstices of which are filled with mud and clay. He believed that the debris, lubricated by the interstitial mud, moves down-slope, the movement aided by freezing and thawing. Kesseli (1941, p. 226–227) similarly thought that rock glaciers are of glacial origin and that many of the smaller surface features result from deformation under the weight of over-riding ice. He thought that the present motion of rock glaciers is due to a remaining core of ice or to a tendency to creep after deposition, but that creep could not have originated the rock glaciers. Richmond (1952) believed rock glaciers to be residual from small glacial advances, retaining the essential configuration and structure of the ice.

The present investigation deals with about 200 rock glaciers in the Alaska Range, all within the Healy 1:250,000 quadrangle, part of which is shown on Figure 1. Two groups of rock glaciers form the major subject of this study. One consists of about 70 rock glaciers on an eastward-trending ridge between the headwaters of the Wood River and the West Fork of the Little Delta River, called the Wood River area (Pl. 1). The other group consists of about 120 rock glaciers on the main ridge of the Alaska Range on either side of the Alaska Railroad (Figs. 4 and 5). The area shown on these figures is called the Windy area. About 10 rock glaciers occurring elsewhere, mainly at the head of Moody and Cody Creeks, near the Wood River glacier, and in the basin of Sheep Creek (Fig. 1) were examined in the field and form a part of this study.

This study is a by-product of a general geologic investigation of the central part of the Alaska Range. The rock glaciers were studied in the field from 1950 to 1957. Commonly only a few hours could be devoted to the field examination of any one rock glacier. Most of the rock glaciers studied on the ground were photographed, many in color. The photographs have proved useful for measurement of height and slope of the fronts of the rock glaciers, the microrelief, average size of fragments on the surface, and for determination of the type and extent of vegetation.

Field observations were supplemented by study of aerial photographs, at scales of 1:20,000 and 1:40,000, taken by the U. S. Air Force in 1946, 1949, and 1952. Outlines of the rock glaciers and patterns of microrelief that could be recognized on the aerial photographs were plotted on enlargements (to a scale of 1:12,000) of the multiplex manuscript of the Healy D-1, D-2, D-3, B-4, B-5, C-4, and C-5 quadrangles. All measurements of altitude, length, breadth, and area, and many of slope and height, were made on these maps at the scale of 1:12,000, contour interval 100 feet.

About three-fourths of the Wood River area rock glaciers were examined on the ground; the others were studied in aerial photographs. In the Windy area, only the rock glacier at the head of Clear Creek was studied on the ground; the others were studied by examination of aerial photographs. Field measurements of the movement of the rock glacier at the head of Clear Creek, the only one whose motion was measured, were made over a period of 8 years. A preliminary study of the rock glaciers (Wahrhaftig, 1954) was based largely on interpretation of aerial photographs. In the course of subsequent field work no major active rock glacier was discovered which had not been identified previously from photographs, and every feature previously identified from photographs as a rock glacier proved to be one when examined in the field.

ACKNOWLEDGMENTS

The marked boulders on the Clear Creek rock glacier were established and surveyed in 1949 by the late Dr. Robert E. Fellows and the senior author. J. H. Birman assisted in re-surveying the boulders in 1952, and R. E. Zartman assisted in 1957.

The manuscript has been reviewed by J. D. Sears, Troy L. Péwé, David M. Hopkins, Arthur H. Lachenbruch, John T. Hack, and Dorothy H. Radbruch of the U. S. Geological Survey, and by Prof. Robert P. Sharp of the California Institute of Technology. We thank all these for calling attention to errors of analysis and obscurities in the prose. George O. Gates, Chief of the Alaskan Geology Branch of the Geological Survey, under whose direction the work was done, encouraged and supported the investigation. Mrs. Pauline Waggoner prepared the text.

Geographic Setting of the Rock Glaciers

The rock glaciers lie in the central part of the Alaska Range (Fig. 1). This part of the Alaska Range consists of several eastward-trending ridges with crests 5000 to 8000 feet in altitude, separated by broad valleys 2000 to 3500 feet in altitude. The northern limit of the Pleistocene cordilleran ice sheet lies in the northern part of the Alaska Range, crossing the middle of the area shown in Figure 1 (Wahrhaftig, 1958, Pl. 2).

Most of the rock glaciers lie in cirques that were occupied by glaciers during the late Wisconsin ice advance. A few, however, are in places that give no sign of having been areas of accumulation of late Wisconsin ice. The maximum extent of the late Wisconsin ice in the Alaska Range is indicated on Figure 1. Carbon 14 determination on a peat sample from a terminal moraine complex near McKinley Park station has dated this advance at slightly before 8600 B.C. (Suess, 1954, p. 471, No. W-49; Wahrhaftig, 1958, p. 46). Glacial ice still covers most of the high mountains of the Alaska Range.

Climate

The Alaska Range separates two parts of Alaska with markedly different climates; climate varies abruptly, and no generalization can be made about the range as a whole. South of the range, summers are fairly cool and cloudy with heavy rainfall; winters are moderately rigorous with heavy snowfall. North of the range, summers are warm and sunny, and rainfall is fairly light; winters are severe with light snowfall. Available temperature data for the central Alaska Range are summarized in the records for McKinley Park Headquarters, Nenana, and Summit (Figs. 2 and 3). Summit lies on the Alaska Railroad south of the Alaska Range at an altitude of 2350 feet, a few hundred feet below timber line (Fig. 1). It represents the cool moist conditions of the south flank of the range. McKinley Park is in the center of the range at an altitude of 2000 feet, about 1000 feet below timber line. It is typical of the intermontane valleys of the central part of the range. Nenana lies in the Tanana flats about 26 miles north of the north edge of the range, at an altitude of 350 feet. Its records are typical of the interior region of Alaska north of the Alaska Range.

The climate of the high mountains where the rock glaciers occur probably differs considerably from that of the stations in the lowlands. The mountains are cooler in summer and probably warmer in winter; the number of cloudy days and amount of precipitation, particularly snowfall, are greater. Cliffs and outcrops in the high mountains probably remain damp during most of the cool, moist summers, particularly during the periods of diurnal freeze and thaw, and much moisture is available for freezing in cracks and joints in the rocks. Temperatures may drop to freezing in any month during the summer. All these factors contribute to greater frost riving in the high mountains of the Alaska Range than in the valleys where the weather stations are located. Weather records indicate that even in the lowlands and valley bottoms, the average annual temperature is at least 5° F. below freezing. Hence the Alaska Range is probably a region of permafrost.

General Description of the Rock Glaciers

The rock glaciers in the Alaska Range are tongue-shaped or lobate masses of poorly sorted angular debris lying at the base of cliffs or talus slopes (Fig. 3 of Pl. 2) or extending down-valley from the lower end of small glaciers (Fig. 1 of Pl. 3). They are a few hundred feet to a little more than a mile long, in the direction of flow, and a few hundred feet to nearly 2 miles wide, measured perpendicular to the direction of flow. The *front* of a rock glacier is a steep face, 30–400 feet high, that marks its down-valley end (Fig. 1 of Pl. 2, Fig. 3 of Pl. 3). The front slopes at an angle that is near the angle of repose for rockfall debris. The upper surface slopes gently toward the front of the rock glacier; it is generally marked by conspicuous microrelief. The sides of many rock glaciers rise as steep embankments, and for this reason the rock glaciers have the general appearance of very viscous lava flows. These steep embankments diminish in height up-valley and are commonly absent from the upper part of the rock glacier. The *head* is taken to be the place where the rock glacier merges with the talus cones that feed it, where it merges with a glacier, or where it ends abruptly at a pit formerly occupied by a glacier.

The rock glaciers in the Alaska Range are grouped into three types on the basis of their map pattern, the ratio of their length to breadth, and their topographic position. These types are intergradational and are believed to

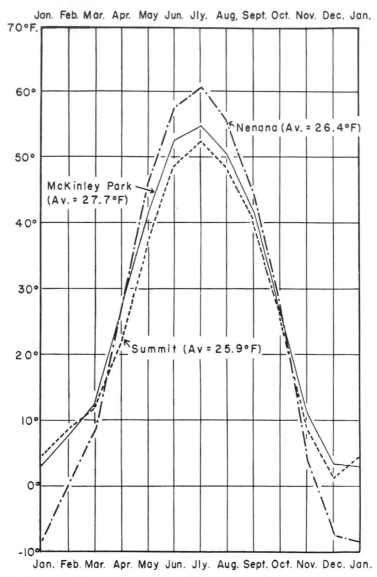

Jan. Feb. Mar. Apr. May Jun. Jly. Aug. Sept. Oct. Nov. Dec. Jan.

Nenana (Av. = 26.4°F)

McKinley Park
(Av. = 27.7°F)

Summit (Av = 25.9°F)

Jan. Feb. Mar. Apr. May Jun. Jly. Aug. Sept. Oct. Nov. Dec. Jan.

FIGURE 2.—AVERAGE MONTHLY AND ANNUAL TEMPERATURES FOR THREE STATIONS NEAR THE
ALASKA RANGE

form a developmental sequence. *Lobate* rock glaciers (Fig. 3 of Pl. 2) are single or multiple lobes extending out from the bases of talus cones or talus aprons. They are characteristically as broad as or broader than they are long. Their lengths range from 200 to 3500 feet, and their widths range from 300 to 10,000 feet. More than 85 per cent of the lobate rock glaciers are less than 1500 feet long. *Tongue-shaped* rock glaciers (Figs. 1 and 2 of Pl. 3, Fig. 1 of Pl. 4) are commonly longer than they

are broad. They are 500–5000 feet long and 200–2500 feet wide. More than 85 per cent of the tongue-shaped rock glaciers are more than 1500 feet long. They completely fill cirques, where they may be fed from three sides, or they extend down-valley from cirques. Tongue-shaped rock glaciers form when lobate rock glaciers from the sides and back of a cirque meet in the center, and the resulting mass flows down-valley from the cirque. *Spatulate* rock glaciers (Pl. 5, left; Fig. 13, lower) widen abruptly near their fronts. In map pattern they are fairly long streams, similar to tongue-shaped rock glaciers, but with rounded bulges at the lower ends. The spatulate rock glaciers

granitic rocks, failed to show any correlation between rock glacier type and rock type. The sequence of events necessary for the formation of tongue-shaped rock glaciers and spatulate

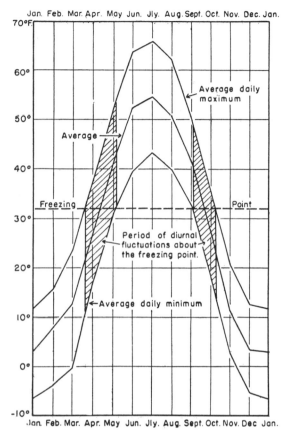

FIGURE 3.—AVERAGE DAILY MAXIMUM AND MINIMUM TEMPERATURES AT McKINLEY PARK, ALASKA, SHOWING PERIODS OF FREQUENT DAILY FLUCTUATIONS ACROSS THE FREEZING POINT

range in length from 3000 to 5500 feet. A spatulate rock glacier forms when a tongue-shaped rock glacier enters a trunk valley from a hanging valley or a side valley, and the debris, formerly confined in the narrow valley, can spread laterally.

Domaradzki (1951, p. 208), who first distinguished between lobate and tongue-shaped rock glaciers, has suggested that tongue-shaped rock glaciers are characteristic of sedimentary terrains, and that lobate rock glaciers are derived from crystalline rocks. However contingency tables of the rock glaciers of the Windy area, part of which is underlain by the sedimentary Cantwell formation and another part of which is underlain by greenstone and

rock glaciers and the probable outcome of continued growth of an apron of lobate rock glaciers lining a cirque wall indicate that these three types are probably stages in the growth of rock glaciers.

Rock glaciers appear to consist of a thin upper layer of rubble of coarse blocks similar to the surface of the talus at the head of the rock glacier and a much thicker lower layer of debris in which similar coarse blocks are mixed with much sandy and silty material (Figs. 1 and 2 of Pl. 2). The debris is derived by rock-fall from cliffs at the head of the rock glacier or at the head of the glacier upvalley from the rock glacier. Interstitial ice cements the blocks in many of the rock glaciers.

Most rock glaciers are connected by smooth, graded talus slopes to the base of their source cliffs; about a third, however, are separated from their source cliffs either by a small glacier a few hundred feet to a mile and a half long, or by a large pit, steep-walled at the lower end, that marks the position of a small recently melted glacier.

All rock glaciers included in this study are above timber line. They are, however, in the zone of low tundralike vegetation, and nearly all are clothed, at least on their upper surfaces, either with nearly continuous turf consisting of low herbaceous plants and grasses or with a continuous mantle of black lichens.

The fronts of nearly half the rock glaciers are completely devoid of vegetation; these fronts are as steep as the angle of repose of their constituent debris, and the angle at the top of each front, where it joins the upper surface, is sharp. These rock glaciers give the impression that they are now moving forward over the valley floor and are called active rock glaciers. The fronts of most other rock glaciers are covered with turf or lichens. These fronts have much gentler slopes, and their tops are gently rounded and merge gradually with the upper surfaces. (Compare Fig. 3 of Pl. 3 and Fig. 3 of Pl. 2.) The fronts of these rock glaciers average about half the height of the fronts of the active rock glaciers. Rock glaciers in this second group are thought to be stable and are called inactive rock glaciers. A few rock glaciers appear to have been inactive for a period after they were formed but seem to have begun to move again recently. These are called reactivated rock glaciers. Plate 1 and Figures 4 and 5, where these three types of rock glaciers are distinguished, show that the active rock glaciers lie upvalley from the inactive rock glaciers or are in higher cirques.

Among the most striking characteristics of rock glaciers is the curious microrelief on their upper surfaces. It is this microrelief that gives the impression of slow plastic or viscous flow, remarked by earlier investigators. Not all rock glaciers show these features, but most do. Especially prominent are the sets of parallel rounded ridges separated by V-shaped furrows, 3–20 feet high and 30–150 feet apart. Some are parallel to the direction of flow of rock glaciers; these are called *longitudinal ridges and furrows* (Figs. 1 and 2 of Pl. 4). Others are approximately perpendicular to the direction of flow and are generally bowed down-valley; these are called *transverse ridges and furrows* (Figs. 4 and

5 of Pl. 4). Another type of microrelief that seems peculiar to rock glaciers appears in the air photographs as a tightly meandering incised trench extending lengthwise down the rock glacier, commonly near the middle. These are called *meandering furrows*. On the ground these were found to be irregular lines of coalescing conical pits, each pit 20–100 feet across. Less common forms of microrelief are crevasses, lobes, and irregular mounds and ridges. The crevasses are straight sharp furrows at large angles to the sets of transverse ridges and furrows.

MOTION OF THE ROCK GLACIERS

Introduction

Each of the theories of the origin of the rock glaciers which have been proposed implies some particular type of motion now or in the past. Much has been learned about the movement of rock glaciers since many of these theories were presented. Direct measurements by surveying methods over considerable intervals of time have established that rock glaciers are now moving. In addition, certain morphological characteristics indicate, indirectly, the way rock glaciers move. All but a very few of the theories of origin can be rejected in the light of present knowledge of the movement of rock glaciers.

Several of the theories indicate that, except perhaps for a certain amount of settling, rock glaciers should not now be moving. The hypothesis that rock glaciers originate as sudden landslides (Howe, 1909), as accumulations of talus that have slid far out onto the floors of cirques over snow banks (Cross and Howe, 1905), or as unusual or subglacial moraines (Cross and Howe, 1905; Kesseli, 1941) all lead to this conclusion and are untenable in view of the evidence that all active rock glaciers for which accurate data are known are now moving at a fairly constant rate.

Remaining are the theory that rock glaciers move down-slope by creep or solifluction, with the movement confined to an unfrozen upper layer (Chaix, 1923), and the theory that rock glaciers are flows of debris-charged ice (Capps, 1910; Tyrrell, 1910; Richmond, 1952).

[*Editor's Note:* Material has been omitted at this point.]

AMOUNT AND RATE OF EROSION REPRESENTED BY THE ROCK GLACIERS

The amount of debris in a rock glacier is the amount of material removed from its source. Divided by the area of the source, it gives the average vertical thickness of bedrock removed from the source area. Thus rock glaciers can be used as a measure of the amount and rate of post-glacial erosion in the areas where they occur.

The vertical thickness of bedrock removed is greater than the thickness normal to the surface by a factor equal to the secant of the average slope angle; where the slope is steep, this factor is large. The vertical thickness removed is given by

$$T_{vs} = T_{rg} \times R \qquad (8)$$

where T_{vs} is the vertical thickness of bedrock worn from the source area; T_{rg} is the thickness of debris in the rock glacier, corrected for increase in volume due to formation of voids, and R is the ratio of the area of the rock glacier to the area of its source. Areas of rock glaciers and source areas were measured on 1:12,000 enlargements of the multiplex manuscript for the topographic map of Healy D-2 quadrangle. The source area is the area upslope from the talus cones at the head of a rock glacier as far as the crest of the ridge; that is, all the area from which any fragment, once dislodged, must eventually find its way to the rock glacier. In Figure 16 the areas of 40 rock glaciers in the Wood River area are plotted against the size of their source areas. The regression line determined for the points in Figure 16 has a slope of 1.36 and passes very nearly through the origin. Hence the average ratio of the source area to the area of the rock glacier is probably close to 1.36, and its reciprocal can be used in equation (8). If the difference in thickness between the active and inactive rock glaciers is assumed to be largely due to the loss of interstitial ice, the average thickness of inactive rock glaciers, approximately 70 feet, would give the thickness of debris with voids. Use of this assumption has probably resulted in an under-

estimation, rather than overestimation, of the amount of debris in active rock glaciers. According to Peele's Handbook of Mining Engineering (1948, p. 3–03) the porosity of gravel is Between 37 and 42 per cent. No figures are available for the porosity of rubble such as that in rock glaciers, and it undoubtedly varies widely from place to place within a rock glacier. If Peele's average value of 0.4 is used, the vertical thickness eroded is no less than 70 × 0.6/1.36, or about 30 feet. This figure indicates little more than the order of magnitude of the erosion involved. The average erosion that produced a particular rock glacier may be one-fifth or 5 times this figure, depending on the size of the rock glacier, its thickness, and how much of it is debris.

A minimum figure for the rate of erosion is given by the time since ice last occupied these cirques, no more than 10,000 years. This gives a rate of erosion of about 3 feet per 1000 years. However, the period of time during which the rock glaciers have been active is probably no more than 3000 years, and many of them have been active for only 1000 or 2000 years. Most of the erosion to produce rock glaciers occurred at the time of formation; hence an average rate of erosion of at least 1 foot per century, and probably 2 or 3 feet per century, is indicated for the relatively short periods of accelerated frost riving and growth of rock glaciers.

Erosion in an individual source area is not uniform. Frost prying of rock fragments, the chief mechanism whereby talus accumulates in the Alaska Range, depends on abundant moisture. It is more rapid in gullies and couloirs, which are kept damp by late-lying snow, than on projecting ridges, spines, and pinnacles, which are generally dry. Fragments falling from a cliff loosen other fragments in their path, and these in turn loosen others. Because the fall of rock fragments is channeled through the couloirs, their erosive effect is concentrated along the couloir walls. The processes that erode the cirque walls therefore tend to accentuate any irregularities that might exist in these walls. As a consequence the source areas of the rock glaciers are masses of sharp ridges and pin-

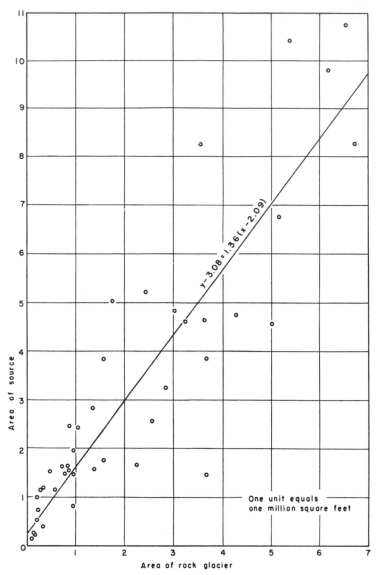

FIGURE 16.—SCATTER DIAGRAM SHOWING AREAS OF ROCK GLACIERS PLOTTED AGAINST SIZE OF THEIR SOURCE AREAS
Regression line derived by least-squares methods from these points.

nacles separated by narrow gullies and steep couloirs. Elsewhere along the cirque and valley walls, where frost action has not significantly worn the cliffs since ice left these valleys, the cliffs are fairly smooth.

The source areas of the rock glaciers are the areas of blocky weathering rocks which are being eroded most actively. Only 7 per cent of the area shown in Plate 1 and 15 per cent of the area shown in Figure 5 are source areas of the rock glaciers, although blocky weathering rocks amount to 50 per cent of the area shown on Plate 1, and the entire area of Figure 5 is blocky weathering rocks. If the rate of erosion for the remainder of these areas of blocky weathering rocks was zero, the average rate of erosion on blocky weathering rocks at the altitudes of the areas shown in Plate 1 and Figures 4 and 5 would be about 1 foot in 1000 years, during periods when rock glaciers are active.

For all post-Wisconsin time it would be about 1 foot in 4000 years. However, erosion does occur elsewhere in these areas. Rock falls are feeding other talus cones at small increments a year, and various forms of solifluction and creep continually move material down-slope, exposing more bedrock to be broken by frost action. Hence, for blocky weathering rocks at these altitudes the average rate of erosion in post-Wisconsin time might lie between 1 foot in 4000 years and 1 foot in 1000 years.

This rate of erosion is a measure oft the rat at which bedrock is being destroyed. I is note directly a measure of the lowering of the land surface. Many rock glaciers are beyond the reach of streams capable of transporting their coarse fragments. Probably, in order for the coarse debris of these rock glaciers to be brought within reach of streams large enough to handle it, the rock glaciers must be incorporated into the till of a glacier and carried to the lower reaches of a large stream system.

ROCK-GLACIER DEBRIS AS A MAJOR COMPONENT OF END AND LATERAL MORAINES

If the climate in a region where rock glaciers abound should become more glacial, the rock glaciers will be overwhelmed by the advancing ice of a new glaciation and will be incorporated into the debris borne along at the front and sides of a glacier, to be deposited as part of the terminal and end moraines. In the southwest part of the area shown in the stereoscopic view of Figure 3 of Plate 7, a typical cirque glacier has advanced over the rock glacier down-valley from it and subsequently retreated (rock glacier 47). The microrelief on the rock glacier has been smeared out into a series of sharp-crested dunelike forms, with short steep lee slopes and long gentle stoss slopes. Longitudinal groove marks scoured by the glacier are also visible. The difference between the overridden surface of this rock glacier and the surfaces of other rock glaciers in this cirque (e.g. rock glaciers 46 and 48) that were not overridden by ice is readily apparent.

A more advanced stage in the transition of a rock glacier to morainal material is represented by the east lateral moraine of the glacier at the head of the northward-flowing tributary of the Wood River about 5 miles west of the Wood River glacier (Fig. 1). This glacier has melted back half a mile from its maximum recent stand. The abandoned terminal moraine in the west half of the valley consists of two low parallel ridges, about 15 feet high and 100 feet apart. The west lateral moraine above them is likewise very low. Behind the terminal moraines is an alluvial flat, which extends back to rolling ground moraine, beyond which is partly till-covered and partly bare ice. A collapse pit at the lower end of the alluvial flat indicates that ice lies beneath the alluvium.

The eastern third of the valley is filled with a long, narrow, flat-topped mound about 60 feet high and about 500 feet wide, which is the east lateral moraine of the glacier. This mound has a front typical of those of rock glaciers and both longitudinal and transverse ridges on its upper surface. This terracelike lateral moraine probably developed from a rock glacier that entered the main valley from the east side, or that formed along the east valley wall. It has been drawn out alongside the glacier to form the massive lateral moraine. Other massive lateral moraines with longitudinal ridges and furrows are near the head of this glacier as well as on the Wood River glacier and on other glaciers in this region.

An apron of lobate rock glaciers around the base of a mountain approximately 1 mile downstream from the foot of the Wood River glacier (Fig. 3 of Pl. 2) was overridden on its upstream side during the nineteenth century by the Wood River glacier, which has since retreated about half a mile. Where the rock glacier was overridden by ice, the rock-glacier topography has been obliterated, and the surface resembles that of till recently exposed by ice and modified slightly by mass wasting.

Lobate rock glacier aprons just downstream from the lower ends of glaciers have been observed on some forks of the Little Delta River. If the onset of glaciation is slow enough, the advance of glaciers down-valley might be preceded by rapid frost spalling of the valley walls and the development of rock glacier aprons downstream from the glacier snout. Such rock-glacier aprons would provide large quantities of debris to the lateral and ground moraines of glaciers and would be a significant factor in valley widening by glacial erosion.

CONCLUSIONS

Origin of Rock Glaciers

Active rock glaciers are masses of debris and interstitial ice and owe their motion to the flow of the ice. Relationships between the velocity, thickness, slope, and surface features of the rock glaciers are consistent with the known physical properties of ice. Rock glaciers, then, are essentially a type of glacier, and hypothetically all gradations between rock glaciers and glaciers would be possible. Because of the manner in which their ice is formed and preserved, the climatic conditions under which rock glaciers occur differ slightly from those under which glacial ice accumulates. Consequently, their geographic distribution is different, and intermediate forms are rare.

The conditions necessary for the growth of rock glaciers are an abundant supply of coarse blocky debris and a climate conducive to the accumulation of ice in its interstices. The amounts of debris necessary for the formation of rock glaciers are commonly found only at the bases of high steep cliffs, such as the walls of cirques and glacial valleys. The average annual temperature must be below freezing in order for interstitial ice to accumulate. Where such temperatures prevail, frost riving of bedrock to form talus is most rapid in the cloudy moist regions around areas of glacial accumulation where bedrock is kept moist during freezing weather; a near-glacial permafrost climate, but without the net accumulation of snow necessary for the formation of glaciers is required.

The talus must have large interconnected interstices. Generally rocks which break into equidimensional blocks, such as granite or greenstone, provide such talus. Highly fissile schistose or platy weathering rocks do not. Ice accumulates within the interstices of the talus, either from snow drifting among the blocks or from freezing of water. When the talus reaches a certain thickness, the weight of the superincumbent debris causes the ice to flow. As the mass flows outward from the base of the cliff, its front is a talus slope 100–300 feet high.

Initially rock glaciers are in the form of lobes at the bases of cliffs or lining the walls of cirques (lobate rock glaciers). When lobate rock glaciers coalesce in the center of a cirque, the resulting mass flows down-valley as a tongue (forming a tongue-shaped rock glacier). If a tongue-shaped rock glacier enters a large valley, it spreads laterally and forms a terminal lobe. In glaciated regions tributary valleys are commonly hanging valleys, and the rock glacier, on reaching the larger valley, cascades down its side as a great talus slide. As the lobe at the base of the talus slide grows, the slope of the talus is reduced until material no longer rolls down it, but flows down the slope as a

rock glacier. Such a rock glacier, with a terminal lobe, is a spatulate rock glacier.

When the climate changes in a direction favoring the formation of glaciers, névé fields and glaciers accumulate at the heads of many rock glaciers. The growth of these glaciers in most cases is slow enough so that the rock glaciers maintain their characteristic form as they move down-valley ahead of the advancing glaciers. The upper surfaces of the glacier and the rock glacier slope continuously from the head of the glacier to the front of the rock glacier, and the two are parts of a single dynamic system. During the last 50 years the climate in these glacier-filled cirques has become less favorable for the accumulation of glacial ice, and the glaciers have melted down and back, leaving large depressions at the heads of many rock glaciers.

In some mountain ranges an interval of rock-glacier formation, separated from other intervals of activity by inactivity and dissection, may be equivalent to a glacial advance elsewhere, where snow accumulated from year to year. If the difference in altitude between the lower limit of rock glaciers and that of glaciers can be established in higher mountain ranges near by, it should be possible to determine, for mountain ranges where only rock glaciers exist, the approximate height of snow line at the time the rock glaciers were formed.

Origin of Microrelief

Most longitudinal ridges and furrows on the surface of a rock glacier are believed to be formed by differential accumulation of talus and snow at the head of a rock glacier. Snow drifts into the swales between talus cones at the head of a rock glacier and may persist from season to season on northward-facing slopes. If these snowbanks fill the swales to the level of the cones, the resulting surface slopes directly away from the base of the cliff, rather than radially outward from the head of the talus cones, as it would if no snow were present. Fragments falling to the heads of the cones would roll outward rather than to the sides of the cones, and the swales would not receive as much debris as they would have if no snow were present. As this snow sinks downward and outward during the growth of the rock glacier, it would be converted to ice, and the rock glacier as a consequence would consist of longitudinal bands of debris with interstitial ice alternating with ice containing embedded debris. It is postulated that the ice-rich bands melt as they leave the shade of the cliff, and a pattern of longitudinal ridges and furrows results. Creep of material from the ridges into the furrows would keep the ridges rounded and the furrows sharp.

Transverse ridges and furrows are believed to result where the motion of the rock glacier is impeded, or where its thickness increases downstream. Downstream decrease in velocity and thickening of the rock glacier both involve shortening in the direction of flow. This appears to be accomplished by overthrusting along closely spaced shear surfaces within the rock glacier. The transverse ridges and furrows are the surface expression of this overthrusting. Because of the manner in which they form, and because of subsequent deformation, transverse ridges and furrows are commonly convex downstream in plan.

Crevasses, which are short straight widely spaced V-shaped furrows trending nearly perpendicular to the trends of transverse ridges and furrows, are believed to be tension cracks, developed either where rock glaciers are spreading laterally, or where there is a convex change in slope. Under these conditions tension cracks develop which are similar to those on glaciers; debris sliding into these cracks converts them to sharp V-shaped trenches.

Conical pits, either single or closely spaced in lines, are formed where running water melts vertical tubes within the ice of the rock glacier. Debris collapses into these tubes as they are formed producing the conical pits. The closely spaced conical pits generally form irregular lines along the lower ends of longitudinal furrows, because these are the natural channels for water running down the surface of the rock glacier and also contain the greatest concentrations of ice.

Lobes on the surface of a rock glacier are thought to result from renewed activity after a period of inactivity. The renewed activity starts at the head of the rock glacier and takes the form of lobes, which are essentially small rock glaciers forming on the back of a large one. Several periods of renewed activity can produce on the surface of a rock glacier a very complex pattern of lobes.

Significance of Rock Glaciers in the Geomorphology of the Alaska Range

Both active and inactive rock glaciers are present in the areas studied and occur in cirques that were filled with ice during Wisconsin time.

The inactive rock glaciers are completely covered with a mantle of turf or lichens, and their fronts are rounded in profile; the fronts of active rock glaciers are barren of vegetation and have a straight profile which intersects the upper surface at a sharp angle. After cessation of activity, some of the inactive rock glaciers were dissected by gullies into which the active rock glaciers have flowed. The active and inactive rock glaciers are therefore interpreted as the products of two different periods of rigorous climate, separated by a period of less rigorous climate, all within the latter part of post-Wisconsin time. The inactive rock glaciers are somewhat lower in average altitude than the active rock glaciers and descend in part from slightly lower mountains. Therefore the earlier cold period was probably somewhat more rigorous than the later one. The rock glaciers indicate the following history for the Alaska Range since Wisconsin time.

(1) General deglaciation.

(2) A period too warm for formation of rock glaciers (the thermal maximum).

(3) A period somewhat colder than the cold period of 1850–1900, lasting at least 1000 and possibly 2000 years. This is tentatively correlated with the cold period of 1000 B.C. to A.D. 1 (Ahlmann, 1953, p. 38).

(4) A period too warm for formation of rock glaciers.

(5) A period cold enough for rock glaciers to form, lasting at least 600 and possibly 1000 years, and continuing to the early twentieth century.

(6) Gradual deglaciation for the last 25 years.

Rock glaciers are a significant factor in erosion of the Alaska Range. Those present today represent an average of 30 feet of bedrock eroded from the areas from which their debris was derived. This material was eroded certainly in the last 10,000 years, and probably during a total of only 2000 or 3000 years. Rock glaciers therefore represent a rate of denudation of bedrock walls amounting to 1–3 feet per century. The average rate of erosion throughout post-Wisconsin time, in the areas that are favorable to rock glaciers, is between 1 foot and 4 feet per 1000 years.

References Cited

Ahlmann, H. W: son, 1953, Glacier variations and climatic fluctuations: Bowman Memorial Lectures, ser. 3, Am. Geog. Soc., 51 p.

Anderson, C. A., 1941, Volcanoes of the Medicine

Lake highland, California: Calif. Univ. Dept. Geol. Sci. Bull., v. 25, no. 7, p. 347–422, 6 pls.

Antevs, Ernst, 1955, Geologic-climatic dating in the west: Am. Antiquity, v. 20, no. 4, p. 317–335

Atwood, W. W., and Mather, K. F., 1932, Physiography and Quaternary geology of the San Juan Mountains, Colorado: U. S. Geol. Survey Prof. Paper 166, 176 p.

Beschel, Roland, 1950, Flechten als Altersmassstab rezenter Moränen: Zeitschr. Gletscherk. Glazialgeol., Bd. 1, H. 2, p. 152–161

—— 1957, Lichenometrie im Gletschervorfeld: Jahrbuch des Ver. zum Schutze der Alpenpflanzen und -Tiere München für 1957, preprint, 22 p.

Brown, W. H., 1925, A probable fossil glacier: Jour. Geology, v. 33, no. 4, p. 464–466

Cailleux, André, 1947, Caractères distinctifs des coulées de blocailles liées au gel intense: Soc. Géol. France, Compte rendu Sommaire des seances, no. 16, p. 323–325

Capps, S. R., 1910, Rock glaciers in Alaska: Jour. Geology, v. 18, p. 359–375

—— 1912, The Bonnifield region, Alaska: U. S. Geol. Survey Bull. 501, 63 p.

Chaix, André, 1919, Coulées de blocs (rock-glaciers, rock-streams) in the Swiss National Park of the lower Engadine: Soc. Physique et d'Histoire Naturelle de Genève, Compte rendu des seances, v. 36, no. 1, p. 12–15

—— 1923, Les coulées de blocs du Parc National Suisse d'Engadine (note preliminaire): Le Globe (Organe de la Societé de Géographie de Genève), v. 62, p. 1–34

—— 1943, Les coulées de blocs du Parc National Suisse, Nouvelles mesures et comparison avec les "rock streams" de la Sierra Nevada de Californie: Le Globe (Organe de la Societé de Géographie de Genève), v. 82, p. 121–128

Cross, Whitman, and Howe, Ernest, 1905, Geography and general geology of the quadrangle in Silverton Folio: U. S. Geol. Survey Geologic Folio no. 120, Silverton Folio, p. 1–25

Deeley, R. M., 1895, The viscous flow of glacier ice: Geol. Mag., v. 32, p. 408–415

—— 1908, The viscosity of ice: Royal Society of London Proc., ser. A, v. 81, p. 250–259

Deeley, R. M., and Parr, P. H., 1913, The viscosity of glacier ice: Philos. Mag., 6th ser., v. 26, p. 85–111

DeMartonne, E., 1920, Le rôle morphologique de la neige en Montagne: La Geographie, v. 34, p. 255–267

Domaradzki, Josef, 1951, Blockströme im Kanton Graubünden: Ergebnisse Wiss. Untersuchungen Schweiz. Nationalparks Bd. 3, no. 23 (Arb. Geogr. Inst. d. Univ. Zurich, ser. A, no. 54, p. 173–235

Emmons, S. F., Irving, J. D., and Loughlin, G. F., 1927, Geology and ore deposits of the Leadville mining district, Colorado: U. S. Geol. Survey Prof. Paper 148, 368 p.

Faegri, Knut, 1933, Ueber die Längenvariationen einiger Gletscher des Jostedalsbre und die dadurch bedingten Pflanzensukzessionen: Bergens Mus. Årb., 1933, Naturvidens. rekke., H. 2, no. 7, 255 p., 47 text figs.

Flint, R. F., and Denny, C. S., 1958, Quaternary geology of Boulder Mountain, Aquarius Plateau, Utah: U. S. Geol. Survey Bull. 1061-D, p. 103–164

Frey, Ed., 1922, Die Vegetationsverhältnisse der Grimselgegend im Gebiet der sukünftigen Stausseen: Mitt. der Naturf. Gesell. in Bern Jahre 1921, p. 85–265

Glen, J. W., 1952, Experiments on the deformation of ice: Jour. Glaciology, v. 2, p. 111–114

Griffiths, Thomas M., 1958, A theory of rock-stream morphology (Abstract); Assoc. Am. Geographers Annals, v. 48, p. 265–266

Haefeli, R., 1951, Some observations on glacier flow: Jour. Glaciology, v. 1, p. 496–500

Hoel, P. G., 1947, Introduction to mathematical statistics: N. Y., John Wiley and Sons, Inc., 258 p.

Hole, A. D., 1912, Glaciation in the Telluride quadrangle, Colorado: Jour. Geology, v. 20, p. 502–529, 605–639, 710–737

Howe, Ernest, 1909, Landslides in the San Juan Mountains, Colorado: U. S. Geol. Survey Prof. Paper 67, 58 p., 20 pls.

Hubbert, M. K., 1951, Mechanical basis for certain familiar geologic structures: Geol. Soc. America Bull., v. 62, p. 355–372

Ives, J. D., and King, Cuchlaine A. M., ♦1954, Glaciological observations on Morsarjökull, S. W. Vatnajökull, Part I: The ogive banding: Jour. Glaciology, v. 2, p. 423–428

Ives, R. L., 1940, Rock glaciers in the Colorado Front Range: Geol. Soc. America Bull., v. 51, p. 1271–1294

Jeffreys, Harold, 1925, The flow of water in an inclined channel of rectangular section: Philos. Mag., 6th ser., v. 49, p. 793–807

Jellinek, H. H. G., and Brill, R., 1956, Viscoelastic properties of ice: Jour. Applied Physics, v. 27, no. 10, p. 1198–1209

Journal of Glaciology, 1953, Conference on terminology: v. 2, p. 229–232

Karlstrom, T. N. V., 1955, Late Pleistocene and Recent glacial chronology of south-central Alaska (Abstract): Geol. Soc. America Bull., v. 66, p. 1581

Kesseli, J. E., 1941, Rock streams in the Sierra Nevada: Geog. Rev., v. 31, p. 203–227

Krauskopf, K. B., 1948, Lava movement at Paricutin volcano, Mexico: Geol. Soc. America Bull., v. 59, p. 1267–1284

Leighton, F. B., 1951, Ogives of the East Twin Glacier, Alaska; their nature and origin: Jour. Geology, v. 59, p. 578–589

Lliboutry, Louis, 1953, Internal moraines and rock glaciers: Jour. Glaciology, v. 2, p. 296

Matthes, F. E., 1941, Rebirth of the glaciers of the Sierra Nevada during late post-Pleistocene time (Abstract): Geol. Soc. America Bull., v. 52, p. 2030

—— 1942, Glaciers, p. 149–219 in Meinzer, O. E., Editor, Physics of the earth, Part 9, Hydrology: N. Y., McGraw Hill Book Co., reprinted 1949, Dover Publications, Inc., New York

Matveev, S. N., 1938, Les coulées de pierres (rock streams): Problemy Fizicheskoi Geografii, v. 6, p. 91–124, Russ. with Fr. Summ., p. 122–124

McCall, J. G., 1952, The internal structure of a cirque glacier, Report on studies of the englacial movements and temperature: Jour. Glaciology, v. 2, p. 122–130

Moffit, F. H., 1954, Geology of the eastern part of the Alaska Range and adjacent area: U. S. Geol. Survey Bull. 989-D, p. 65–218

Moxham, R. M., Eckhart, R. A., Cobb, E. H.,

West, W. S., and Nelson, A. E., 1953, Geology and cement raw materials of the Windy Creek area, Alaska: U. S. Geol. Survey open-file rept., 49 p.

Muller, S. W., 1945, Permafrost or permanently frozen ground and related engineering problems: U. S. Army Engineers, Strategic Eng. Study, Special Rept. no. 62, 231 p. (Reprinted by J. W. Edwards, Inc., Ann Arbor, Mich., 1947)

Nangeroni, Giuseppe, 1954, Neve-acqua-ghiaccio; fenomeni crionivali delle regioni periglaciali nelle Alpi Italiane: Como, Italy, Antonio Noseda, publisher, 43 p.

Nye, J. F., 1951, The flow of glaciers and ice sheets as a problem in plasticity: Royal Soc. London Proc., v. 207, p. 554–572

—— 1952, The mechanics of glacier flow: Jour. Glaciology, v. 2, no. 12, p. 82–93

—— 1953, The flow law of ice from measurements in glacier tunnels, laboratory experiments, and the Jungfraufirn borehole experiments: Royal Soc. London Proc., v. 219, p. 477–489

Parsons, W. H., 1939, Glacial Geology of the Sunlight area, Park County, Wyoming: Jour. Geology, v. 47, p. 737–748

Patton, H. B., 1910, Rock streams of Veta Peak, Colorado: Geol. Soc. America Bull., v. 21, p. 663–676, pls. 48–52

Peele, R., 1948, Mining engineers' handbook: 3d ed., v. 1, N. Y., John Wiley and Sons, Inc.

Perutz, M. F., and Gerrard, J. A. F., 1951, Determination of the velocity distribution in a vertical profile through a glacier: Internat. Geod. Geophys. Union, Assoc. Sci. Hydrology, General Assembly Brussels Proc., 1951, v. 1, p. 214

Richmond, G. M., 1952, Comparison of rock glaciers and block streams in the La Sal Mountains, Utah (Abstract): Geol. Soc. America Bull., v. 63, p. 1292–1293

Roots, E. F., 1954, Geology and mineral deposits of Aiken Lake map-area, British Columbia: Geol. Survey Canada Mem. 274, 246 p.

Rubey, William W., 1952, Geology and mineral resources of the Hardin and Brussels quadrangles (in Illinois): U. S. Geol. Survey Prof. Paper 218, 179 p.

Salomon, Wilhelm, 1929, Arktische bodenformen in den Alpen: Heidelberger Akad. Wiss., Math-natur. Kl., Sitzungsber. Jahrg. 1929, Abh. 5, p. 1–51

Sharp, Robert P., 1942, Mudflow levees: Jour. Geomorphology, v. 5, p. 222–227

—— 1954, Glacier flow, a review: Geol. Soc. America Bull., v. 65, p. 821–838

Sigafoos, R. S., and Hopkins, D. M., 1952, Soil instability in regions of perennially frozen ground, p. 176–192 in Frost action in soils; a symposium: Natl. Research Council, Highway Research Board Special Rept. No. 2

Spencer, A. C., 1900, A peculiar form of talus (Abstract): Science, v. 11, p. 188

Suess, Hans, 1954, U. S. Geological Survey radiocarbon dates I: Science, v. 120, no. 3117, p. 467–473

Tyrrell, J. B., 1910, "Rock glaciers" or chrystocrenes: Jour. Geology, v. 28, p. 549–553

Wahrhaftig, Clyde, 1949, The frost-moved rubbles of Jumbo Dome and their significance in the Pleistocene chronology of Alaska: Jour. Geology, v. 57, p. 216–231

—— 1954, Observations on rock glaciers in the Alaska Range (Abstract): Geol. Soc. America Bull., v. 65, p. 1353

—— 1958, Quaternary geology of the Nenana River and adjacent parts of the Alaska Range, Alaska: U. S. Geol. Survey Prof. Paper 293A, 68 p.

Washburn, Bradford, 1935, Morainic banding of Malaspina and other Alaskan glaciers: Geol. Soc. America Bull., v. 46, p. 1879–1890

Williams, Howel, 1944, Volcanoes of the Three Sisters region, Oregon Cascades: Calif. Univ. Dept. Geol. Sci. Bull., v. 27, no. 3, p. 37–84

Zumberge, J. H., and Potzger, J. E., 1955, Late Wisconsin chronology of the Lake Michigan basin correlated with pollen studies (Abstract): Geol. Soc. America Bull., v. 66, p. 1640

Alaskan Geology Branch, U. S. Geological Survey, Menlo Park, California; Bacon Hall, University of California, Berkeley, California

Manuscript Received by the Secretary of the Society, May 20, 1958

Publication Authorized by the Director, U. S. Geological Survey

18

Reprinted trom *Amer. Jour. Sci.*, **265**, 762–772 (Nov. 1967)

PROTALUS RAMPARTS ON NAVAJO MOUNTAIN, SOUTHERN UTAH

JOHN W. BLAGBROUGH and WILLIAM J. BREED

Museum of Northern Arizona, Flagstaff, Arizona 86001

[*Editor's Note:* In the original, material precedes this excerpt.]

PROTALUS RAMPARTS

Protalus ramparts are most numerous along the base of the upper slopes of Navajo Mountain on the southeastern flank. They have also been observed on the northern and western slopes. The ridges are separated from other protalus ramparts and talus aprons on the flanks of the mountain by ditches that are either free of or partially filled with talus debris. At many localities protalus ramparts and ditches constitute platforms forming distinct breaks in talus slopes near the base of the upper flanks of the mountain.

PLATE 1

A. Younger talus platform 5 on southeastern slope of Navajo Mountain. Large blocks are on crest of outer ridge. Arrow identifies middle protalus rampart. Smaller debris is in the ditches. Large blocks comprising talus apron on the upper slope of the mountain are in the foreground.

B. Older talus platform E on southern flank of Navajo Mountain. Man to left is on outer protalus rampart. Most of the talus is covered by a soil developed on sand and silt of probable eolian origin. Man to right is near the center of the platform.

[*Editor's Note:* Certain figures have been omitted owing to limitations of space.]

A.

B.

235

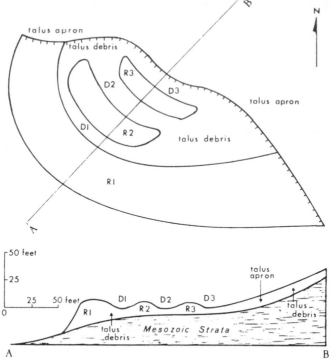

Fig. 3. Map and profile of younger talus platform 6 on southern slope of Navajo Mountain. Protalus ramparts and ditches are delineated: R1, outer protalus rampart; R2, middle protalus rampart; R3, inner protalus rampart; D1, outer ditch; D2, middle ditch; D3, inner ditch.

Talus platforms of two ages are distinguished by the extent to which their constructional topography is preserved and the amount of soil and vegetation cover. The younger platforms have pronounced physiographic expression characterized by well-defined ramparts and distinct ditches. They lack a soil cover and contain a sparse growth of trees and shrubs (pl. 1-A). The talus debris is slightly weathered and has a moderate coat of desert varnish. Talus blocks have a moderate to heavy growth of lichen on their exposed surfaces and are completely stabilized. No apparent variation in the intensity of lichen growth and desert varnish was observed on the debris forming the younger platforms and that of the talus aprons.

Older talus platforms have a subdued constructional topography distinguished by a moderate expression of the ramparts and scant delineation of the ditches. They are partially covered by about 6 inches of sand and silt of probable eolian origin and a growth of trees and shrubs (pl. 1-B). A dark brown soil about half an inch thick containing much organic material has developed on the sand and silt. Talus debris is only moderately well exposed on the ridges and is nearly always obscured over the rest of the platform. The blocks are covered with lichen and desert varnish and are moderately to slightly weathered.

Younger talus platforms.—Nine younger talus platforms have been mapped on Navajo Mountain (fig. 2). They have elevations ranging between 8400 and 9220 feet, and their surfaces slope from the flanks of the mountain with angles varying between 6 and 16 degrees. The platforms face in all major quadrants of the compass, but southern and southeastern exposures are most prevalent. They are both arcuate and parabolic in form and are composed of one, two, or three protalus ramparts.

The protalus rampart furthest from the slope of the mountain is generally the largest on compound talus platforms and forms the outer wall (figs. 3 and 4). This outer ridge surrounds the smaller protalus ramparts and debris in the ditches and abuts against the flank of the mountain on its two termini. The inner ridge is usually the smallest, and the middle protalus rampart is intermediate in size. Three ditches or depressions are on talus platforms containing three protalus ramparts, and the outer separates the outer ridge from the middle. The middle ditch is between the middle and inner protalus rampart, and the inner depres-

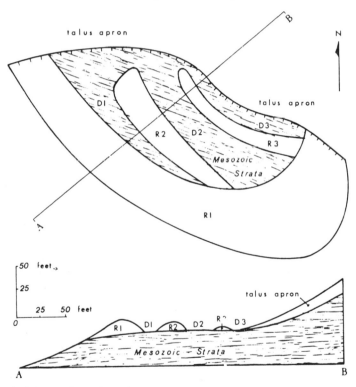

Fig. 4. Map and profile of younger platform 8 on southern flank of Navajo Mountain showing protalus ramparts and ditches (see pl. 2-A). R1, outer protalus rampart; R2, middle protalus rampart; R3, inner protalus rampart; D1, outer ditch; D2, middle ditch; D3, inner ditch.

sion divides the inner ridge from the talus aprons on the flanks of the mountain. Middle and inner protalus ramparts are either separated from the talus aprons by small depressions contiguous with and at the same level as the floors of the ditches or abut against the outer ridges (figs. 3 and 4).

Considerable variation exists in the size of the younger talus platforms and the ramparts and ditches that compose them. Average length measured along the base of the mountain is about 300 feet, and the width determined at right angles to the length is approximately 200 feet. Outer protalus ramparts are between 10 and 80 feet high and have breadths measured at their bases from 45 to 120 feet. Crests of the middle and inner ridges are between 5 and 30 feet above the floor of the depressions, and their widths are considerably less than those of the outer protalus ramparts. Ditches have breadths ranging between 10 and 50 feet. They are arcuate shaped and generally parallel the protalus ramparts.

Three types of younger talus platforms are distinguished by the extent to which debris covers the floors of the ditches. Seven are characterized by a veneer of talus in the depressions and around the middle and inner protalus ramparts (pl. 1-A and fig. 3). Thickness of the debris is between 5 and 10 feet, and the protalus ramparts are from 10 to 30 feet above the debris-covered floor. Platform 8 on the southern side of the mountain is a second type distinguished by an almost complete absence of talus between the ramparts (pl. 2-A and fig. 4). The ditches are broad, flat depressions underlain by bedrock, and the protalus ramparts stand out as distinct ridges. A third type is represented by platform 9 on the western flank and is characterized by a partial cover of debris on the floors of the outer and middle ditches (pl. 2-B). Talus extends into the depressions from the middle and inner protalus ramparts. Bedrock crops out on the floor in small arcuate-shaped areas directly behind the outer and middle ridges.

Debris forming the younger talus platforms and talus aprons is composed of angular blocks of sandstone derived from the Mesozoic strata exposed on the upper slopes of the mountain. Many of the fragments are rectangular in shape, the result of the intersection of joint surfaces and bedding planes. The blocks have maximum diameters ranging between 10 and 15 feet and minimum diameters of about 3 feet. Subangular to subrounded fragments have dimensions varying from 6 inches to 1 foot and fill the spaces between the larger blocks. In some compound talus platforms the outer ridge is composed of slightly larger fragments than the middle and inner protalus ramparts. In others there is no apparent difference in size. Boulders forming the ramparts and partially filling the ditches are randomly distributed with no orientation imbricate to the slope of the platforms.

Talus debris on the floors of the ditches and on the flanks of the protalus ramparts is smaller than on the crest of the ridges and on the surfaces of the talus aprons (pl. 1-A). This relationship suggests that the larger blocks are segregated on the crests of the ramparts and may cover

A.

B.

PLATE 2

A. Middle protalus rampart of younger talus platform 8 on southern side of Navajo Mountain. Blocks in foreground are part of talus apron. Ditch between talus apron and middle protalus rampart is almost free of debris (see fig. 4).

B. Younger talus platform 9 on western slope of Navajo Mountain. Man to right is on outer protalus rampart. Debris in foreground extends from middle protalus rampart and has partially filled the outer ditch.

smaller debris.. On many of the talus platforms the average diameter of the smaller fragments is about 3 feet as opposed to 6 feet for the larger blocks. On others only a slight difference in size was observed between the larger and smaller debris.

Older talus platforms.—Six older talus platforms are on the southern and eastern flanks of Navajo Mountain along War God Bench. They are more difficult to recognize than the younger because of their degree of modification and soil cover. Other platforms not shown in figure 2 may be in the heavily forested areas along the base of the upper slopes. The older talus platforms are between 20 and 200 feet below the younger and, in some instances, are separated from them by talus-covered slopes (fig. 5). The talus debris was probably deposited during formation of the younger protalus ramparts and talus aprons and encroaches upon the floor of the older platforms.

Older talus platforms resemble the younger in size and form and are composed of angular blocks of sandstone with average diameters of about 6 feet. They contain one or two recognizable protalus ramparts behind and around which talus debris has accumulated. The older talus platforms have average elevations of 8740 feet. Their surfaces slope from the cliffs rising above War God Bench and are not dissected by the intermittent streams eroded in the upper flanks of the mountain. The outer ridges are the larger and constitute walls containing the inner protalus ramparts and debris on the floors of the ditches. They abut against the slopes of the mountain on their two termini and have average heights of about 25 feet.

ORIGIN AND CLIMATIC IMPLICATIONS

Protalus ramparts are ridges of coarse blocks and boulders that accumulated in front of snow banks or temporary firn fields. The boulders rolled, slid, or bounced across the snow or firn and accumulated at the base. No fine material moved with the coarse talus because the fine debris cannot roll, slide, or bounce across snow or firn (Sidney E. White, written communication, 1967). Protalus ramparts are easily confused with

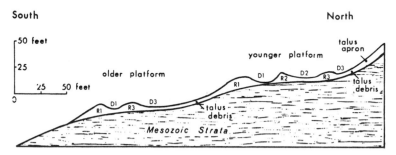

Fig. 5. Profile of older talus platform E and younger talus platform 7 on southern flank of Navajo Mountain. Protalus ramparts and ditches are delineated (see pl. 1-B). R1, outer protalus rampart; R2, middle protalus rampart; R3, inner protalus rampart; D1, outer ditch; D2, middle ditch; D3, inner ditch.

protalus lobes defined by Richmond (1962, p. 20) as "a tonguelike or lobate mass of rubble or debris that is the product of creep or solifluction of the debris toe of a talus". Protalus lobes may contain one or more arcuate ridges in which the blocks tend to be oriented imbricate to the slope (Richmond, 1962, p. 61). The lobes are composed of all the various sizes of fragments that constitute the parent talus and may contain fine as well as coarse debris (Sidney E. White, written communication, 1967). The ridges on Navajo Mountain are protalus ramparts rather than protalus lobes because they are well-defined and separated from the talus-covered slopes of the mountain and other protalus ramparts by ditches, some of which are free of debris. They are also distinguished by their lack of fine debris which along with coarse blocks composes the talus aprons on the flanks of the mountain.

The protalus ramparts on Navajo Mountain are believed to be of Wisconsin age for the reasons given below. They were first observed by Gregory (1917, p. 82) who attributed them to "incipient ice work" and "nivation operating at a time when a perennial snow cap occupied the highland". This origin implies that the ridges formed above the orographic snow line because ice and permanent snow fields are assumed. An evaluation of the altitudes of the Wisconsin regional and orographic snow lines in southwestern United States indicates that the protalus ramparts probably accumulated near the orographic snow line, and the more extreme climatic conditions suggested by Gregory were not necessary for their development.

The regional snow line is defined by Matthes (1940) as "the level above which snow accumulates from year to year to generate ice bodies over a large part or all of the land depending upon latitude, altitude, and topography". It is approximated by the lower limit of ice caps (Richmond, 1965, p. 228). Orographic snow line is the lower limit of small, isolated snow banks and patches of névé which owe their preservation to favorable mountain surroundings (Ray, 1940, p. 1911).

The elevation of the Wisconsin regional snow line on the Aquarius Plateau 80 miles northwest of Navajo Mountain in south-central Utah was somewhat below 11,000 feet (Flint and Denny, 1958, p. 160). Its altitude over Navajo Mountain was probably slightly higher because the regional snowline rises to the southeast in the western United States (Ray, 1940, p. 1911), and it is estimated to have been at about 11,000 feet. At the present time the orographic snow line may impress itself 2500 feet below the regional snow line in the Sangre de Cristo Mountains of southern Colorado and northern New Mexico (Richmond, 1965, p. 225, fig. 3). If today there is such a difference between the orographic and regional snowlines at these latitudes, it is assumed that this was true in the Pleistocene, and the Wisconsin orographic snow line on Navajo Mountain may have been at about 8500 feet.

The protalus ramparts on Navajo Mountain formed between 8400 and 9200 feet or close to the postulated Wisconsin orographic snow line. Thus the snow banks responsible for their development probably melted considerably or disappeared completely at the end of each summer. This would have been particularly true for those on the southern and western flanks of the mountain. The spring and early summer were opportune times for the dislodgment of blocks by frost wedging because of diurnal freezing and thawing and saturation from water supplied by melting snow. During this season the snow banks maintained a fairly constant size since the debris accumulated as ridges at their bases. As the snow banks began to melt later in the summer, frost wedging diminished, and the supply of blocks dwindled; otherwise the ditches would have become filled during their retreat. This balance between the supply of debris and persistence of the snow bank must have existed each spring and early summer for a considerable number of years in order for the ridges to develop.

The segregation of larger blocks on the crest of many of the protalus ramparts on Navajo Mountain gives an indication as to their origin and the climatic conditions under which they developed. Many of the boulder fields in the Driftless Area of Wisconsin show this same relationship (Smith, 1949, p. 212). The smaller debris is thought to have been derived by frost action from a weathered zone developed during an interglacial stage when chemical weathering was prevalent. The coarser debris formed after the fine material had been removed by gravity and frost induced slope processes. It was derived from deeper bedrock in which there had been less preparation by previous chemical weathering. Hence, only the more persistent and through-going joints offered points of attack and only large blocks were dislocated.

If the same reasoning is applied to the ridges on Navajo Mountain, it can be postulated that many protalus ramparts are composed of talus formed during two stages of weathering. The smaller debris was deposited at the beginning of a periglacial stade and was derived from terrain that had undergone chemical weathering in an interperiglacial stade. The larger blocks were dislodged by frost wedging and deposited a considerable time after the onset of periglacial conditions. This interpretation suggests that several of the younger talus platforms may have formed during three distinct periglacial stades separated by two interperiglacial stades-because they contain three well-developed protalus ramparts.

The excellent development of protalus ramparts on Navajo Mountain may be attributed in part to the lithology and structure of the Mesozoic strata, which form the upper slopes and crest of the mountain. These stratigraphic units are especially susceptible to weathering by frost wedging because they contain porous, well-cemented sandstone beds with well-developed joint surfaces and bedding planes. After the blocks were dislodged, gravity carried them down the upper flanks of the mountain, and they accumulated near the break in slope at the foot. The base of the upper slopes was a favorable area for the formation and preservation of snowbanks during periglacial stades. Irregular cliffs rising above the topographic benches acted as a windbreak, causing the development of large snowdrifts in the winter, and provided shade in the warmer seasons enabling the snowbanks to persist well into the summer.

AGE AND CORRELATION

The talus platforms are believed to be of Wisconsin age because the degree of weathering of the talus debris and the physiographic expression of the ridges and ditches preclude them from being older, since the chief characteristics of pre-Wisconsin deposits in the Cordilleran region of North American are the development of a' very thick, mature soil with little evidence of constructional topography remaining (Flint, 1957, p. 329). Weathering of the talus on Navajo Mountain has been slight to moderate, and the boulders lying near the surface show little decomposition and chemical alteration. Soil on the older talus platforms is only about half an inch thick and has developed upon sand and silt of probable eolian origin rather than on material formed from decomposition of the talus. The talus platforms are little dissected and still maintain their constructional topography, even though they are at the base of steep slopes where erosion is intense. The younger talus platforms are modified very little, and the older still maintain well-defined ridges in spite of considerable change. No other protalus ramparts, till sheets, or moraines of any glaciation have been observed below or above the protalus ramparts here described. This suggests that the protalus ramparts are not either pre-Wisconsin or Neoglaciation in age.

The periglacial deposits on Navajo Mountain probably formed during glacial substages of the Wisconsin because the same climatic changes that caused glaciers to develop in the higher mountain ranges of the Colorado Plateau and Rocky Mountain region would also produce periglacial environments on the mountain. The older and younger talus platforms probably developed during separate Wisconsin substages since the degree of modification of the older indicates a considerable time lapse between their formation and that of the younger.

The talus platforms may be correlated with the Wisconsin glacial deposits of the Colorado Plateau and Rocky Mountain region using physiographic similarities and degree of weathering as criteria. Two substages of Wisconsin glaciation are recognized in the La Sal Mountains of eastern Utah (Richmond, 1962), on the Aquarius Plateau in southwestern Utah (Flint and Denny, 1958), and on the San Francisco Mountains of northern Arizona (Sharp, 1942). These uplands are on the Colorado Plateau and are the three glaciated areas nearest to Navajo Mountain (fig. 1). The two substages are correlative with the Bull Lake and Pinedale glaciations in the Rocky Mountain region, and minor end moraines, protalus ramparts, and rock glaciers of Recent age are assigned to the Neoglaciation (Flint, 1957, p. 329; and Richmond, 1965, p. 222-223).

Throughout the Rocky Mountain region and Colorado Pleateau, the Bull Lake drift has subdued constructional topography and is moderately weathered. A soil separates it from overlying glacial deposits at some localities. The Pinedale drift is less extensive than the Bull Lake. It has pronounced constructional topography and shows little alteration by weathering. The Neoglacial deposits are very little affected by weathering and have exceedingly well-developed constructional topography. The older talus platforms on Navajo Mountain are probably correlative with the Bull Lake Glaciation because they have moderate expression of the protalus ramparts and ditches and are partially obscured by a thin soil. The younger talus platforms may have formed during the Pinedale Glaciation since they have well-developed ridges and depressions, lack a soil cover, and have undergone little weathering. Some of the debris forming the talus aprons on the flanks of the mountain probably accumulated during the Neoglaciation after the formation of the younger protalus ramparts.

REFERENCES

Baker, A. A., 1936, Geology of Monument Valley-Navajo Mountain region, San Juan County, Utah: U. S. Geol. Survey Bull. 865, 106 p.

Behre, C. H., Jr., 1933, Talus behavior above timber line in the Rocky Mountains: Jour. Geology, v. 41, p. 622-635.

Bryan, Kirk, 1934, Geomorphic processes at high altitudes: Geog. Rev., v. 24, p. 655-656.

Daly, R. A., 1912, Geology of the North American Cordilleran at the forty-ninth parallel: Canada Geol. Survey Mem. 38, 857 p.

Flint, R. F., 1957, Glacial and Pleistocene geology: New York, John Wiley and Sons, 553 p.

Flint, R. F., and Denny, C. S., 1958, Quaternary geology of Boulder Mountain, Aquarius Plateau, Utah: U. S. Geol. Survey Bull. 1061-D, p. 103-164.

Gregory, H. E., 1917, Geology of the Navajo country; a reconnaissance of parts of Arizona, New Mexico, and Utah: U. S. Geol. Survey Prof. Paper 93, 161 p.

Matthes, F. E., 1940, Report of the committee on glaciers: Am. Geophys. Union Trans., 21st Ann. Mtg., pt. 1, p. 396-405.

Ray, L. L., 1940, Glacial chronology of the southern Rocky Mountains: Geol. Soc. America Bull., v. 51, p. 1851-1918.

Richmond, G. M., 1962, Quaternary stratigraphy of the La Sal Mountains, Utah: U. S. Geol. Survey Prof. Paper 324, 135 p.

————, 1965, Glaciation of the Rocky Mountains, *in* Wright, H. E., Jr., and Frey, D. G., eds., The Quaternary of the United States: Princeton, New Jersey, Princeton Univ. Press, p. 217-230.

Sharp, R. P., 1942, Multiple Pleistocene glaciation on San Francisco Mountain, Arizona: Jour. Geology, v. 50, p. 481-503.

Smith, H. T. U., 1949, Periglacial features in the Driftless Area of southern Wisconsin: Jour. Geology, v. 57, p. 196-215.

19

Reprinted from *Geog. Ann.*, **48**(2), Ser. A, 55, 57–58, 70–73, 84–85 (1966)

BLOCK FIELDS, WEATHERING PITS AND TOR-LIKE FORMS IN THE NARVIK MOUNTAINS, NORDLAND, NORWAY

BY RAGNAR DAHL

Department of Physical Geography, University of Uppsala

ABSTRACT. The block-field investigation in the Narvik-Skjomen district has shown that the largest areas with flat continuous block fields are to be found in the higher parts of the mountains. There is a marked but variable lower limit (not conditioned by the gradient) for these block fields. The lowest values observed in the whole district lie within a zone between 1140 and 1220 m a.s.l.

The depth of the block fields displays great local variations. The maximum depth measured is about 2 m. The blocks are underlain by finer material. Even today in the summit area of the district the frost is transporting blocks to the surface and sorting of the material is taking place.

The well-developed state of the block fields around large snow-fields on scarcely weathered rock surfaces, the high local variability of the development of the block fields and the result of comparative X-ray analyses of the mineral distribution yield the conclusion that the block-field material in the district consists of postglacial weathering products and isolated erratics. Thus the block fields cannot be cited as indications of ice-free sites during the Würm glacial period.

The highest rock surfaces near the edges of vertical precipices have most of the weathering pits in the district, which is probably due to the fact that these surfaces are subject to a large number of freeze-thaw cycles and high temperatures. The conditions favour both local microgelivation and chemical decomposition. Weathering pits also occur on block surfaces in the block-field material formed under recent periglacial conditions. Thus the weathering pits cannot be used as a proof that an area was not glaciated during the Würm glacial period.

Finally the origin of the tor-like weathering stacks is discussed. On account of rapid drainage and evaporation the macrogelivation may be diminished locally in the higher parts of periglacial areas. On the other hand, microgelivation and chemical decomposition are locally favoured, especially along the lines of structural weakness. The result may be tor-like forms even in areas glaciated during the Würm glacial period.

CONTENTS

GENERAL INTRODUCTION

The present paper deals with the weathering material and weathering forms—block fields, weathering pits and tor-like weathering stacks—investigated in the Narvik–Skjomen district of Nordland in Norway. The greater part of the field work was done by students at the Department of Physical Geography at Uppsala University under the author's direction. Preliminary reports from the respective sub-areas were presented in the form of duplicated essays by these students: B. Calmborg (area B), B. Henriksson and B. Nilsson (area F), A. Holmelid (area C), L. Jacobsson (area A), G. Lindqvist (area C), R. Nordlander (area D), U. Olsson (area G),

[*Editor's Note:* Certain figures and tables have been omitted owing to limitations of space.]

R. Storkull (area E) and L. Strömquist (area B). See also Fig. 1.

The aim was to obtain a picture of the character and location of the different weathering phenomena by means of a field investigation planned on broad lines and, if possible, to correlate this picture with the results of extensive glaciomorphological investigations in the same district.

The bedrock of the north-western part of the Narvik–Skjomen district is predominantly mica schist (Narvik schist) and that of the south and south-east granite (Skjomen granite). The boundary between these bedrocks is often very well marked in the terrain, because of vegetation differences. The granite mountains are often bare and have a marked sheet structure (Fig. 2 and R. Dahl, 1963). Near the Norwegian–Swedish frontier the granite window is bounded by other types of mica schist and further to the south-east there is amphibolite.

Since block fields exist in a number of summit areas in the Narvik–Skjomen district that are of importance in the discussion on refuges, investigations were carried out with a view to assessing the value of these block fields as indicators of areas free from ice during the Würm glacial period. The summit areas investigated are as follows (see Fig. 1): Rundtind and Skjellingfjell (area B), Stortind and Tverfjell (area C), Elvegårdstind, Haugbakktind and Gamnestind (area E), Sandvikfjell and Rapisflåget (area F) and Sandvikfjell and Kongsbaktind (area G). All the areas are the highest flat parts of peaks bounded by glacially sculptured precipices or deeply indented glacial cirques. As regards the situations and topography of the areas in other respects, see R. Dahl, 1963, Fig. 2, and Figs. 1, 5, 13 and 14 in the present paper.

REVIEW OF EARLIER LITERATURE

Block fields

Like most of the high parts of the Scandinavian mountains, the peaks of the Skjomen district are usually covered by accumulations of blocks, known as block fields (cf. R. Dahl, 1963). In spite of the differing views as to their age and origin, these block fields have come to play an important part in the discussion as to the extent of the Quaternary glaciations in

Fig. 2. The post-glacial exfoliation along sheet fractures at Sukkertoppen ("Sugar-loaf Mountain") in the Tverdalen valley has resulted in considerable talus accumulations at the foot of the mountain. Photograph: T. Romsloe.

Scandinavia. Block fields and block streams arise on slopes and mountain ridges where the gradient is not so steep ($< 25°$) as to cause the material to fall down and form a talus slope (cf. Fig. 4).

Location and limits

According to the literature, block fields occur principally in flat mountain areas but also in boggy forest areas at lower levels. Foslie (1941), E. Dahl (1961) and others have found that in Norway the block fields are mainly located in the summit areas. In the Enontekiö district of Finnish Lappland, however, B. Ohlson (1964, p. 108) has recorded that there are no block fields at altitudes above 970 m a.s.l., while they are common in lower, boggy areas.

Opinions are still greatly divided as regards the existence of any marked, general, lower limit for block fields. According to E. Dahl (1961) and O. Löken (1962), the limit is very conspicuous and is situated at a steadily increasing altitude as one moves inland from the coast. E. Dahl even states (p. 88) that under favourable conditions the altitude of the limit can be determined to the nearest 10 m. According to Rudberg (1954, p. 207), the lower limit of the block fields in Västerbotten is determined partly by the terrain and partly by the climate. Rapp and Rudberg (1960, pp. 145 ff.) consider that the lower limit of the block-field zone may be very diffuse and difficult to follow.

In spite of the difficulty in determining the lower limit of the block fields there are some approximate data in the literature. Högbom (1914, p. 280) gives the following limit values: on Spitsbergen at sea-level, on the Kebnekajse mountains at about 1300 m, on the Jämtland and Härjedalen mountains at above 1400 m and in the Dovre area at 1600 m. According to Foslie (1941, p. 273), the coast of Tysfjord, SW. of Narvik, has a minimum value of about 750 m and to the east of this point, on the frontier between Sweden and Norway, the corresponding value is about 1250 m a.s.l. In the Snöhätta Massif E. Dahl (1961, p. 88) found the limit at 1900 m a.s.l.

The above values indicate that the altitude of the limit zone declines from the mountain chain westwards and from south to north. According to J. Lundqvist's profile along the Swedish Caledonides (1962, Fig. 8), the decline in the approximate lower limit of the block-field zone is greatest between the 62° and 63° parallels of latitude. The figure lends no support to the idea that the limit steadily declines towards the north.

The origin of the material in the block fields

Svenonius (1909, p. 171), Högbom (1914, p. 280) and G. Lundqvist (1948, pp. 379 and 400) consider that the block fields may have been formed from material loosened by the frost from the underlying solid bedrock. These autochthonous block fields are characterized, according to Högbom, by a close connection between the blocks and the local bedrock. Foslie (1941, p. 286) defines them as autochthonous block fields widespread in arctic regions and in high-lying or boggy zones in sub-arctic regions.

According to E. Dahl (1961, p. 86), every large collection of blocks is a block field, whether it is autochthonous or allochthonous. For material consisting of large blocks or fine-grained components which has been formed by being weathered from solid bedrock on high mountains with fairly flat summit plateaux, he has introduced the appellation *mountain-top detritus*. This appellation is also used by O. Löken (1962). Both these authors consider that mountain-top detritus is clearly autochthonous and therefore not composed of erratics.

According to Högbom (p. 280), the block-field material may also be derived from till, in which several kinds of rock appear. In the boggy Swedish forest region he considers (pp. 244 ff.) that there are both allochthonous and autochthonous block fields. Rapp and Rudberg (1960, pp. 145 ff.) consider that it is not uncommon to find many erratics which are not frost-weathered in the block-field zone. If they are numerous, it is possible that the block field consists of glacial drift.

According to Rudberg (1964, p. 193), the results of two detailed investigations, one in the Norra Storfjäll area (Werner) and the other on Mt. Nipfjället in Dalarna (Strandvik), showed a strong predominance (normally > 90 %) of stones and boulders of local origin in the block fields. But this does not prove that the material has originated from postglacial weathering, as almost the same percentage is known from ordinary till in these districts.

Markgren (1963, p. 56) maintains that in the northern Scandes loose material is encountered *in situ* interspersed with allochthonous material. He considers that the block fields in the higher parts of the Scandes generally consist of predominantly local weathering material with an admixture of erratics. B. Ohlson (1964, pp. 110 ff.) points out that on account of the re-sorting of the block-field material by frost it is difficult to decide whether it is autochthonous, allochthonous or polymict.

[*Editor's Note:* Material has been omitted at this point.]

DISCUSSION

The location and limits of the block fields

The block-field investigations have shown that the largest areas with flat continuous block fields in the Narvik–Skjomen district are to be found in the higher parts of the mountains (see Fig. 5). It is true that relatively large block fields have also been found at lower levels, but there they appear more sporadically, partly in flat areas with poor drainage and partly in passes situated at 800–1000 m a.s.l. and oriented more or less transversely to the main direction of the ice. In both cases the blocks are considerably rounder than in the real block fields in the summit areas. The genesis of the high and low block fields may have been essentially different, though the result has been the same.

The distribution of the block fields in the Narvik-Skjomen district forms an interesting contrast to what B. Ohlson (1964) found in the Enontekiö district of Finnish Lappland, where flat block fields do not exist above 970 m a.s.l. Ohlson explains this by the rapid run-off of water from the broken terrain at the higher levels, where instead he observed dry bedrock (*Felsendürre*) on the highest and often glacially polished mountain ridges.

From Ohlson's information we may conclude that the topographical conditions play a greater part in the genesis of the block fields than the time available. Where favourable terrain, drainage and bedrock conditions made it possible for the water to be retained, block fields could probably have been formed after the last ice culmination. The possibilities of this also increase substantially in areas with an early initial sinking of the ice surface and a consequent prolongation of the time for which the periglacial processes were active.

The existence of a marked lower limit for the block field is called in question by certain research workers, while others consider that it can be determined to the nearest 10 m (p. 58). The results of the field investigations in the Narvik–Skjomen district indicate that, for large block fields in the summit areas, there is a lower limit (not conditioned by the gradient). The lowest values in the whole district fall within a zone between 1140 and 1220 m a.s.l. Within the zone the limit is highly variable, a circumstance which is probably bound up with several factors.

The lower limit zone of the block fields in north-western Scandinavia may to a certain extent be related to a glacial evacuation below this zone during the period after the initial sinking of the ice surface, but it is in no sense whatever a gauge of the maximum extent of the last glaciation.

The origin and age of the block-field material

The material which after the last phase in its evolution constitutes a block field may be autochthonous, allochthonous or polymict. With few exceptions, authors from Högbom (1926) to Ohlson (1964) have found the question of the origin of block fields a very difficult one.

O. Löken (1962) uses the term *mountain-top detritus* to mean surface material that has been derived from the subjacent bedrock by weathering *in situ*. He thinks that a profile of such material should show a continuous transition from the loose detritus on top to the unweathered bedrock below. In spite of the fact that he has not recorded such a profile, he considers that the block field of the higher zone in an area in north-eastern Labrador (Ungava) can properly be called mountain-top detritus.

As is clear from the brief account given above, there was a substantial proportion of fine-grained material under the blocks in all the pits the students dug. Even today a freezing up of blocks and a sorting of the material is taking place. In the periglacial environment creep movements contribute to the upward transport of blocks, even on slopes with modest gradients.[1] Since all the present summit areas in Europe and North America have or at any rate had periglacial conditions, the profile assumed by Löken would seem to be extremely rare and difficult to explain.

The frost-conditioned sorting of loose material in the mountain-top areas greatly reduces the possibilities of deciding in the field whether the area in question has contained a covering of till or of weathering products formed *in situ*. Only in cases in which, in spite of the re-

---

[1] Laboratory experiments are in progress with a view to ascertaining in more detail the influence of these processes.

sorting, traces of special features in the underlying bedrock are found at the surface can the material be classified to a certain extent as a weathering product. As regards the Narvik–Skjomen district, clear traces of these features have been observed in the norite field in the NW., in the schist field and (less clearly) in the granite field.

A detailed examination of the block fields on both sides of the marked granite–schist limit on Sandvikfjell (Fig. 13) tells immediately against the idea of large elements of till, with the exception of certain erratics. However, this by no means excludes the possibility that till may form a large proportion of the original material on the higher and flatter summits, where the content of fine-grained material has proved to be greater. Most of the diggings in the block fields in the Skjomen district were terminated on account of the hole caving in, frost in the ground or ground water in relatively fine material at somewhat varying depths. A hole dug by Storkull on Gamnestind to a depth of 1.7 m yielded interesting information. At this locality and depth the material, which had become finer and finer with increasing depth, was suddenly replaced by coarse blocks. This state of affairs may be interpreted in three ways:

(a) The material in this locality was formed *in situ* throughout. The disintegration of blocks takes place under the cover of blocks and soil, which is about 2 m deep. The blocks are then transported up to the surface by the further action of the frost.

(b) The material was originally till, which was sorted down to a certain depth by the action of the frost.

(c) The material is a mixture of till and weathering products, which has taken on its present profile through the action of the frost.

The absence of large blocks in the middle part of the vertical profile, i.e. between the block layer at the surface and the bottom of the block field, may perhaps be explained by saying that the upward transport of blocks from the bottom has now ceased. The reason for this may be changes in climate and drainage or that the upward transport ended when the thickness of the block field exceeded a certain value.

However, it is quite conceivable that frost-heaving of blocks occurs, even at the present day, from a depth of 2 m in the usually relatively snow-free areas, if the drainage is poor. But the continued frost-bursting in the underlying bedrock may cause cracks which suddenly improve the ground-water drainage, which in its turn reduces the frost-heaving effect..

The fact that most of the blocks of the block fields investigated are more angular than established erratics may argue that there is a substantial element of frost-weathering products. However, in this respect a comparison between schist blocks and granite erratics is extremely hazardous, as is a comparison between granite blocks and, for example, amphibolite erratics from the est. In many cases it is difficult to decide whether a granite block in the granite-block field is an erratic or not. Rothers (1965) argues that the rounding of the block should be interpreted with the greatest caution.

Grain-size analyses of the fine-grained material (Table 3) have not yielded any clear information either, as weathering products and till may produce similar curves and the analyses do not eliminate the re-sorting due to the frost.

The striking difference, as regards minerals, between the material from the block-field zone and the neighbouring till (see Table 2 and p. 68) proves that the greater part of the block-field material in the district consists of weathering products and not till. The most probable explanation of the mineral character of the fine-grained material in the Rundtind block field is that the material as a whole is of no great age and that mineral transformation is still going on. The ratio of fresh minerals to transformation products is probably dependent on the extent to which favourable conditions for the transformation processes exist in the respective localities.

The fact that clay minerals with the above-mentioned (p. 68) Ångström units are also present in one of the samples from the till ridges between Kobbvatnet and Baatvatnet and that traces of clay minerals are indicated in the analytical results from other till samples strengthens the hypothesis of a relatively extensive postglacial mineral transformation. Recent clay-mineral formation in the actual till ridges at a depth of 1–2 m appears far more likely than the late-glacial ice transport of old weathering products. Apart from small, scattered, late-glacial forms, there are no accumulations of loose material even in the fjord district. The ice flowing to the NW. seems to have had a high transporting capacity[1] and deposits would seem to have been mainly made in the deep waters off the coast. The actual till ridges were deposited in a period when the ice surface had sunk several hundred metres below its level at the time of the ice culmination in the district. Proximally of the till ridges, there is scarcely any part of the mountains, with the possible exception of the Kebnekajse Massif, which can be considered as having been a nunatak during the culmination of the last glaciation.

Assuming that the clay minerals in question were not already present in the matrix fissure, it therefore seems very far-fetched to introduce ice transport into the interpretation of their appearance in the till ridges. *The clay minerals in all probability originated* in situ *in the till ridges during the postglacial period.*

The conclusion from E. Dahl's reasoning as to the great age of the block-field material must be that the denudation transport from the much-discussed western part of the mountains was minimal, in spite of the fact that they undoubtedly had periods of markedly periglacial conditions.

The question of the magnitude of the periglacial denudation transport during the postglacial period may be said to be of fundamental importance in several respects. A recent example of this is Ragg's and Bibby's (1966) interesting report on the conditions in southern Scotland, which may lead to widely differing conclusions, depending on one's conception of the periglacial transport capacity after the culmination of the last glaciation in the area.

In spite of thorough discussions (reported in the "Slope Report" of 1960, 1963 and 1964), opinions are still divided as to the intensity of the periglacial denudation processes.

Lakes have not been encountered in the block-field zone in the Narvik–Skjomen district and, with the exception of erratics, the destruction of glacial features has been total here (cf. Büdel, 1960, p. 82, Rapp, 1960, p. 178 note, Rudberg, 1964, pp. 201–2, and Häggblom, 1963). However, the present author reckons that the summit parts concerned were laid bare of

ice as early as 20–25000 years ago on acount of shifting ice culminations.

As regards the periglacial destruction of traces of glaciation, the contrast between the western Norwegian summit areas and northern Lappland may be due not only to differences in climate and bedrock but also to a substantially longer period of strictly periglacial conditions in the west. The data and observations from the Narvik–Skjomen district indicate that the periglacial transport is rather rapid. For that reason it is also improbable that Tertiary deep-weathering material remains on slightly arched mountain ridges at any rate in the block-field zone of western Scandinavia. The presence of the clay minerals in question requires another explanation.

Hoff-Ostwald's law implies that a temperature drop of 10°C results in the rate of chemical reactions being reduced to half. Temperature measurements have shown (p. 65) that there may be very high temperatures locally in the block-field zone. If the temperature in a diabase block in the Antarctic can reach +20°C (J. H. Mercer, 1963, p. 9) and chemical weathering there can be considered to contribute to the genesis of weathering pits, the postglacial temperature conditions in the special localities in Norway discussed here would not seem to be any serious obstacle to the creation of the transformation products mentioned above. If, in addition, we take into account Tuytuynov's argument (1964, p. 19) about the chemical reactions in frozen bedrock, it would seem to be quite conceivable to reckon on the postglacial formation of the clay minerals which appear in the block-field material.

As regards the nature of the material, nothing has emerged from the investigations in the Narvik–Skjomen district which argues against the hypothesis that the greater part of the block-field material originated through weathering in postglacial times. As was pointed out earlier (R. Dahl 1963), the areas were probably covered with ice also during the early phase of the last glacial period. The topographical conidtions are favourable for the retention of water and therefore for frost bursting. The water supply is also reinforced by the contributions from neighbouring snow-fields and snow patches.

The marked contrast on Elvegårdstind (Fig.

---

[1] The dearth of loose deposits in the district will be dealt with in more detail in a forthcoming paper.

5), for example, between the central parts which have no block fields and which during normal summer periods are protected by a permanent snow cover and the border of well-developed block fields round these parts (cf. p. 62) proves that the Elvegårdstind block fields were formed under climatic conditions of the type that prevails now. The main part of the material is thus weathering products of recent date.

The striking local variation at the corresponding levels, as regards the block-field thickness, the component sizes and the block-field development in all the block fields investigated, may be connected neither with ancient deep-weathering processes nor with till but may require the recent formation of loose material as an explanation.

The postglacial climatic conditions in the district were favourable for frost-bursting, as is clear from the unusually large quantities of talus (cf. R. Dahl, 1963, p. 128). Traces of recent block and gravel formation were observed in several cases at precipices and certain weathering stacks.

The fine-grained material in the block fields may be products of mechanical micro-bursting and chemical weathering. These processes are probably active even today at the level concerned (see also p. 78), since at that level there may be the optimal possibilities of interplay between the disintegration due to freeze–thaw temperatures and chemical action.

Conclusions

The following conclusions may be drawn from the above account, as regards the block fields in the summit areas:

1. The lower limit is locally very variable.

2. The constant profile, showing blocks on the surface and fine material below it, is a consequence of the re-layering of the material by frost.

3. The blocks heaved up to the surface are preserved on account of the lack of moisture.

4. The block-field material in the district consists of postglacial weathering products and isolated erratics.

5. Block fields cannot be cited as indications of unglaciated areas during the Würm period.

[*Editor's Note:* Material has been omitted at this point.]

References

Büdel, J., 1960: Die Frostschutt-Zone Südost-Spitzbergens. Coll. Geogr., 6. Bonn.

Calmborg, B. (Dahl, R.), 1965: Nedisnings- och isrörelsespår samt postglacial vittring vid Rundtind, Nordland, Norge. Stenc. report. Dept. Phys. Geogr. Univ. Uppsala.

Dahl, E., 1961: Refugieproblemet og de kvartaergeologiske metodene. Svensk Naturkunskap. Stockholm.

Dahl, R., 1963: Shifting ice culmination, alternating ice covering and ambulant refugee organisms? Geogr. Annaler, Vol. XLV, No. 2—3, Stockholm.

Foslie, S., 1941: Tysfjords geologi. Norges geol. Unders. 149. Oslo.

Henriksson, B. and Nilsson, B. (Dahl, R.), 1964: Blockhavsundersökning och glacialmorfologiska studier i området Sandvikfjellet-Rapisflaaget, Skjomen, Nordland, Norge. Stenc. report. Dept. Phys. Geogr. Univ. Uppsala.

Holmelid, A. (Dahl, R.), 1963: Studier av vittringsförhållanden, nedisnings- och isrörelsespår i trakten av Elvegård. Stenc. report. Dept. Phys. Geogr. Univ Uppsala.

Häggblom, A., 1963: Sjöar på Spetsbergens Nordostland. Om sjöarna som indikatorer på den postglaciala utvecklingen inom Murchisonfjorden-området på Spetsbergens Nordostland. Ymer. Stockholm.

Högbom, B., 1914: Über die geologische Bedeutung des Frostes. Bull. Geol. Inst., XII. Uppsala.

— 1926: Beobachtungen aus Nordschweden über den Frost als geologischer Faktor. Bull. Geol. Inst. XX. Uppsala.

Jacobsson, L. (Dahl, R.), 1963: Inventering av glaciala formelement på Gangnesaksla i Nordland, Norge. Stenc. report. Dept. Phys. Geogr. Univ. Uppsala.

Lundqvist, G., 1948: De svenska fjällens natur. Sv. Turistf, Förl. Stockholm.

Lundqvist, J., 1962: Patterned ground and related frost phenomena in Sweden. SGU, Ser. C. N:o 583. Stockholm.

Löken, O., 1962: On the vertical extent of glaciation in north-eastern Labrador-Ungava. Canadian Geographer, VI (3—4). Toronto.

Markgren, M., 1962: Detaljmorfologiska studier i fast berg och blockmaterial. I. Vittringsprocesser. Svensk Geogr. Årsbok. Lund.

— 1963: II. Residuala blockfält och talus samt polygena blockfält. Svensk Geogr. Årsbok. Lund.

Mercer, J. H., 1963: Glacial geology of Ohio Range, Central Horlick mountains, Antarctica. Inst. of Polar Studies. Report 8.

Nordlander, R. (Dahl, R.), 1963: Postglacial vittring och glaciala formelement inom Tverdalen-Gamnestind. Stenc. report. Dept. Geogr. Univ. Uppsala.

Ohlson, B., 1964: Frostaktivität, Vervitterung und Bodenbildung in den Fjeldgegenden von Enontekiö, Finnisch-Lappland. Fennia 86. Helsingfors.

Olsson, U. (Dahl, R.), 1964: Blockhavsundersökning jämte studier av vittringsformer och glaciala formelement inom området Sandvikfjell-Kongsbaktind, Norge. Stenc. Report. Dept. Phys. Geogr. Univ. Uppsala.

Ragg, J. M. and Bibby, J. S., 1966: Frost weathering and solifluction products in southern Scotland. Geogr. Annaler. Vol. 48. Ser. A, 1. Stockholm.

Rapp, A., 1960: Recent development of mountain slopes in Kärkevagge and surroundings, northern Scandinavia. Geogr. Annaler, Vol. XLIII, No. 2—3. Stockholm.

Rapp, A. and Rudberg, S., 1960: Recent periglacial phenomena in Sweden. Biuletyn Peryglacjalny 8. Łódź

Rudberg, S., 1960: See Rapp, A. and Rudberg, S., 1960.

— 1964: Slow mass movement processes and slope development in the Norra Storfjäll area, southern Swedish Lappland. Zeitschrift für Geomorphologie. Supplementbd 5. Berlin.

Storkull, R. (Dahl, R.), 1964: Blockhavsundersökning i området Gamnes- Haugbakk- och Elvegårdstind, Nordland, Norge. Stenc. Report. Dept. Phys. Geogr. Univ. Uppsala.

Strömquist, L., (Dahl, R.), 1965: Blockhavsundersökningar på fjället Rundtind, Nordland, Norge. Stenc. report. Dept. Phys. Geogr. Univ. Uppsala.

Svenonius, F., 1909: Om skärf- eller blockhaven på våra högfjäll. GFF Bd 31. Stockholm.

Tyutyunov, I. A., 1964: An introduction to the theory of the formation of frozen rocks. Intern. Series of Monographs on Earth Sciences. Vol 6.. Pergamon Press. Oxford.

20

Reprinted from *Jour. Geol.*, **14**, 94–97, 104–106, 110, 112 (1906)

SOLIFLUCTION: A COMPONENT OF SUBAERIAL DENUDATION

J. G. Andersson

[*Editor's Note:* In the original, material precedes this excerpt.]

SOLIFLUCTION IN BEAR ISLAND

This islet situated in $74\frac{1}{2}°$ N. L., in the Atlantic part of the Arctic Ocean, is a small remainder of a once larger stretch of land. Sea cliffs, in some places rising to the height of 400 meters and fringing the island almost all around, tell us of the destructive action of the abrasion. The northern and larger part of the island forms a lowland plateau rising very slowly from the recent coastal outline to the interior and cut out through slightly dislocated Devonian and Carboniferous beds. Several reasons, among others the occurrence of old, somewhat obliterated coastal cliffs at the inner margin of this plateau, incline me to consider it as worked out by marine abrasion.

The southern part of the island is mountainous, with tops rising to 460–539 meters. In this region the ground is composed of a great variety of sedimentary rocks from Silurian to Triassic age, and the older strata are cut up by vertical dislocations into a few parallel blocks. In some cases the denudation has laid bare parts of the precipitous surfaces of the faults, but as a rule the land-sculpturing agents show a marked tendency to smooth the tilted blocks to rounded hills and broad river valleys with gentle slopes. Thus the land forms of the interior of the island are in a striking contrast to the perpendicular cliffs on the sea side.

On the hill-slopes and valley sides of Bear Island there are almost everywhere clear indications of a moving of the waste from higher to lower ground. Many of the small hills show a very marked streakiness of surface, which is due to the peculiar arrangement of the detritus; sometimes the flowing soil forms real streams which have much likeness to a glacier in miniature, and often in the depressions between the hills there are numerous small circular or semi-

circular walls composed of rock fragments and including a patch
of muddy material, often hardened through desiccation. Evidently
all these phenomena only represent different facies of the displacement of the waste.

It is not very difficult to find out the mode of formation of
the "mud-glaciers" mentioned above. When in summer time the
melting of the snow has reached an advanced stage, often the bot-

Fɪɢ. 2.—Scenery from the southern part of Bear Island, showing the contrast
between the rounded land-forms caused by subaërial denudation and the perpendicular coastal cliffs caused by marine abrasion, dissecting the island.

tom of the valleys are free from snow, while big masses still rest in
sheltered places on the valley sides. Every warm and sunny day
new quantities of water trickle from these melting drifts into the
rock-waste at their lower edge. As the masses of detritus are composed not only of coarser rock fragments such as blocks, slabs, and
gravel, but also of finer particles filling the interspaces between the
coarser material, they are able to absorb considerable quantities
of water. When once saturated, they form a semifluid substance
that starts moving slowly down-hill. This process, the slow
flowing from higher to lower ground of masses of waste saturated with water (this may come from snow-melting or rain), I

propose to name *solifluction* (derived from *solum*, "soil," and *fluere*, "to flow").

As the snowdrifts of Bear Island diminish in the course of the summer, new ground is exposed, its waste is thawed, and then, saturated with water, it follows the slow downward movement. The flowing detritus does not generally move as a "sheet-flood" with a broad front, but more often flows in some slight depression of the slope, taking the form of a narrow tongue, offering a most striking parallel to a glacier. The névé region is represented by the area of water-saturated detritus at the lower edge of the melting snow-drift, and the flowing tongue of mud is the glacier proper that moves down the valley. The terminal moraine even is often to be seen in the shape of slabs and pieces of rock that the mud-stream has pushed together in front of its lower end.

These mud-streams do not consist of finer particles only, but also of coarse material, gravel and blocks, frequently intermixed with, and also carried on the top of the muddy substance.

Small mud-glaciers of the type just described are very common in Bear Island, and sometimes they reach noticeable dimensions. On the slope of Oswald Hill I measured such a tongue of detritus that had a breadth of 35 meters and a depth (thickness) of at least 2.1 meters. In this case "the terminal moraine" was represented by a zone of sandstone plates, pushed together so that they were all standing edgewise beautifully concentric to the rounded front of the detritus tongue, and this piled-up zone had the rather start-ling width of 17 meters! This example may give some idea of the considerable quantity of detritus that can be removed in a single stream, as well as of the energy which such a mud-glacier can develop before its movement is arrested.

My companion in the voyage to Bear Island, Dr. G. Swenander, during his botanical researches made an observation that was highly illustrative for a proper understanding of the extent and importance of the solifluction.

Bear Island looks very barren also when compared with more northern Arctic islands, and its sterility is proved not only by the small number of species of higher plants, but also by the complete barrenness of large areas

Swenander gave an explanation of this scantiness of the flora: All the phanerogamic plants of Bear Island—with the exception of a single species—are perennial, and these forms cannot endure upon the flowing soil, where they are easily drowned in the moving mud. The sterility of the areas of solifluction is quite a striking feature of the island. On many slopes small hills of the solid rock crop out of the covering of moving waste, and, though the soil is very scant in the small fissures and crevices of the rock, here the plants grow more willingly than upon the perfidious flowing ground. Thus these small rocky hills form cheerful verdant patches scattered over the barren slopes. In the rare cases when Swenander found single plants growing upon the moving waste, he could demonstrate a remarkable adaptation to this mode of life. The root system of these specimens was exceptionally developed in order to keep the plant afloat on the moving medium. These roots were stretched out in the direction of the movement of the soil, and in some cases where the slope was very steep it was noticed that the proximal part of the root system, carried by the more rapid flowing of the superficial layer of the mud-stream, had reached a *lower* level than the distal branches, which, joining the slower movement of deeper parts of the stream, were left behind in the displacement down-hill. These facts, which I owe to my friend Dr. Swenander, illustrate better than anything else the mode of progress and the importance of solifluction in Bear Island. Evidently, then, this process is a chief agent of the denudation. Every summer large masses of detritus in this way flow down-hill into the bottom of the valleys, where the streamlets undertake the work of transport. It is characteristic for this hitherto undervalued component of subaërial denudation, which I have named "solifluction," that— in contrast to the wind and the running water—it does not lift its material, but like the glacier-ice—with which it has been already compared—it removes blocks and gravel almost as well as the fine mud.

[*Editor's Note:* Material has been omitted at this point.]

REGIONAL EXTENSION OF THE SOLIFLUCTION

From Bear Island in the arctic and from the Falkland Islands in the subantarctic regions I have described in the previous pages two occurrences of solifluction—one recent, the other fossil—where in both cases the phenomenon is developed in rather exceptional dimensions. But, if often only in a moderate scale, this mode of waste transport works in many parts of the world. In the geological, geographical, and botanical literature there are to be found many notes on mud-streams and flowing slopes, illustrating the process here in question. In the following pages I have collected, partly with the assistance of helpful friends, some descriptions of of soil-flowing, but the list of them is given without any pretentions to completeness. Surely many a valuable note on processes of this kind has escaped me, and I hope that colleagues in different lands, after having read this article, will kindly call my attention to observations still unknown to me.

South Georgia.—This subantarctic island is a ridge of folded land, deeply indented by transverse fiords between which the mountains rise steeply to 1,000–2,000 meters. Small glacier caps of varied type occupy a large part of the mountains, and in many of the valleys splendid glacier streams extend into the fiords. But the narrow lowland strips and much of the mountain slopes were free from ice-covering, and here frost-weathering, solifluction, and running water are the chief agents of destruction. Slowly moving slopes and small mud-streams were frequently seen, and the forms of the detrital masses offered many a striking parallel to my earlier experiences on Bear Island.

Graham Land.—The ice-covering of the antarctic lands is all too complete to leave any place for solifluction on a large scale; but on the very limited areas where the land was exposed we could trace almost everywhere this mode of transport working in the summer. On the plateau above the winter station on Snow Hill, Nordenskjöld kindly pointed out to me a locality where the land surface showed the same streakiness that I had seen before in Bear Island, beautifully developed, and which was evidently due to displacements of the soil.

North America.—I have had no time for looking into the immense American literature for notes indicating observations on solifluction, and shall be most thankful for any information on this subject relative to that continent where the land forms are so splendidly developed and so admirably surveyed.

FIG. 3.—Fox Bay, West Falkland Scenery showing the vast plains and the low, rounded mountains.

The only observation from the United States that I am able to mention here is cited by James Geikie in *The Great Ice Age,* and is taken from Hayden, who has given a very illustrative description of solifluction in some valleys in the Rocky Mountains. These valleys are

covered thickly with earth, filled with more or less worn rocks of every size, from that of a pea to several feet in diameter. The snow melting upon the crests of the mountains saturates these superficial earths with water, and they slowly move down the gulch much like a glacier.[1]

[1]*Geological and Geographical Survey of Colorado,* 1873, p. 46.

A note on "flowing soils" is also cited by James Geikie[1] from the arctic part of the American continent. The observation was made by Sir Edward Belcher on Buckingham Island in 77° N. L. He ascended a hill 200 feet above the sea to make observations, and found the soil well frozen and firm and covered with a slight crust of snow. But as the power of the sun increased toward noon, the snow disappeared.

As noon passed the soil in all the hollows or small watercourses became semifluid, and very uncomfortable to walk on or sink into. At the edge of the southern bank the mud could be seen actually flowing, reminding one more of an asphalt bank in a tropical region than our position in 77¹ 10′ N. The entire slope, in consequence of the thaw, had become a fluid moving chute of débris for at least one foot in depth.[2]

Spitzbergen.—During the arctic expedition of 1898, under the command of Professor A. G. Nathorst, when I made my first acquaintance with the solifluction in Bear Island, I also had opportunity to see some phenomena of this kind in Spitzbergen, where they were studied somewhat more in detail in the neighborhood of Hecla Cove, in Treusenberg Bay by my companions G. Andersson, H. Hesselman, and A. Hamberg. The following year, 1899, Dr. Th. Wulff made some observations on the same phenomenon in the first-named place and in Green Harbor in the Ice Fiord.[3]

Recently Professor DeGeer has given a brief account of his researches on moving slopes made during several expeditions to Spitzbergen, partly long before the observations mentioned above.[4] DeGeer has noticed the phenomenon in several localities in the Ice Fiord region, as at the Wahlenberg glacier, at Temple Bay, where the cover of vegetation had got dispersed by a mass of moving soil, and at Cape Thordsen. At the last-mentioned place a small railroad was built many years ago (1872) for mining purposes. Ten years later the railroad could still be used, but in 1896 De Geer found it to be greatly distorted by the effect of the moving soil.

De Geer points out that the waste-sheet generally moves over a bedding of frozen soil, and that also regelation may act in the displacement of the moving masses.

[1] *Op. cit.*, pp. 387, 388.
[2] *The Last of the Arctic Voyages*, Vol. I, p. 306.
[3] Th. Wulff, *Botanische Beobachtungen aus Spitzbergen* (Lund, 1902), pp. 85, 86.
[4] *Geologiska Föreningen i Stockholm Förhandlingar*, 1904, pp. 465–66.

[*Editor's Note:* Material has been omitted at this point.]

CONCLUSIONS

From the evidence given in the preceding pages it is easy to recognize the climatic features which form the optimum of the solifluction. In the polar and subpolar regions, where the ground is not covered with ice, we find this process working with more or less intensity almost everywhere, and in the same manner, the alpine tracts of lower latitudes are favorable for the development of this phenomenon. In these regions, characterized by a "subglacial" climate with heavy deposits of winter snow melting in summer, solifluction is a chief agent of destruction. The unceasing succession, summer after summer, of mud-streams and moving slopes indicates that here the removal of the waste runs on at a rate that may be unsurpassed in other parts of the earth's surface—except in the deserts.

But this effective removal of the waste must be balanced on the one hand by a rapid production of new waste, and on the other, by an effective clearing out of valley bottoms by running water. The "subglacial" climate is most favorable for both these processes. In the extra-glacial regions the frost-weathering works with an intensity well known to every student of polar lands and alpine mountains, and the violent summer floods caused by the melting of the winter's snow give to the rivers of these regions an erosive power that is quite surprising.

When all these conditions are taken into consideration, the "subglacial" climate must be looked on as an optimum of destructive action. In fact, the high mountains, the large folded ranges, where all the agents mentioned work in high intensity, are found to be short-lived features in the earth's surface. In fact, one might be tempted to raise the question whether complete peneplanation is possible in regions where the action of running water is not supported by an effective removal of the waste, such as is produced by solifluction. With this premise, every peneplain found a climatic token of the same kind as the stone-rivers of the Falkland Islands and the limestone breccias of Gibraltar. Still I think that such a presumption hides a dangerous exaggeration. It is very possible that in many cases, under other climatic conditions of less erosive power, the process has had time to reach, though at a very slow rate, the end of the cycle of erosion.

But, on the other hand, it seems certain that solifluction has often been an important agent toward peneplanation. The importance of the process, because of its humbleness, has been much undervalued. But I feel sure that it is not until we get a knowledge of all the contributive agents that we can reach a full understanding of all the varied and complicated forms of land-sculpture.

21

Reprinted from *Biul. Peryglac.*, No. 11, 89–101 (1962)

ALTIPLANATION TERRACES AND SLOPE DEVELOPMENT IN VEST-SPITSBERGEN AND SOUTH-WEST ENGLAND

Ronald S. Waters [*]

Exeter

Abstract

Conspicuous among periglacial modifications of recently de-glacierized dolerite terrain in Ekmanfjord, Vest-Spitsbergen, are rock benches and terraces on moderately inclined slopes (7—25°) lying between 50 and 100 metres above sea-level. Field observations suggest that the benches are fashioned primarily by frost heaving, frost shattering and the gravitative transfer of debris. Their continuing development is converting formerly ice-smoothed, convex-upward slopes into stepped, but overall rectilinear, slopes which retain their steepness as they retreat.

A similar origin is suggested for terraces hillslopes displaying features akin to altiplanation terraces which are very characteristic of the higher parts of south-west England. The demonstration that these landforms are periglacial features not only supports the conclusion that many parts of Southern England exhibit relict periglacial landscapes. It also suggests that in some cases cryergic processes have created rather than smoothed out irregularities of slope. The mature or equilibrium slope of cryoplanation is indeed smooth and compounded of accordant elements the inclination of each of which represents the slope limit appropriate to the calibre of the debris that is moved across it. But the achievement of cryoplanation on certain moderately inclined initial slopes is effected via the production and development of altiplanation terraces or congelifraction benches. These are characteristic of those parts of Southern England of which for one reason or another the complete periglacial metamorphosis was never realized.

Altiplanation terraces were first recognised in Britain by Guilcher (1950). In 1950 he interpreted as relict periglacial features terraced hillsides developed on indurated arenaceous sediments (Middle Devonian *Hangman Grits*) near the north Devon coast. Six years later Te Punga (1956) described similar terraces from Cox Tor, a 435 m hill on the western margin of the Dartmoor granite upland. Both authors suggested that such forms might be found elsewhere on the upland terrain of south-west England; in fact, Te Punga mentioned five other sites which exhibit at least superficially similar morphological features.

Benched hillsides are indeed common on both Exmoor and Dartmoor; but they are not confined to the Palaeozoic oldland. It would appear that many, perhaps most, of the hillside „terraces" that are so widespread on the more coherent members of the Jurassic and Cretaceous formations of the English Plain are but man-made modifications of pre-existing features genetically related to the wholly natural, high-level benches of south--west England.

The following results of a comparative study of relict „terraces" on

---

[*] Department of Geography, The University, Exeter, Devon, England.

Dartmoor and currently-developing forms in Vest-Spitsbergen are offered as a contribution towards the elucidation of the origins and evolution of these landforms which are so characteristic of English landscapes.

HIGH-LEVEL BENCHES ON DARTMOOR

The upland of Dartmoor with an area of some 390 km² between 300 and 600 m above sea-level is developed on the' outcrop of the easternmost

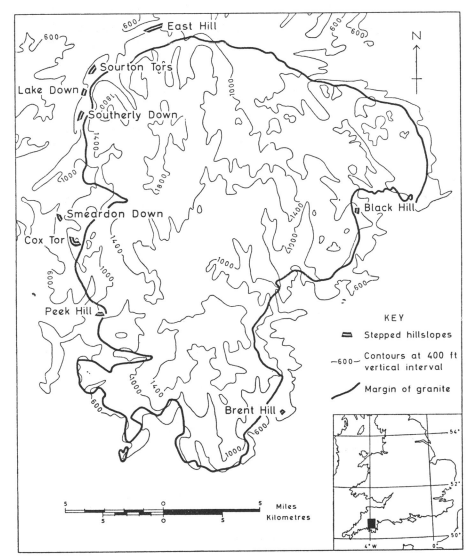

Fig. 1. Sites of altiplanation terraces on Dartmoor

and largest of the five major post-tectonic Armorican granite masses of south-west England and its aureole of altered Palaeozoic sediments and basic intrusions. Its bounding slopes rise distinctly and, in places, abruptly from the upper margins of the surrounding plateau-lands on unaltered country rock which decline gently from a little below 300 m to c. 120 m at the coast. It is these marginal slopes on the metamorphic aureole that are differentiated by benches and terraces, particularly between the 300 and 450 m contours.

Suites of well-developed benches have been mapped at nine localities on the western, northern and eastern flanks of the upland (fig. 1) and incipient or degraded forms can be seen at many other sites.

MORPHOLOGY

The stepped hillsides have mean slopes of from 8° to 14° (fig. 2). Their constituent steps possess gently-sloping treads of between 3° and 8° inclination and moderately steep risers of from 15° to 22°. The width

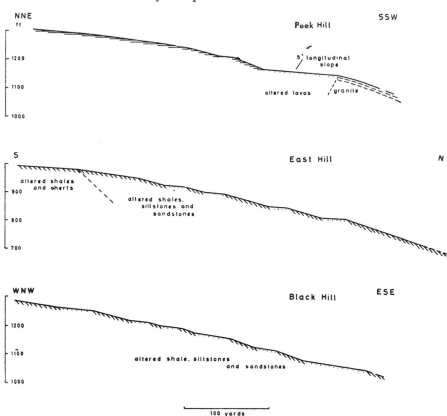

Fig. 2. Profiles of benched hillsides on Dartmoor

of the steps varies from 10 to 90 m, and the height of the backs ranges from 2 to 12 m. The longitudinal dimension of each step always exceeds its transverse dimension; indeed, some benches may be traced along the slope for over 800 m. But although the successive treads and risers appear to follow the contours very few of them are, in fact, longitudinally horizontal. Normally they display a longitudinal slope that is less than, but occasionally equal to, the transverse slope of the treads (pl. 1).

Although the hillsides are for the most part covered with a moorland vegetation of grass and/or heather, buttresses of bare rock diversify some of the risers and near the tops of the slopes on basic intrusives they may form free faces up to 100 m in length. Thus the uppermost benches on Cox Tor and Sourton Tors are backed by vertical rock cliffs instead of by steep, vegetated slopes (pl. 2). Even the isolated buttresses on the metasediments commonly extend to the top of the riser but rarely do they project above the tread of the superjacent step (pl. 3).

STRUCTURAL RELATIONS

It would appear from the geological evidence provided by these buttresses and free faces, by occasional quarries in the steeper back slopes and by adjacent stream sections that there is little lithological control over the distribution of the benches within the aureole. They occur on the more or less contact-altered representatives of interlaminated shales and fine siltstones; interbedded shales, thin limestones and radiolarian cherts; shales interbedded with siltstones and alternating with beds of medium-grained sandstone up to 0,5 m thick; fine- to medium-grained sills of oligoclase-dolerite; fine-grained albite-biotite dyke rock, and vesicular lava. Near to the granite margin the shales are represented by hornfelses, the calcareous rocks by calc-flintas, the siltstones by quartzites, the dolerite by a tough diabase-hornfels and the lava, the vesicles of which were filled by calcite, by a banded rock of the calc-flinta type (Reid *et al.* 1912; Dearman & Butcher 1959). Many of these metamorphic rocks also exhibit the effects of fissure metasomatism by means of which they were further hardened through silicification and tourmalinisation.

These rocks, apparently so diverse lithologically and hardly less so petrologically, are nonetheless possessed of one common attribute: they are all well bedded and/or well jointed, and between their well-defined divisional planes they are uniformly hard and relatively dense. With their joints and foliation these indurated rocks are less massive than the coarse-grained, biotite granite, and less laminated and friable than the non-metamorphosed shales. But within these broad limits the density of parting planes is by no means uniform. Variations in the spacing of both joints and foliation planes occur in each lithological group, and they are reflected in the calibre of the debris yielded by the different outcrops. Consequently any attempted explanation of the origin of the benched hillsides on the flanks of Dartmoor must take account of this variable structural attribute which might be expected to encourage the differential action of any erosional process.

A more obvious, but equally indirect and subtle expression of structural influence is seen in the longitudinal attitude of the benches. They are horizontal where they extend along a hillside parallel with the strike of the subjacent rocks, but if they are aligned in any other direction they assume a gentle longitudinal slope. This slope is always in the same sense but rarely as steep as the apparent dip of the bedding (normal or inverted); and it is always in a direction away from the granite [1].

DEBRIS-MANTLE

It is obvious from the evidence furnished by various exposures that the benches are cut in solid rock. Although the steps carry variable thicknesses of debris — up to 2 m on some treads and 20 cm on the risers — they are not built features. Consequently the morphology of the stepped hillsides cannot be explained in terms of the disposition of superficial material.

The mantle consists of a mélange of coarse and fine debris in variable proportions. On the altered sediments and lavas the coarse fraction (>2 mm) comprises flat, angular rock fragments of gravel and stone up to 20 cm in length (hornfels and calc-flintas < 10 cm, quartzites < 20 cm); on the altered basic intrusives it includes large boulders up to 1,0 m long. The boulders are concentrated at the base of the diabase cliffs but are also scattered on the treads of subjacent steps. By contrast, the stone and gravel grades of the coarse fraction are fairly evenly distributed. However, across the steps there appears to be some variation in the proportions of coarse and fine material in the mantle. The bulk of the debris is coarse near the backs (upper margins) of the treads; and the proportion of fine

---

[1] The longitudinal slope of the hillside benches on the Jurassic and Cretaceous rocks of the English Plain is approximately equal to the apparent dip of the bedding. Therefore it is not unreasonable to suggest that on the margins of the Dartmoor batholith the structural controls are not the bedding planes but planes of weakness produced during the initial emplacement (late-Armorican) and subsequent uplift (Miocene). Recumbent folds on the northern and western margins of Dartmoor were certainly tilted outwards by the granite and considerable movement occurred on pre-existing fault-planes during the Miocene (Dearman & Butcher 1959).

earth increases   progressively towards their fronts where it frequently
exceeds 50% (by volume). In the uppermost 1 m of this heterogeneous
debris near the front of the terrace it is usual for the stones to be disposed
vertically, and sometimes the material is particularly well sorted (pl. 4).

On the steps that are developed on or adjacent to the outcrops of the
altered basic intrusions the texture profile of the soil exhibits features which
are unlikely to be due solely to pedological processes. It comprises a virtually stoneless silty loam surface horizon over a sandy loam with angular
rock fragments of all sizes. The large boulders that are more or less densely
scattered over these steps protrude through the soil which itself is disposed
in closely packed, grass-covered mounds. The circular mounds, first noted
by Te Punga (1956) on Cox Tor, are up to 1 m across and 30 cm high.
Their origin and preservation would appear to be unrelated to the presence
of the boulders. The most perfect forms occur above 390 m on or near to
dolerite and diabase outcrops; subjacent steps on calc-flints exhibit degraded heather (*Calluna*) — clad mounds which less clearly surmount intervening grassy furrows.

HYDROLOGY

On nine days out of ten there is no surface water flow on the stepped
hillsides, though at certain spots near to the upper margins of the treads
the ground may be damper underfoot than elsewhere. But after a period
of continuous rain springs emerge at the erstwhile damp spots and braided
streamlets follow the steepest slope towards the front of the treads (pl. 5).
The streamlets have no channels; the water merely oozes up through the
grass sod, flows over it in barely perceptible depressions and sinks into
it again before reaching the lower edge of the terrace. There is consequently
no obvious surface erosion; yet little depressions up to 1 m deep and 3 m
broad head at the feet of some risers and extend for a few metres across
the steps, becoming shallower downslope and finally dying out altogether.
These depressions are indicative of a transfer of fine earth along narrow
belts from the backs to the fronts of the treads. The effects of this transfer
of material are well seen on steps with earth mounds. Within the relatively
narrow zone of surface water movement, the mounds are absent from the
back of the step, form well-defined islands in the middle and are partially
buried towards the front. It is possible that the transfer of fines is effected
by a kind of sub-cutaneous „flow" or mass movement which is more effective
than the soil creep which operates elsewhere. If not, then it may reflect
the more vigorous operation of existing processes either during abnormal
weather conditions or during slightly different bioclimatic conditions in the

very recent past. If, as seems likely, these relatively insignificant depressions are currently-developing forms they are an indication not only of the ineffectiveness of existing processes so far as the modification of pre-existing slope forms is concerned; they also show that the steps themselves are relict features, the results of processes which are no longer operating.

STEPPED SLOPES IN VEST-SPITSBERGEN

Comparable benched slopes have been studied in Vest-Spitsbergen — on Blomesletta, the south western part of the peninsula between Ekmanfjörden and Dicksonfjörden (fig. 3). They are best developed on the outcrop of a discordant dolerite sill which forms the backbone of, and the highest ground within, the northern half of the low-lying (100 m), glaciated peninsula built largely of Permo-Carboniferous cherts, limestones, dolomites

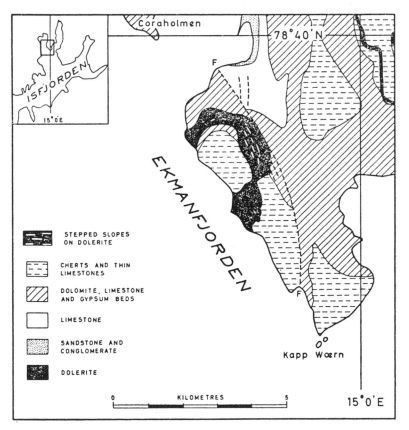

Fig. 3. Geology of Blomesletta, Vest-Spitsbergen (after Bates and Swarchzacher)

and gypsum beds (Bates & Swarchzacher 1958). The dolerite outcrop is marked by an approximate semi-circle of ice-smoothed hills which rise above low, shingle-covered plateaux and meet the coast in prominent head-lands. The ridge of hills is broadly asymmetrical in cross-section, with steep outward-facing slopes (free faces and scree slopes) and more mode-rately inclined, inward-facing slopes which descend to a kind of strike vale between the ridge and the chert plateau. It is these moderately sloping hillsides that are diversified by the rock-cut benches (pl. 6). The hillsi-des are rarely rectilinear; more commonly they exhibit convex upper segments and concave lower parts which may be separated by more or less extensive, uniformly sloping elements.

MORPHOLOGY

The benches are not limited to one particular part of the slope, though their size and morphology do appear to be related to their position on the hillside. On the steeper middle portions with average slopes of from 12° to 15° the steps tend to be narrower (c. 5 to 10 m) and flatter (1° to 2°) than they are lower down. In fact, it is clear that the overall concavity of the lower parts of the hillsides has been produced by the more extensive development of their constituent terraces, the treads of which may have

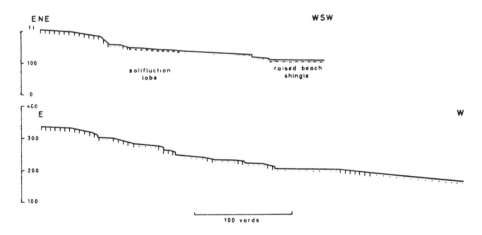

Fig. 4. Profiles of stepped hillslopes on Blomesletta

widths of up to 60 m and mean slopes of between 2° and 5°. The backs or risers are most commonly vertical and only 1 or 2 m high (fig. 4). On the upper parts of the hillsides where the steps are mere nicks in the slopes, the vertical backs pass upwards into convex segments of the pre-existing,

269

Pl. 1. Longitudinally sloping benches on Cox Tor

Pl. 2. A gently sloping terrace tread, with turf-covered mounds, backed by a steep riser with free—faces and *ravins de gélivation*, on the metadolerite of Cox Tor

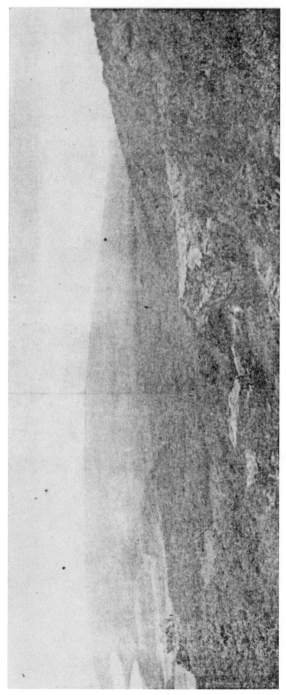

Pl. 3. View downslope across a broad bench on Peek Hill showing a „tor" in metasediments on the steep backslope and one in granite on the slope below the bench

Pl. 4. Section in debris on the front of a bench on Southerly Down showing sorting in the congelifluction material

Pl. 5. Braided streamlets flowing obliquely across a terrace tread on Cox Tor

Pl. 6. Stepped hillside on dolerite, Blomesletta, Vest-Spitsbergen

glacially-smoothed surface. Most of the benches exhibit longitudinal slopes, the longer ones showing a greater departure from the horizontal ($< 5°$) than the shorter. It would appear that the shorter slopes are controlled, initially at least, by the very slight inclination of the quasi-horizontal jointing or pseudo-bedding in the dolerite.

DEBRIS-MANTLE

The dolerite ridge carries a thin, discontinuous debris cover. On the relatively flat summits and broadly convex upper parts of the slopes unmodified portions of the initial surface exhibit ice-smoothed, -grooved and -striated exposures of bare rock separated by very thin patches of red, shelly drift. Lower down the hillsides the pre-existing surface is more or less replaced by the currently-developing stairways with risers of bare rock and treads which carry, at most, thin skins (<20 cm) of moving gelifractate. The proportion of bare rock at the surface decreases towards the bottom of the slopes where continuous sheets of soliflual debris up to 50 cm in thickness extend across the broader terraces and obliquely down from one terrace to the next thereby breaking the longitudinal continuity of the intervening riser. Small, tor-like masses of dolerite project above the debris-mantled, rock-cut surfaces of the lowest and broadest terraces. They are of two kinds: one represents the more massive remnants of a former riser which has disappeared with the coalescence of two terraces; the other is the result of the vertical heaving of joint-bounded blocks by ground ice.

THE MORPHOLOGICAL EVOLUTION OF STEPPED HILLSLOPES

CURRENTLY-DEVELOPING FORMS

It is obvious from the fresh, unweathered appearance of the rock cliffs and the angular blocks at their bases that the small benches on the higher parts of the hillsides on Blomesletta are developing by the retreat of the vertical risers. It would appear that this is affected primarily by basal frost-sapping. Even in July the upper parts of most of the treads were moist underfoot while their lower frontal portions were quite dry. This suggests that the top of the saturated zone in the mollisol on the well--jointed dolerite is less accidented than is the ground surface. Consequently frost-shattering is likely to be particularly effective at the base of the riser where the saturated zone approaches sufficiently near to the surface to be affected by diurnal freeze and thaw, especially during Spring and Autumn. There it is that large gelifracts shed from the riser remain until sufficient fines have been produced for solifluction to occur. The soliflual debris then streams down the steepest slope of the tread; i.e., not directly across it as a broad sheet but either obliquely across or along it and down to the next bench. Thus the backs of lower steps are not overwhelmed all along their lengths by debris from above; rather does their independent retreat continue save at localized points where it is precluded by the concentrated soliflual stream.

The proportion of the surface that is covered with the thin layer of solifluction debris increases progressively in a downslope direction and breaks in the continuity of the risers become increasingly wide until on the lowest parts of the hillside the benches finally lose their individuality in an ever-broadening footslope. This is an almost wholly debris-covered, rock-cut, concave-upwards, solifluction slope of less than 6°, diversified by lines of ,,tors'' which mark the positions of vanished risers. It grades imperceptibly into the flatter surface of the heavily-aggraded, strike vale.

Although it is clear that each individual step develops more or less independently as a solifluction slope of debris-removal by the retreat of its superjacent vertical free face, it is not known how effective the thin moving streams or spreads of soliflual material may be in degrading the terrace treads. It is conceivable that a certain amount of rock-robbing from the treads does occur through ablation of frost-heaved, joint-bounded blocks, particularly on the lower parts of the slope. But the ubiquity of rock-cliffs on the steeper parts of the hillsides suggests that downwearing is subordinate to backwearing, at least during the earlier stages of slope development.

The effects of slowly-melting snow-beds (Lewis 1939) may, however, be relatively important during these early stages. Alternate freeze and thaw beneath an elongated snow-patch in the basal angle between the tread and riser encourages basal sapping of the back-slope and the continued fragmentation of the gelifracts shed from it. As silt-sized particles are produced, meltwater carries them farther down the tread and a relatively flat wash-slope of limited but variable width may then be seen between the narrow belt of coarse debris at the back of the tread and the position where soliflual movement supervenes. This zone of surface washing retreats with the slope elements above it and the area over which solifluction operates increases progressively. A melting snow-bank may thus encourage the parallel retreat of the face against which it is lying, particularly in the early stages of step development, but its presence is not necessary for the continued operation of the processes whereby the backwearing is effected.

Undoubtedly the steps originated in slight depressions in the initial, glacially-moulded slopes. These depressions, which through the presence of snow-beds or by virtue of their position relative to the saturated zone in the mollisol would be moister than their surroundings, would encourage the cryergic processes to begin their differential action. The disposition of the initial hollows was probably joint-controlled; in any event, during their subsequent conversion into, and development as, treads and risers divisional planes in the dolerite exercised an important control. And the morphological effects of structural control of one kind or another are still everywhere to be seen on the stepped hillsides. Even the inclination of their most evolved portions — the foot-slopes — is a function of structure, as is the disposition of the residual rock masses which rise above these foot-slopes.

RELICT FORMS

On the benched hillsides of Dartmoor the adjustment of surface forms to underlying structure is as striking as it is on comparably inclined slopes in semi-arid and periglacial landscapes. It arises where, in the absence of a continuous vegetation cover, the alternation of more „mobilisable" (Tricart 1950, p. 221) and less „mobilisable" rock outcrops promotes differential fragmentation and ablation. That the Dartmoor hillslopes are, in fact, relict periglacial forms, as Te Punga believed, is confirmed by the character and disposition of their debris mantle and by the morphological resemblances between them and the stepped slopes in Vest-Spitsbergen.

In both areas steps or benches occur on hillsides with mean slopes (8°—15°) that are less steep than direct gravity-controlled slopes but are steeper than the minimal slopes over which soliflual movement may occur. As on Blomesletta so on Dartmoor these moderately inclined hillsides have been broken up into a series of treads and risers; but the process would seem to have gone farther in the latter area where virtually no part of the pre-existing surface remains. Moreover, on Dartmoor the treads are broader and the backs higher. This may be a reflection of structure or of the fact that south-west England experienced several periglacial phases during the Pleistocene. If the amount of recent, post-periglacial slope modification is in any way indicative of the extent of inter--periglacial modification, then we may interpret the existing benched hillslopes as pure periglacial forms representing the cumulative effects of several periglacial phases. In any event most of the Dartmoor benches are more mature than the Blomesletta steps.

In one further respect the morphology of the benched hillslopes exhibits the ever-present influence of structural differences. Risers on dolerite in Vest-Spitsbergen and on equally massive basic intrusives on Dartmoor are vertical or nearly so, while those on the metasediments slope at between 15° and 22°. Insofar as the inclination of a basally-sapped cliff is related to the strength of the bedrock some variation in slope is to be expected. Subsequent differential downwearing of the relict free faces initially developed on rocks of different kinds may also be significant.

SUMMARY AND CONCLUSIONS

Of the several processes which have been collectivelly responsible during the last 10 000 years for converting moderately-inclined, glacially--smoothed, dolerite hillsides on Blomesletta into irregularly-stepped slopes, frost weathering — including frost-heaving, -thrusting and -shattering — and solifluction would appear to have been by far the most important. However, ablation of fine material by snow meltwater and of coarse material by free fall may have been significant contributory processes.

The differential action of both zonal and azonal processes which was occasioned by slight surface irregularities and has been subsequently encouraged by structural attributes, by the disposition of snow-beds and by the manner of disposal of surface water during the season of thaw has produced a series of individual free faces (risers) or waste-production slopes whose continued parallel retreat has, in turn, created an equal number of foot-slopes (treads) or waste-removal slopes. The inclination of the basally-sapped free faces depends ultimately on the strength of the bedrock and that of the foot-slopes is related to the calibre of the debris that moved over them. The removal of debris by solifluction along rather than — or as well as — across the foot-slopes normally keeps pace with its production, consequently intermediate debris-slopes or waste-accumulation slopes are rarely represented.

By virtue of their morphology and the nature and disposition of their debris mantle, the benched slopes on Dartmoor are interpreted as relict examples of more mature, stepped hillsides fashioned by cryergic processes during successive Pleistocene periglacial phases. This interpretation also accounts most satisfactorily for their marked adjustment to structure.

The demonstration that these landforms are periglacial features not only supports the already widely-accepted conclusion that many parts of Southern England exhibit relict periglacial landscapes. It also suggests that in some areas cryergic processes created rather that smoothed out irregularities of slope. The mature or equilibrium slope of cryoturbation (geliturbation) is indeed smooth and compounded of accordant elements the inclination of each of which represents the slope limit (Tricart 1952, p. 232—233) appropriate to the calibre of the debris that is moved across it. But the achievement of cryoplanation on certain moderately-inclined, initial slopes is effected *via* the production of altiplanation steps and benches. These are characteristic of those parts of Southern England of which, for one reason or another, the complete periglacial metamorphosis was never realized.

ACKNOWLEDGEMENTS

This paper is based in part on observations made while the writer was a member of the Birmingham and Exeter Universities 1958 Spitsbergen Expedition. Grateful acknowledgement is made of the financial assistance that was provided by the Univerisities of Birmingham and Exeter and by the Royal Geographical Society to make the expedition possible. The author's thanks are also due to A. Horton for his valuable help in Vest-Spitsbergen.

References

Bates, D. E. B., Swarchzacher, W. 1958 — The geology of the land between Ekmanfjörden and Dicksonfjörden in Central Vestspitsbergen. *Geol. Magazine*, vol. 95; p. 219—233.

Dearman, W. R., Butcher, N. E. 1959 — The geology of the Devonian and Carboniferous rocks of the north-west border of the Dartmoor granite, Devonshire. *Proc. Geol. Assoc.*, vol. 70; p. 51—92.

Guilcher, A. 1950 — Nivation, cryoplanation et solifluction quaternaires dans les collines de Bretagne Occidentale et du Nord du Devonshire. *Rev. Géomorph. Dyn.*, vol. 1; p. 53—78.

Lewis, W. V. 1939 — Snow-patch erosion in Iceland. *Geogr. Jour.*, vol. 94; p. 153—161.

Reid, C. *et al.* 1912 — The geology of Dartmoor. *Mem. Geol. Survey U.K.*

Te Punga, M. T. 1956 — Altiplanation terraces in southern England. *Biuletyn Peryglacjalny*, nr 4; p. 331—338.

Tricart, J. 1950 — Le modelé des pays froids, fasc. 1: Le modelé périglaciaire. Cours de géomorphologie, 2ᵉ partie, fasc. I. CDU, Paris.

Tricart, J. 1952 — La partie orientale du Bassin de Paris. Paris.

22

Reprinted from *Biul. Peryglac.*, No. 22, 117–119, 124–130 (1973)

THE VALLEY CRYOPEDIMENTS IN EASTERN SIBERIA

T. Czudek and J. Demek

Brno

Abstract

The paper concerns the problems of pedimentation in the subnival environment of Eastern Siberia (mainly n the area of the Stanovoj Range, the Elginskoye Plateau and the Čerski Range). A concave erosion surface at the foot of valley sides occurs here, gently sloping to the valley bottoms. The erosion surface attains up to 3 km in width with an inclination of usually 3 to 10°. In some places the piedmont surface is bound directly on the valley bottom, in other places on low river terraces of Middle up to Upper Pleistocene age. The gentle erosion surface usually extends headwards from the main valleys into the side valleys. The studied surface is cut in solid rocks and covered with a thin layer of slope deposits (usually 1–2 m only).

The erosion surface develops by backwearing of steep valley sides due to Pleistocene and present-day cryogenic slope processes. On steep slopes mainly frost weathering, frost creep and suffosion manifest themselves. These processes are mainly active in shallow dells. Sapping of the steep slope by nivation together with the dells activity are the main reasons for parallel slope recession. The foot is exposed by further transportation of material mainly due to frost heaving and solifluction. Linear transport on the gentle erosion surface is also concentrated in shallow dells.

The authors consider the gentle erosion piedmont surfaces a subnival variety of the pediments cf arid and semi-arid regions and call them cryopediments.

INTRODUCTION

In the years 1966 and 1969 we dealt with the problems of slope development in the present-day subnival climate of Eastern Siberia. We simultaneously studied pedimentation under conditions of permafrost occurrence. Our investigations were concentrated mainly on the regions of the upland and mountain relief of the Stanovoj Range, the Elginskoye Plateau and the Čerski Range.

In these regions valleys are deep but widely open. On their bottoms misfit streams underfit to the valleys occur. Free meanders are common in valley bottoms. A gentle concave surface often rises from the valley bottoms with an inclination of dominantly 3 to 10° its width ranging from several hundreds of metres to 1.5 km and/or even 3 km. In places it is in direct contact with the valley floor, elsewhere it is separated by a step from the valley bottom. The piedmont surface passes over in a break of slope into steeper valley sides. The gentle surface usually extends headwards from the main valley into the side valleys.

[*Editor's Note:* Certain figures have been omitted owing to limitations of space.]

The localities investigated occur in the taiga zone. The valley floors are often muddy and covered with peat bogs, grass, shrubs and trees. The piedmont surfaces are usually overgrown with taiga consisting predominantly of *Larix dahurica*. Sometimes, muddy places are found on them. The steep valley sides are partly covered with taiga and partly with debris and block fields. The bedrock often outcrops directly on them. In the north, the upper slope sections are already above the timber line and extend into the zone of mountain tundra and/or frost desert with cryoplanation terraces.

The piedmont surfaces are developed in various geological structures. They occur in both the hard rocks of the Siberian Platform (granites, gneisses) and in soft Mesozoic sandstones and shales of the Verhkoyansk complex in the young mountain ranges of NE Siberia. It is evident at first sight that they are no structural surfaces.

In former papers the gentle concave piedmont surfaces were usually considered erosion or accumulation river terraces. Subsequent works (Timofeev, 1962, 1963; Kašmenskaya & Khvorostova, 1965; Simonov, 1966; Panasenko, 1968; Keyda, 1970; etc.) point to the share of pedimentation prosesses in their development.

THE CLIMATE

The area investigated belongs climatically to the continental type of subnival climate with severely cold, dry winters and short, warm summers with the maximum of precipitation. The mean annual temperatures are lower than 0°C (Aldan: −6.2°C, Yakutsk: −10.2°C, Khandyga: −11.7°C, Deputatskij: −15.2°C, Ust-Nera: −15.3°C, Verkhoyansk: −15.6°C). The maximum July temperatures exceed 30°C. The mean July temperatures amount e.g. in Verkhoyansk to 15.3°C, Ust-Nera to 15.5°C, in Aldan to 16.8°C, in Yakutsk to 18.8°C. The absolute minimum in Verkhoyansk in the Yana River valley is −67.8°C and in Ojmyakon in the Indigirka River valley −77.8°C. The amount of precipitation is very low. The mean annual precipitation amounts in Verkhoyansk to 154 mm, in Ojmyakon to 194 mm, in Yakutsk to 202 mm, in Deputatskij to 218 mm, in Ust-Nera to 225 mm, in Khandyga to 430 mm, in Aldan to 564 mm. The present-day climate resembles that of the last glacial period.

PERMAFROST

Owing to the negative thermal balance continuous permafrost is developed in the area investigated. On the whole the permafrost is in thermal equilibrium with the present-day climate of Eastern Siberia. Its thickness varies and ranges from several tens of metres to several hundreds of metres. In the Arkagala River valley this thickness is from 120 to 200 m. In some regions of Eastern Siberia the thickness of permafrost attains as much as 1500 m. Open or closed taliks are developed beneath lakes and larger water courses. In the regions investigated the thickness of the active layer ranges from about 0.5 m in the north to 3 m in the south.

In connection with the climatic conditions and the presence of permafrost a complex of cryogenic processes are operating on slopes. Several facts are characteristic of the intensity of their effects, such as:

(a) considerable and relatively rapid oscillation of air temperatures,

(b) mean annual temperatures below 0°C and extremely low winter temperatures,

(c) short transition seasons (spring and autumn) with temperatures ranging usually around 0°C and the occurrence of numerous freeze–thaw cycles,

(d) considerable aridity and exceedingly small quantities of snow,

(e) temperature inversions in mountain basins,

(f) presence of permafrost of very low temperatures (mainly in the north – as low as −12°C).

[*Editor's Note:* Material has been omitted at this point.]

The Arkagala River displays near the village of Arkagala (some 150 km SE of the town of Artyk) in the Čerski Range a wide and in its cross profile asymmetrical valley. The right valley side facing NE is steeper, the left one exposed SW is gentler. The NE slope has in its lower section an inclination of 16 to 22°, in its middle and upper sections of 17 to 33°. In the lowest slope section there is a dense moss growth moving slowly downslope and creating moss terracettes as much as 1 m high. In cracks, water occurs above the permafrost. Higher upslope angular fragments of sandstone mixed with sandy loam with numerous traces of downslope movements are found forming small solifluction terraces in the uppermost part of the slope. Bedrock often crops out on this slope.

At the foot of the left valley side a gentle piedmont surface is developed (Pl. 1). The inclination of the surface is 5 to 10° and decreases successively down to 3°. In the place of the profile (Fig. 5) its width is 540 m. In the cross profile it is slightly undulated by shallow dells. The piedmont surface is separated by a 15 m high step from the Arkagala River floodplain. On this step terrace gravels occur. The terrace gravels are found in non-sorted circles of 0.5 m in diameter at a distance of 200 m from the step edge. In higher

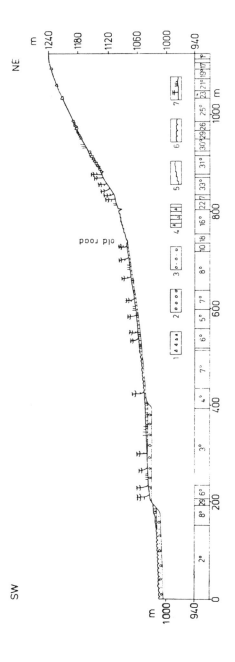

Fig. 5. Profile through the cryopediment in the Arkagala River valley near the village of Arkagala in the Čerski Range

1. angular sandstone fragments; 2. terrace gravel; 3. floodplain gravel; 4. slope deposits (angular sandstone and shale fragments with sandy loam); 5. moss terracettes; 6. non-sorted circles and polygons; 7. vegetation (*Larix dahurica*, grass)

Surveyed by the authors, June 18, 1969

sections of the piedmont surface the gravels do not occur any more. On the basis of the cut of the old road in the uppermost part of the piedmont surface, the thickness of the mantle can be estimated to have 1 to 2 m. Bedrock (sandstone) occurs beneath the mantle. On the basis of the facts mentioned the upper part of the piedmont surface may be regarded as an erosion surface, and the lower one as a river terrace partly covered with slope material (Fig. 5).

The piedmont surface is overgrown with taiga (*Larix dahurica*) with bent tree trunks. Downslope, intensive processes of frost heaving are taking place. Mainly on bottoms of flat dells traces of present-day frost heaving and solifluction can be found manifesting themselves in the development of small solifluction terraces and the elongation of non-sorted circles in the direction of the slope. The non-sorted circles are formed of brown loam with small fragments, predominantly shales of Triassic age. The movements are indicated also by cracks in the moss and turf cover which were open and water-filled during the observations on June 18, 1969.

LOCALITY NEAR THE PLACER OF POLEVOJ

About 40 km east of the town of Susuman the Svetlaya River valley near the placer of Polevoj in the Čerski Range runs in E–W direction. The valley is asymmetrical in its cross profile. The slope facing south is gentler and has in its lower section an inclination of 22° (Fig. 6). There are fragments of sandstones of the Verkhoyansk complex on it and the bedrock outcrops often directly to the surface. At the slope foot begins a gentle surface (angle of slope 6 to 7°). This piedmont surface is 360 m wide and falls with a 5 m

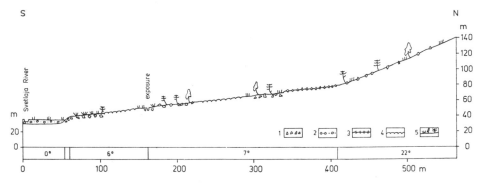

Fig. 6. Profile through the cryopediment in the Svetlaya River valley near the placer of Polevoj in the Čerski Range

1. angular sandstone fragments; 2. floodplain gravel; 3. sorted stone polygons; 4. non-sorted circles; 5. vegetation (birch, alder, larch, grass)

Surveyed by the authors, June 10, 1969

Note: Vertical scale is not related to sea level

high step down to the Svetlaya River floodplain (Fig. 6). The step is covered with coarse debris. A long exposure dug across the piedmont surface at a distance of 100 m from the step showed that there are slightly weathered sandstones of the Verkhoyansk complex below 1.40 up to 1.50 m thick loams mixed with sandstone fragments (Pl. 2). This surface can be therefore considered an erosion surface. It is overgrown with birch and alder shrubs and larches with bent tree trunks. On the surface there are non-sorted circles (with diameters of 1.5 m) and sorted stone polygons with diameters of as much as 3 m.

THE GENESIS OF THE PIEDMONT SURFACE

The analysis of the localities mentioned shows that gentle erosion surfaces have developed under permafrost conditions in many valleys of Eastern Siberia. They can be found on one or both sides of the valley bottoms. In some places low river terraces occur along the valley floor. In those cases the erosion surface passes into accumulation terraces which constitute the local erosion base level of its development.

The sides of the wide valleys are steep in the area described (inclined often above 20°). But they do not bear evidence of lateral sapping due to meandering of rivers. A typical feature is that the rivers are underfit to the width of the valleys. The valley gradient is small and the rivers usually flow in free meanders in the middle stripe of very wide floodplains. The development of the wide valleys can only difficultly be explained merely by river activity. The valley sides are mostly not within direct reach of the effects of river erosion. The present shape of the valleys may be assumed to be a result of slope processes on valley sides.

The studied test pits and exposures have shown that the piedmont rock surface is covered with a thin layer of slope deposits. Its thickness ranges usually from 1 to 2 m in the area investigated. It consists of fragments of local rocks mixed with sandy loam and loamy sand. Their structure shows that these deposits developed owing to the action of a complex of cryogenic processes, such as frost weathering, frost heaving, frost creep, solifluction and sheet wash. The piedmont surface shows a number of cryogenic forms. These are mostly non-sorted circles, non-sorted polygons, non-sorted steps, sorted stone polygons, solifluction terraces, etc. The presence of these forms together with that of bent trees (drunken forest) point to present-day development of the piedmont surface.

The main denudation lines on the forms described are the shallow dells. Dells are very common on the piedmont surface. Due to a higher quantity of water, intensive cryogenic processes take place within them. Movements occur not only in the direction of the longer axis of the dells, but even lateral movements can be observed. On the bottoms of the dells shallow furrows come into being during snow melt quickly filled up with the material sliding from the sides.

Many dells occur also on the steep rocky valley sides partly covered with angular debris. On these slopes frost creep acts in the area investigated, i.e. slow movement of debris due to sliding on ice crusts developing at the inferior surface of the fragments in debris mantles. The effects of needle ice are also important. Sometimes solifluction terraces develop. During snow melt suffosion acts too. The main denudation lines are even here the dells cutting continuously parallely into the slope. In the area investigated no accumulation of the material removed from the steep slope at its foot was observed. Although the transition between the steep slope and the piedmont surface is gentler than in arid regions the material supplied from the steep scarp is carried away. This is evidently promoted by snow accumulation on the break of slope and the nivation connected with this accumulation. With regard to the dry cold climate in the area studied the snow accumulation and its later thawing are an important source of water and consequently of the increased intensity of cryogenic processes. We could observe that the foot of the steep slope was even in summer wetter than the other parts of the piedmont surface. J. G. Simonov (1966) explains this fact by issues of shallow ground waters above the permafrost. The removal of the material supplied from the steep slope causes the sapping of its foot. The sapping due to nivation is together with the effects of the dells the main cause of the retreat of the steep slope.

The removal of the material on the piedmont surface is promoted by suffosion washing out the fine material of the block fields and debris mantles on the steep slope. The fine soil is transported on the piedmont surface assisting the development of frost heaving and solifluction.

It follows from the above mentioned facts that gentle erosion-denudation surfaces develop in the valleys of Eastern Siberia by the retreat of the steep valley side due to the effects of a complex of geomorphological processes connected with present-day climatic conditions. This is why these surfaces can be considered a subnival variety of the pediments of arid and semi--arid regions. The pediments developing in the subnival area under conditions of permafrost are called cryopediments (pediments of the Siberian type according to D. A. Timofeev, 1963).

The extent and the intensity of development of the valley cryopediments depend on both the tectonic stability of the territory and the climatically controlled intensity of cryogenic processes. The valley pediments develop in regions where lateral widening of the valleys prevailed over vertical erosion during the Pleistocene and in the Holocene due to slope processes.

As to the age of the cryopediments of the area investigated they can be considered young with regard to their relation to low terraces and/or the valley floor (see profiles). J. G. Simonov (1966) states that there are extensive valley pediments linked with the so-called main terrace of Middle to Upper Pleistocene age in the valleys of the large rivers, such as Šilka, Olona, Ingoda, Argun. In some places the pediments in the valleys of the said rivers display a width of several kilometres. This fact points to a relatively rapid development of the cryopediments.

The valley pediments penetrate from the main valleys into the side valleys. Under favourable conditions the cryopediments of two opposite valleys merge together and pediment passes develop. Thus a relief of isolated mountain groups and inselbergs comes into existence which was seen in the Stanovoj Range. In this region there are at least two pediment steps one above the other. The erosion base level for their development were evidently river terraces of various heights.

The pediments were observed in some places of Eastern Siberia even above the upper edge of deep incised valleys. An example is the Nera River valley (right-side affluent of the Indigirka) bordered with as much as several kilometres wide pediments at heights of 170 to 200 m above the valley bottom. The deep valleys of the Nera River and of its tributaries are incised into the pediments which are probably older and evidently developed before the Quaternary.

CONCLUSION

A number of authors (Baulig, 1952; Dylik, 1957; Khvorostova, 1970; Baranova & Čemekov, 1970; etc.) have already pointed out the common features of slope development in arid and semi-arid regions on the one hand and in subnival regions on the other. The analysis of the localities in the valleys in the continental subnival climate of Eastern Siberia confirms this opinion. In the dissected relief of upland and mountain regions built of solid rocks a parallel retreat of valley sides and the development of valley cryopediments take place owing to cryogenic processes operating in the presence of permafrost. On steep slopes, mainly frost weathering, frost creep and suffosion manifest themselves. These processes are concentrated chiefly in shallow dells with a high gradient (*Hangdellen*) which, together with the sapping of the foot by nivation are the main causes of the parallel slope retreat. The material supplied is removed from the foot mainly by frost heaving and solifluction. At the foot of the steep slope a pediment develops most often with an inclination of 3 to 10°. Even on the pediment, the most effective transportation is concentrated in shallow dells.

The authors believe that the pediments investigated are a subnival variety of the pediments of arid and semi-arid regions and call them cryopediments.

References

Baranova, J. P., Čemekov, J. F., 1970 – Penepleny, pedipleny i poverkhnosti vyravnivaniya skladčatykh oblastej (Peneplains, pediplains and planation surfaces of orogenetic regions). *Problemy poverkhnostej vyravnivaniya*, no. 1; p. 13–15; Akad. Nauk SSSR, Irkutsk.

Baulig, H., 1952 – Surfaces d'aplanissement. *Annales Géogr.*, No. 325: p. 161–183; No. 326: p. 245–262.

Dylik, J., 1957 – Tentative comparison of planation surfaces occurring under warm and under cold semi-arid climatic conditions. *Biuletyn Peryglacjalny*, No. 5; p. 37-49.

Kašmenskaya, O. V., Khvorostova, Z. M., 1965 – Geomorfologičeskij analiz pri poiskakh rossypej na primere Elginskogo zolotonosnogo rajona v verkhovyakh reki Indigirki (Geemorphological analysis in the research of mineral deposits in the gold-bearing Elginskij region, the Indigirka river basin). Ed. Akad. Nauk SSSR; 166 p., Novosibirsk.

Keyda, E. P., 1970 – Sovremennoye razvitiye dolinnogo pedimenta v nižnem tečeniji Irtyša (Present-day development of the valley pediment in the lower course of the Irtyš River). *Problemy poverkhnostej vyravnivaniya*, No. 1; p. 43-44; Akad. Nauk SSSR, Irkutsk.

Khvorostova, Z. M., 1970 – Problema obrazovanija pedimentov (Problem of the origin of pediments). *Problemy poverkhnostej vyravnivanija*, No. 1; p. 90-94; Akad. Nauk SSSR, Irkutsk.

Panasenko, V. I., 1968 – O četvertičnykh pedimentakh v rykhlykh porodakh Amuro-Zejskoj depressii (On Quaternary pediments in loose rocks of the Amur-Zeya depression). *Problemy izučeniya četvertičnogo perioda*; p. 39–41, Khabarovsk.

Simonov, J. G., 1966 – Dolinnye pedimenty lesnoj zony vostočnogo Zabajkala i jikh mesto v analize relefa (Valley pediments of the forest zone of Eastern Transbaikal and their role for an analysis of the relief). *Vestnik naučnoj informacii Zabajkalskogo filiala Geografičeskogo Obščestva SSSR*, No. 6; p. 26–33, Čita.

Timofeev, D. A., 1962 – K probleme proiskhoždeniya formy rečnykh dolin (na primere rečnykh dolin Južnoj Yakutii) (On the origin of river valley forms as exemplified by the river valleys of South Yakutya). *Izvestiya Akad. Nauk SSSR, ser. geogr.*, no. 3; p. 82–89, Moscow.

Timofeev, D. A., 1963 – Poverkhnosti vyravnivaniya Aldano-Oljokminskogo meždurečya, Južnaya Yakutiya (Surfaces of planation of the watershed between the Aldan and the Oljokma Rivers, South Yakutya). *Zemlevedenije*, t. 6; p. 169–183, Moscow.

23

Reprinted from *Biul. Peryglac.*, No. 21, 51–54, 62, 64–73 (1972)

ASYMMETRICAL SLOPE DEVELOPMENT
IN THE CHILTERN HILLS

H. M. French *

Ottawa

Abstract

The asymmetrical nature of Chalk dry valleys in Southern England is discussed with reference to some of the Chiltern valleys. The periglacial origin of the asymmetry is not disputed. Instead, a model of asymmetrical slope and valley development is proposed in which the importance of fluvial processes in periglacial environments is stressed. Some implications with respect to the origin of chalk dry valleys are outlined. It is argued that while the major valleys may very well be explained in terms of a ground.water hypothesis in accordance with a progressively falling watertable throughout the Pleistocene, the periods of periglacial conditions promoted: (a) the asymmetrical modification of the slopes of already existing valleys and (b) the initiation of the periglacial valleys and gullies – les vallons en berceau – upon the frozen subsoils existing at that time.

INTRODUCTION

It has been widely assumed that the Chalk landscape of Southern England is essentially a "fossil" periglacial landscape, fashioned by processes of a cryonival and niveo-fluvial nature. Williams (1968) has recently, in this journal, reviewed much of this evidence. Additional support for such an interpretation has been provided by the recognition of the widespread existence of periglacially modified asymmetrical valleys (Ollier & Thomasson, 1957; French, 1967). Such an asymmetry expresses itself in a steeper slope orientated towards the SW or W and is suggested to be directly comparable to that which characterises the dry valleys of North France (Gloriod & Tricart, 1952), of Belgium (Grimberieux, 1955), and of the Muschelkalk Plateau of Germany (Helbig, 1965). The asymmetry is thought to be the result of differential insolation and freeze-thaw processes operating upon slopes of varying orientations during the fluctuating climates of the Pleistocene. Similar asymmetry also exists upon other lithologies in France (Taillefer, 1944; Cavaillé, 1953), Belgium (Alexandre, 1958), the Netherlands (Geukens, 1947), Germany (Büdel, 1953), Czechoslovakia (Czudek, 1964) and Poland

---

* Department of Geography, University of Ottawa, Canada.

[*Editor's Note:* Certain figures have been omitted owing to limitations of space.]

(Dylik, 1956). This asymmetry is regarded as the "normal" asymmetry for W-Europe.

Areas in Southern England where this asymmetry is particularly well developed are not only the Chiltern Hills, as Ollier and Thomasson have described, but also the North Wiltshire Chalklands and the North Dorset Downs (French, 1967). In these three areas, the well developed but simple NW–SE alignment of the valley patterns has been favourable for the development of the asymmetry. In other areas, structural and lithological considerations, together with rather more complex valley pattern have operated to disguise or negate the development of this asymmetry.

In the Chiltern Hills, the simple asymmetry pattern is complicated by the fact that there would appear to be at least two different types of W-facing asymmetrical valleys existing in close juxtaposition to symmetrical valleys. For this reason, the area offered special problems of explanation as regards asymmetry. This paper is an attempt to establish the relationship between these various forms as a contribution towards an understanding of the origin of such valleys. Emphasis in this paper is upon the forms of these valleys since the significance of the deposits has already been discussed by Ollier and Thomasson (1957), Avery (1959) and others. The resultant model of asymmetrical slope development is suggested to be applicable to the other areas of the Chalklands possessing the "normal" asymmetry.

THE SETTING OF THE PROBLEM

The majority of the valleys which dissect the dip slopes of the Chalklands of S. England do not posses surface streams. It is possible to distinguish two types of dip-slope dry valleys: (1) the large valleys which constitute the major drainage lines and valley patterns and (2) the numerous small tributaries and gullies which feed into the large valleys and which often produce asymmetrical tributary patterns. Such a distinction is similar to that made by Klatkowa (1965) in Poland between the *vallons en berceau* (niecki denudacyjne) and the *vallées sèches* (doliny denudacyjne). This study is primarily concerned with the asymmetrical form of the large valleys – les vallées sèches. In the Chiltern Hills at least four different types of valley forms can be recognised:

(1) Small shallow symmetrical valleys and gullies, often elongated or "paddle--shaped". These are found at the extreme heads of the larger valleys and on the gentle slopes of the asymmetrical valleys. These are analogous to the *vallons en berceau*.

(2) Normal Asymmetrical Valleys, Type I. On a much larger scale, valleys exist possessing a broadly U-shaped, yet asymmetrical, cross-profile.

Examples are the Radnage, Little Hampden and Callow Downs valleys (pl. 1). This form of asymmetry is regarded as the Type I asymmetry in the Chiltern Hills and may be contrasted with the Type II asymmetry existing in the same area.

(3) Normal Asymmetrical Valleys, Type II. This type of asymmetry is rather more striking than Type I since the valley is rather more clearly defined. The cross-profile is asymmetrical, but basically V-shaped. Such valley forms are found in the Chesham and Hughendon areas (pl. 2).

(4) Symmetrical U-shaped Valleys. These are broad flat valleys such as the Great Missenden valley and the "through" valleys of the region. The three latter types constitute the *vallées sèches* or *doliny denudacyjne* of the English Chalklands.

Because both symmetrical and asymmetrical valleys are present in the Chiltern Hills, the problem is not merely to explain asymmetry but also it is to establish why some valleys are not asymmetrical. Any hypothesis must be of general applicability to the whole area if it is to be anything but special pleading.

The problems posed by symmetrical valleys are exemplified by examining the broad "through" valleys. They lack any noticeable slope asymmetry throughout their lengths. This could be the result of either the size of the valley limiting any cross-valley variations conducive to asymmetry formation or the eradication of asymmetry by glacial meltwater modification from the Vale of Aylesbury (Brown, 1964). However, the nature of these valleys is rather more complex than this, for Ollier and Thomasson (1957) note that the asymmetrical pattern of soils found in all the asymmetrical valleys is also present in these large "through" valleys, (see soil survey map "Aylesbury" for many examples, e.g. Gt. Missenden Valley). If meltwater had indeed acted to modify the slope forms of these valleys one would have expected it to have also destroyed the much more delicate soil asymmetry. Furthermore, the very presence of an asymmetrical development of soil types implies differential cross-valley variations of some sort. For several reasons, therefore, it is clear that the symmetrical nature of the broad valleys poses rather more difficult problems of explanation than are, at first, appreciated.

The existence of "paddle-shaped" symmetrical valley heads is also problematical. Is there, for example, a "threshold" of some sort at which such a form changes to an asymmetrical one? If so, the establishing of this is essential to understanding the development of asymmetry. Helbig (1965), for example, has suggested a depth as small as 15′ (4.5 m) for the initiation of asymmetry. Alternatively, one can ask: what conditions are exclusive to these extreme valley heads?

Thus, in examining the asymmetrical nature of the Chiltern valleys, one is inevitably committed to a consideration of the symmetrical valleys also. If asymmetry is a climatic phenomena, the location of the areas of no asymmetry may be just as revealing as those with asymmetry. Finally, a given sequence of asymmetrical slope development must be able to incorporate symmetrical forms if it is to be considered as representing any sort of reality for the Chiltern Hills.

ASYMMETRICAL SLOPE DEVELOPMENT

In the past, many slope studies have been conceived of within a cyclic or stage dependent conceptual framework. Thus, a downvalley slope sequence may be thought to be an indirect reflection of stage. However, it is also possible to examine a downvalley sequence of slopes without any implication of stage. Instead, the various downvalley profiles are considered in relation to such variables as slope height, distance from source, and valley dimensions in general.

Slopes profiles were measured by the author in dry valleys on the Chalk possessing the "normal" asymmetry[1]. The fieldwork was carried out in the Chiltern Hills, the Marlborough Downs, the N. Dorset Downs, the South Downs and the Southern Hampshire Chalklands.

Distinctive slope angles were found to characterise the valleys. Maximum slope angles had major peaks of 7–11° (41%) and of 19–23° (15%) which correspond to the gentler and steeper slopes respectively of these valleys. Secondly, the gentle slopes were characterised by long R. segments of a constant angle between 5–9°.

The interaction of these characteristic angles to produce the asymmetrical forms is clearly of importance. Consequently, a detailed analysis of the slopes of two valleys in the Chiltern Hills is outlined. The two valleys chosen were those of Little Hampden and Bryant's Bottom which correspond, respectively, to the two types of asymmetrical valley already recogn sed.

---

[1] The slopes were measured by means of an Abney level, a tape and a pair of ranging poles. Analysis was made first, of the maximum slope angles measured and secondly, of the frequency of rectilinear segments within the profiles. The following terminology was decided upon: a *measured length* – the distance over which each slope measurement was taken (a constant interval of 30′); a *rectilinear segment* (R) – at least 3 measured lengths having an interval variation of no more than 1° (plus or minus ½°); a *maximum angle* – the highest slope angle recorded over one measured length for one complete profile. The index of asymmetry, Id, as defined by Tricart (1947) was used to describe the degree of asymmetry present. However, it was defined with respect to maximum angles and the resultant index was given the title of the *Asymmetry Index*, A/I.

[*Editor's Note:* Material has been omitted at this point.]

A MODEL OF ASYMMETRICAL SLOPE DEVELOPMENT

Using the evidence which has been presented above, it is possible to develop an inductive model within which all of these observations can be incorporated.

The association between asymmetry development and the existence of R-segments of 5.0–8.9° upon the gentle slope would appear to be crucial. The evidence suggests that asymmetry develops in response to the increasing

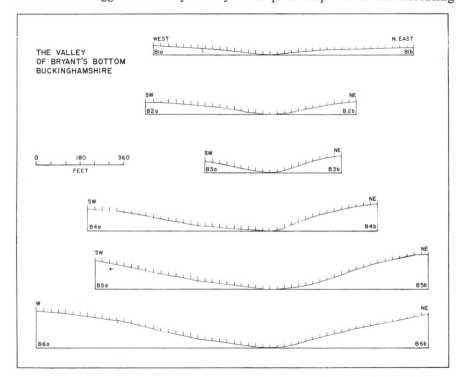

Fig. 5. Surveyed slope profiles: The Bryant's Bottom Valley, Buckinghamshire

extent of such slope angles upon the gentle slope. It seems probable that the major asymmetry forming process in the Chiltern operates most effectively upon the gentle slope at that angle. Thus, not only may 5.0–8.9° be a characteristic angle of the Chalk but it may also be a limiting angle for asymmetry processes.

It may therefore be possible to postulate a threshold angle, rather than a threshold depth, below which asymmetry does not develop. The evidence indicates that this angle is approximately 7–9°. On the Chalk, it would appear that a depth of encasement of approximately 40–50′ is necessary for the normal attainment of this angle. Upon reaching 7–9°, the asymmetry forming processes begin to operate upon the slopes; to smooth out, to lower slightly in angle, and to extend the gentle slope at this constant angle of 5.0–8.9°. With the continued downcutting of the basal channel asymmetry subsequently begins to develop. This is because the gentle slope extends itself constantly at 5.0–8.9° and there is, therefore, an inevitable lateral movement of the basal channel to the foot of the other, now steeper, slope. It must be emphasised that this latter slope begins to develop in response to two factors; firstly, the lateral movement of the basal channel to the foot of the slope, and secondly, the continued downcutting of the valley. Lateral movement and downcutting are thus synonymous at this stage of development. On the steeper slope, the R-segments extend downslope and the maximum angle, as well as increasing, does likewise.

Continued downcutting affects the steeper slope even more, however, and the R segments of this slope increase to perhaps, a maximum of 70% of the total slope length. At the same time a maximum angle is reached of about 19–22°. Contemporaneously, the gentle slope still extends itself at its constant angle and, as downcutting is still operating in the valley, the extension of this slope and the steepening of the SW facing slope result in a continued lateral movement of the basal channel. At about this stage, the maximum asymmetrical development will be reached for the steeper slope will have obtained its maximum angle. An A/I of between 2.0–3.0° will be typical at this point. With the development of this stage of maximum asymmetry, subsequent slope development will depend to a large extent upon the efficiency of the basal channel to cut downwards. It would appear that it is this criteria which differientiates the Little Hampden Valley from Bryant's Bottom, and thus the form of the Type I asymmetrical valleys from that of the Type II asymmetrical valleys.

There are many reasons why the downcutting of a valley may become less important. While many of them are connected with increasing valley dimensions (e.g. lowering of channel gradients, increasing proximity to base level) some are associated with decreases in the debris *production : removal* ratio. In the latter case, an increase in the volume of weathered material arriving at the basal channel will leave the stream with less available energy to erode, either laterally or vertically. Moreover, any increase in slope length as the valley deepens will provide more debris even if the slope angle remains constant. It may be, therefore, that as the steeper slope increases to 19–22° in angle, the amount of weathered material arriving at the foot of this slope becomes sufficient to reduce, and finally negate, any further downcutting of the basal channel. If this is the case, the asymmetrical valley form, at its maximum

Pl. 1. Chiltern Hills asymmetry, type I. The Callow Downs Valley, looking SE from Slough Hill

Pl. 2. Chiltern Hills asymmetry, type II. The Bryants' Bottom Valley, looking N

asymmetrical development, suffers modification. The gentle slope remains constant in angle and the basal channel does not move laterally in either direction since a state of adjustment or balance has been attained between the stream and the debris being supplied by the two slopes. The steeper slope, having attained its maximum angle, continues to retreat parallel to itself. This results in the development of a lower angled debris slope between the maximum angle and the basal channel (fig. 3).

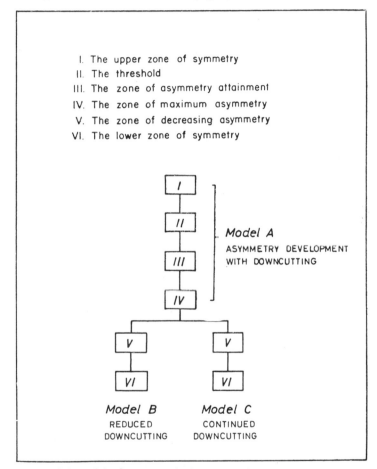

I.   The upper zone of symmetry
II. The threshold
III. The zone of asymmetry attainment
IV. The zone of maximum asymmetry
V.   The zone of decreasing asymmetry
VI. The lower zone of symmetry

Model A
ASYMMETRY DEVELOPMENT
WITH DOWNCUTTING

Model B
REDUCED
DOWNCUTTING

Model C
CONTINUED
DOWNCUTTING

Fig. 6. Sequential model of asymmetrical valley development in the Chiltern Hills

If, however, downcutting is not impeded to the extent which has been implied, the slope development is slightly different. Asymmetry decreases owing to an increase in the angle of the gentle slope and, perhaps, a slight decline of the steeper slope. It is suggested that this decline in angle of the steep slope is associated with impeded basal removal, for any further undercutting of the maximum angled steep slope (22°) will produce a totally disproportionate amount of debris to be removed from the base. Only an exceptionally large stream would have the power to do this and, at the same time, have excess energy available for lateral erosion. If downcutting continues, therefore, the gentle slope will begin to suffer modification and steepen in angle for the simple reason that it cannot extend itself at a constant angle because the position of the basal channel is relatively fixed. With the increase in angle of the gentle slope, the asymmetry decreases. Moreover, it may well be that, with the disappearance of the R segments of 5.0–8.9° the asymmetry-forming processes would also diminish in importance.

In the case of those asymmetrical valleys where downcutting is unimportant, and the steep slope is beginning to retreat parallel to itself, the asymmetry also decreases after a time and eventually disappears. The lack of any surveyed profiles at this point makes the following evolution rather conjectural, but it may be tentatively proposed that continued parallel retreat of the steep slope and a diminution in the extent of the maximum angle results, ultimately, in the 'steeper' slope being composed of a lower angled debris slope and/or footslope. The asymmetrical distribution of deposits within the large symmetrical valleys of the Chiltern Hills rules out the alternative possibilities of either increasing valley dimensions reducing crossvalley variations or increasing discharge acting to modify the slopes.

It is now possible to summarise the manner in which the normal asymmetry develops, maintains itself and then diminishes, by recognising a sequence of stages of downvalley development of an asymmetrical valley (fig. 6):

Stage I: The Upper Zone of Symmetry. In the shallow weakly incised heads of valleys where the slope angles have not reached the 'threshold' angles of 7–9°, there is a symmetrical development of both slopes of the valley.

Stage II: The 'Threshold'. This is reached when both slopes attain angles of 7–9°. This usually coincides with a depth of incasement of approximately 40–50′.

Stage III: The Zone of Asymmetry Attainment. After reaching the threshold angle, the gentle slope suffers very little modification, remaining constant in angle with R segments of 5.0–8.9°. Continued downcutting and a subsequent lateral movement of the basal channel to the foot of the steep slope promotes an increase in angle length and rectilinearity of that slope. Asymmetry, therefore, begins to appear in the cross-profile.

Stage IV: The Zone of Maximum Asymmetry. The steep slope reaches a maximum angle of 19–22° in response to downcutting and a lateral movement of the basal channel. The gentle slope has a uniform angle of 5.0––8.9°. The A/I reaches a maximum of 2.75–3.00. Asymmetry is maintained f (1) downcutting ceases or is reduced in importance, for the steep slope then begins to retreat parallel to itself producing a lower angled debris slope (e.g. Little Hampden) or (2) the basal channel is particularly strong in its erosive power to undercut the steep slope even further and to transport material away at the same time. This produces 24–25° meander bluffs' as the steeper

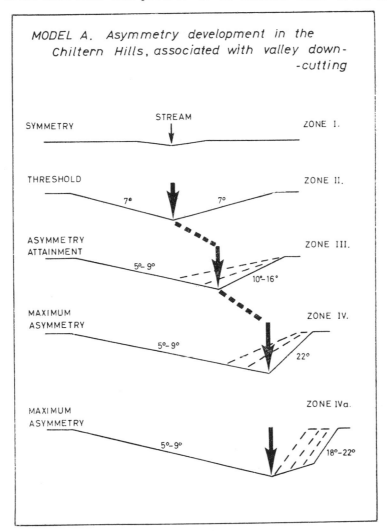

Fig 7. Model A. Asymmetry development contemporaneous to downcutting

slope. (This feature is not present in the Chiltern Hills but is found in Dorset where powerful scarp foot springs cross the dip-slope, e.g. the Devil's Brook, Dorset).

Stage V: The Zone of Decreasing Asymmetry. Asymmetry decreases owing to (1) the continued retreat of the maximum angle if downcutting has ceased or (2) an increase in the angle of the gentle slope if there is a continuance of downcutting (e.g. Bryant's Bottom).

Stage VI. The Lower Zone of Symmetry. Symmetry again becomes established in the valley (e.g. the large 'through' valleys). In these cases, the retreating steeper slope would have ultimately 'consumed' its free face. This

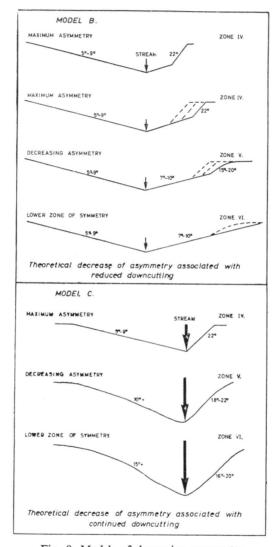

Fig. 8. Models of decreasing asymmetry

slope would consist of the debris and/or footslope components which would approximate in angle to the gentle slope.

There are certain aspects of this sequence which warrant further comment.

It would appear that, in many of the Chiltern valleys, slope development has proceded to the stage of M a x i m u m A s y m m e t r y. The development towards this stage is conceived of as model A (fig. 7). The Z o n e o f M a x i m u m A s y m-m e t r y reflects a change in the environmental slope conditions from those of downcutting and lateral migration of the stream to those of an equilibrium between the debris arriving from both slopes and the erosive powers of the basal stream. The manner in which asymmetry disappears suggests two further models. Asymmetry may decrease in one of two ways: (1) a continued retreat

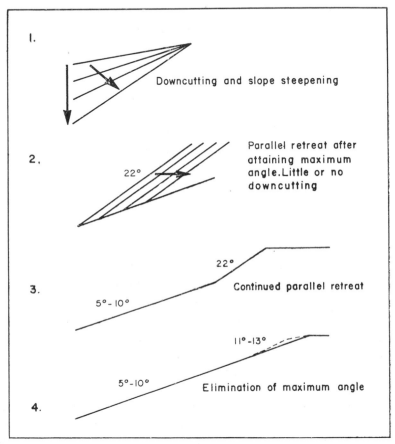

Fig 9. The evolution of the steeper west-facing slope in the Chiltern Hills

of the steep slope – Model B (fig. 8) and (2) a renewal of downcutting – Model C (fig. 8). In model B a broad shallow valley would be the end result which would be approximately symmetrical in cross-profile and in model C a deeply incised V-shaped valley would develop.

It may very well be that the form of the large 'through' valleys of the Chiltern Hills, such as the Gt. Missenden Valley with the symmetrical cross-

-profiles but asymmetrical pattern of soils and deposits is the end result of this slope sequence. How else, it must be argued, can an asymmetrical pattern of soils and deposits be present in both symmetrical and asymmetrical valleys? Thus, it can be visualised that the three rather different valley forms of, for example, Bryant's Bottom, Little Hampden and the Great Missenden Valley are reflections of but one sequence of asymmetrical slope development (fig. 9) which has become arrested and fossilised with the drying up of the valleys and the eradiction of conditions conducive to asymmetry formation. It is suggested that this sequence is applicable to those areas of the Chalk of Southern England which possess the 'normal', W-facing asymmetry. However, there are many instances where it is inapplicable. For example, the E-facing asymmetry of the valley NW of Brighton, the randomly orientated asymmetry of S. Hampshire, and the predominantly N-facing asymmetry of Salisbury Plain (French, 1967) are of a different nature. Finally, the rather unique, 'abnormal' asymmetry of Clatford Bottom, N. Wiltshire (Clark, *et al.*, 1967) and "The Valley of Stones" in S. Dorset finds no place within this model. To all of these, other explanations of a lithological, structural or different climatic nature are necessary (see French, 1967; Williams, 1968).

CONCLUSIONS

There are several observations of general relevance to periglacial problems which may be drawn from this study. Firstly, the morphologically more active slope is the SW-facing slope and it appears to undergo a process of parallel retreat as Ollier and Thomasson (1957) concluded. The recognition of the existence of debris slopes beneath the free face operating as slopes of transport for material from the free face is further confirmation of this mechanism. Secondly, the importance attached to the gentle slopes in the development of the asymmetrical valleys is considerable. Although little morphological modification takes place upon this slope, it would appear that the extension of this slope often at an angle of between 5–8° is instrumental in the lateral movement of the basal channel and the promotion of the steeping of the other slope. A 'threshold' angle of about 7° may very well be a valid concept for the initiation of asymmetrical valley development.

Thirdly, the importance of linear erosion and its relationship to the production of weathered material is basis to this model of asymmetry development. The efficacy of erosion by running water in periglacial regions has often been underestimated in the past. Peltier (1950), for example, does not give it the mportance which it deserves. However, recent work in the

Canadian Arctic is indicating that stream erosion is probably more effective in a periglacial region than in any other morphoclimatic region (St--Onge, 1964) since the stream flow is highly concentrated and there is an absence of a continuous vegetation cover (Robitaille, 1960). Running water in the valley bottom and solifluction and creep acting upon the gentle slope with freeze-thaw processes upon the steeper slope were the most likely combination of processes operating to, produce the asymmetry.

Finally, there are the implications of this study to the problematic origin of the dip-slope dry valleys of the Chalk. The interpretation given here to the existence of the 'normal' asymmetry implies that a surface water hypothesis may be valid (i.e. higher surface run-off in the past due to e.g. frozen sub-soil, Bull, 1940). However, many of these valleys are undoubtedly polycyclic, since different stages of evolution of the steeper slope are found in different valleys within the same area. Furthermore, their long profiles give evidence of successive rejuvenation (Culling, 1956). A surface water hypothesis is not completely adequate, therefore, and it is probable that the major dry valleys – les vallées sèches – were cut by stream action resulting from the existence of a higher water table at the time (i.e. a groundwater hypothesis, Fagg, 1954). In all probability, the valleys became successively drier from the headwater downwards, as the water table fell during the Pleistocene. If this were so, the downstream sections could have been formed and modified while upstream parts were dry and fossilised. The influence of the periglacial asymmetry-forming processes upon the Chalk landscape may therefore have been twofold; firstly, to modify the slope forms of already existing valleys and secondly, to initiate true periglacial valleys and gullies – les vallons en berceau – upon the frozen subsoils existing at that time. Such periglacial conditions probably recurred several times throughout the fluctuating climate of the Pleistocene and the Chalk landscape of today is, essentially, a reflection of the latest of these periods, preserved by the progressive desiccation of the Chalk.

ACKNOWLEDGEMENT

The author is indebted to Dr. R. J. Small, University of Southampton, and Mr. C. D. Ollier, University of Papua and New Guinea, for comments on a draft of this paper; however, they bear no responsability for the ideas expressed.

References

Bull, A. J., 1940 – Cold conditions and landforms in the South Downs. *Proceedings Geologists Association*, vol. 51; pp. 63–71.

Cavaillé, A., 1953 – Les vallées dissymétriques dans les pays de la moyenne Garonne. *Bull. Soc. Géogr., Comité Trav. et Scient.*, 1953; pp. 51–68.

Clark, M. J., Lewin, J., & Small, R. J., 1967 – The Sarsen stones of the Marlborough Downs and their geomorphological significance. *Southampton Research series in Geography*, 4; pp. 3–40. University of Southampton.

Culling, E. W. H., 1956 – The Upper Reaches of the Chiltern Valleys. *Proceedings Geologists Association*, vol. 67; pp. 346–68.

Czudek, T., 1964 – Periglacial slope development in the area of the Bohemian Massif in Northern Moravia. *Biuletyn Peryglacjalny*, No. 14; pp. 169–195.

Dylik, J., 1956 – Esquisse des problèmes périglaciaires en Pologne. *Biuletyn Peryglacjalny*, No. 4; pp. 57–71.

Fagg, C. C., 1954 – The coombes and embayments of the Chalk Escarpment. *Trans. Croy. Nat. Hist. Soc.* vol. 9; pp. 117–31.

French, H. M., 1967 – The asymmetrical nature of Chalk dry valleys in Southern England. Unpublished Ph. D. Dissertation, University of Southampton.

Geukens, F., 1947 – De asymmetric der drage dalen van Haspengower. *Natuurwet. Tidschr.*, 1, pp. 13–18.

Gloriod, A., & Tricart, J., 1952 – Etude statistique des vallés asymmétriques de la feuille St. Pol au 1 : 500000. *Rev. Géom. Dyn.* 3, pp. 88–98.

Grimbérieux, J., 1955 – Origine et asymmétrie des vallées sèches de Hesbaye. *Ann. Géol. Soc. de Belgique*, vol. 78; pp. 267–286.

Helbig, K., 1965 – Asymmetrische Eiszeittäler in Suddeutschland und Östereich. *Würzburger Geog. Arbeiten*, H. 14; 103 p.

Ollier, C. D., & Thomasson, A. J., 1957 – Asymmetrical valleys of the Chiltern Hills. *Geog. Jour.*, vol. 123; pp. 71–80.

Peltier, L. C., 1950 – The geographic cycle in periglacial regions as it is related to climatic geomorphology. *Ann. Assoc. Amer. Geog.*, vol. 40 (3); pp. 214–236.

St – Onge, D. A., 1964 – Géomorphologie de 1' Ile Ellef Ringes, T. du Nord-Ouest. *Etude Géographique*, 38; Ottawa.

Taillefer, F., 1944 – La disymmétrie des vallées gasconnes. *Rev. Géog. des Pyr. et du Sud-Ouest*, vol. 15, pp. 153–181.

Williams, R. G. B., 1968 – Periglacial erosion in Southern and Eastern England. *Biuletyn Peryglacjalny*, No. 17; pp. 311–35.

24

THE *GRÈZES LITÉES* OF CHARENTE

Y. Guillien

*This condensed translation was prepared expressly for this
Benchmark volume by Cuchlaine A. M. King, The University
of Nottingham, from "Les Grèzes litées de Charente,"
Rev. Géogr. Pyrénées Sud-Ouest, 22, 152–162 (1951)*

From the point of view of size distribution, the Vindelle deposit has an upper limit of gravel (2.5 cm); locally sand is characteristic, and *grèze* is also present. The particles of *grèzes* are not too small and bear indisputable marks of frost action; they show shattering, cracking, and wear that are not necessarily due to transport. The material is arranged in narrow beds that differ in size; the slope rarely exceeds 40 percent. The beds lie parallel and some are cellular, where fine material is lacking. These strata are called stratified *grèze,* or freeze–thaw deposits, and they lie at the base of long slopes in Vindelle. Others occur in midslope or at the head of nivation hollows. Most of the deposits are found at a break of the slope that they partially hide. They do not occur on slopes of less than 7 to 8 percent. In the larger occurrences the beds can exceed 1 m in thickness near the lower part of the slope, thinning upslope. In character the deposits differ from scree deposits of the same material. Scree deposits form by individual, free-falling particle accumulation, making sorting by size impossible. The *grèze* deposit is an exogenic material on the rock on which it rests.

Thin deposits of sandy material are laid down by wash over the slope mainly in winter. The deposits of *grèze* consist of alternating layers of material deposited by sheet wash or sheet flood, and *ruissellement en nappe* and *écoulement en lave,* a type of deposit similar to a lava flow. The distinction is between concentrated flow, and creep or wash over the whole slope.

Ground ice plays a part in the formation of *grèze litée.* The deposits are concentrated on southeast facing slopes and none occur facing northwest. This orientation is related to the pattern of snow accumulation. Each deposit of *grèze* corresponds to a semipermanent snowfield. Snow melt, thaw–freeze, and nivation processes are responsible for producing the material that makes the beds of *grèze* and for its deposition. The *grèze* started to form in the Mousterian. Accumulation took place in cold, dry phases when cryoturbation occurred and the ground was permanently frozen to a depth of 8 m. The features are an indication of a subarctic climate, not a cool or even a fairly cold one.

The fairly coarse, heterogenous strata occur on appreciable slopes, where rill wash carries away fine particles. The absence of gullying is due to permafrost or permanent snow cover. The formation of the cellular beds may be affected by snow pressure and melt water flow, and only operates on appreciable slopes, which are ordinarily 30 to 45 percent, although occasionally as low as 15 percent. Another process is needle-ice formation, which is more effective with coarser fragments, and which creates sorting at the upper limit of the snow.

Part III
SNOW ACTION

Editor's Comments
on Papers 25 Through 28

Snow plays an important part in the periglacial environment. Indeed, many of the processes considered in Part II are related to the action of snow or are facilitated by its presence. The interstitial ice of some rock glaciers, for example, probably was initially snow, and protalus ramparts owe their distinctive form essentially to the presence of semipermanent snow patches across which the blocks move to build the rampart. Snow, as it gradually melts, also helps to keep solifluction going for a longer period in the summer by releasing a continuous supply of moisture to the ground beneath it and the slope below it. Altiplanation terraces are also intimately linked with the presence of elongated snow patches, which are important in the widening of the terraces. Some forms of slope asymmetry have also been associated with uneven snowfall on either side of the valley. The part played by snow in the formation of stratified slope deposits is not so clear, but again snow has been suggested as one agent to be considered.

The processes associated with snow can be broadly divided into two major types according to whether the snow is lying passively on the surface or whether it is moving. The processes associated with passive snow are usually referred to as "nivation processes." The term "nivation" was originally introduced by

Matthes in 1900 in his study of the Big Horn Mountains of Wyoming. The processes associated with rapidly moving snow are mainly related to the downslope movement of snow in avalanches, which have been shown to produce recognizable geomorphic features of both erosion and deposition. At very low temperatures individual crystals of falling snow may be blown at considerable velocity by the wind. These sharp crystals, with a hardness of 6, can produce a large amount of erosion, especially on exposed, relatively soft rocks, such as some sandstones. The forms created in this way resemble those due to wind erosion by sand grains in arid areas; polishing of the rocks and etching away of the softer parts of the strata produce the effect described by the term "tafoni."

The processes associated with nivation will be considered first and then the forms attributed to avalanches will be described. Two papers have been selected to represent each type of snow process. Paper 25, by W. V. Lewis, on nivation in Iceland is a very valuable detailed study of the process in the field; Paper 26, by E. Watson, describes the features resulting from nivation from examples in north Wales. Paper 27, by A. Rapp, describes the avalanche boulder tongues of northern Scandinavia and is a pioneer study of these features; in Paper 28, N. Caine discusses the topic from a more quantitative point of view in his ideas concerning a model representation of the effect of avalanche on talus slopes.

NIVATION

The process of nivation depends on the nature of both the snow and of the ground on which it lies. The snow can be either thick or thin, and the ground can be either frozen at depth with permafrost or unfrozen. However, if the snow cover is prolonged and especially if it is thick, it will inhibit the formation of permafrost, because thick snow is a very good insulating material. It prevents the freeze–thaw cycles from penetrating into the ground very far.

Because of its insulating effect, the nivation processes associated with freezing and thawing of the snow will be most effective around the margins of the snow patch. The area of the snow patch, however, varies constantly during the melting season as it gradually diminishes in size. Thus a relatively large area will be covered by the thaw–freeze process, which is one of the most important associated with nivation. Even thick snow patches, which turn into relatively hard nevé near their base, may be traversed by tunnels created by flowing melt water; these may give access to the cold, and thus allow some thaw–freeze activity to take place beneath the snow.

Because thaw–freeze processes are the most important in connection with nivation, the process will operate most effectively in those areas where the thaw–freeze cycles are most numerous and effective in rapidity of temperature change. These will be in the high-altitude mountain areas, rather than the high-latitude polar areas of periglacial processes. The melt water produced by thawing snow may be sufficient in some circumstances to produce some erosion, especially where it can become concentrated, usually below the snow patch.

Chemical action has also been associated with melting snow by some authorities. Water near 0°C absorbs carbon dioxide and oxygen most efficiently, which suggests that snow meltwater may be particularly effective in the solution of limestone. The process is not probably effective on other than calcareous rocks, and is probably not generally nearly so important as the mechanical effects of the snow associated with thaw–freeze and the operation of the melt water in facilitating solifluction and other slope processes associated with mass movement.

Snow has been observed to move slowly across a slope, as it can creep as it undergoes modification and settlement. This slow movement of snow over bedrock has been observed by means of colored stones placed in marked positions beneath the snow. After the snow melted, the stones were found to have moved downslope; striations were also observed to have been formed by this process. Nevertheless, snow is generally not an effective agent of erosion itself unless it is moving rapidly. The nivation processes can, nevertheless, produce features of erosion, as well as facilitate the depositional forms already considered.

The major feature associated with nivational erosion are nivation hollows. At one end these form very shallow hollows on the hillside; at the other end of their range they grade into true cirque basins. True nivation hollows do not, however, have the reversed slope characteristic of a well-developed cirque basin, although it should be noted that not all true cirque basins necessarily have the reversed slope. A well-developed nivation hollow is a hemispherical hollow that varies in size, according to Lewis, from a few hundred meters to over 1 km across. Their form varies according to the rock structure, and Lewis recognized three types: (1) transverse hollows, which may initiate altiplanation terraces, are elongated parallel to the contours; (2) longitudinal hollows are elongated parallel to the maximum slope, and often develop in gullies in rocks in which vertical joints are well developed, whereas transverse hollows are more common on rocks with horizontal structures and varying resistance; and (3) circular hollows, which develop where the rocks are homogeneous. The longitudinal type may be modified by short, steep streams during the melt season, which may also link up with avalanche activity. Some of these features are probably modified water-eroded gullies, such as those occupied by the small niche glaciers described by Groom (1959) in Spitzbergen. In these funnel-shaped features the snow has thickened enough to flow as small glaciers.

The most typical nivation hollows are circular, as they develop independent of structure in the rocks and fluvial or avalanche processes. They develop best on fairly gentle slopes; there are many good examples on the flatter, high-level interfluves of the Front Range of the Colorado Rocky Mountains. Fossil forms can be seen in the British mountains, for example, on the Isle of Man and those described by Watson near Aberystwyth in Wales. Where permafrost is present, the nivational process can operate most effectively because the meltwater from the snow is confined to the upper active layer, thus increasing the water content and the movement of material.

AVALANCHE ACTIVITY

Avalanches are common phenomena in all mountain areas that have steep slopes and heavy snowfalls. An avalanche will take place when the weight of snow exceeds the frictional resistance holding it on the slope. Four factors influence avalanching: (1) the internal cohesion of the snow, (2) the thickness and density of the snow, (3) the character of the underlying surface, and (4) the slope. The steeper the slope, the thinner the snow need be to avalanche, although the character of the snow and its underlying surface play an important part. A number of classifications of avalanches have been suggested, depending on the nature of the breakaway, the position of the sliding surface, and the character of snow.

The only avalanches that are significant from the point of view of the landforms created are the "dirty" avalanches, which involve the whole snow layer that slides downward over the ground surface. These are also called "ground avalanches." They can break away from one point or in the form of a slab covering a considerable area, and they can consist of powder snow or con-solidated snow. The amount of water incorporated in the snow also plays an important part, and can be used to distinguish dry-snow avalanches from slush avalanches. The latter occur mainly in spring during thawing and, because of their great density and mass, can often be potent geomorphic agents. The character of the snow plays an important part in the initiation of avalanches. Newly fallen snow may undergo *constructive* metamorphism, whereby depth hoar, in the form of very unstable cup crystals, forms. This process renders the snow mass very unstable and avalanching is more likely. The *destructive* form of metamorphism involves the breakdown of the crystal structure, usually leading to the consolidation and stabilization of the snow mass.

Avalanche activity is related to the position of the snow line and the tree line in Alpine areas where mountain slopes are forested at lower levels. In the European Alps the tree line is at 1500 m and the snow line at about 3000 m; 4 percent (394) of the avalanches took place below the snow line, 28 percent (2632) between 1500 and 2000 m, 42 percent (3806) between 2000 and 2500 m, and 24 percent (2210) between 2500 and 3000 m. Only 326 or 3 percent took place above 3000 m. Spring is the most common time of the year for avalanching to occur. Dense forest is the best method of preventing avalanching. In the Alps avalanches tend to follow the same tracks year after year, and can thus in time pro-duce geomorphic forms. In the tundra areas of higher latitudes, avalanches tend to be small and more widespread, but again they tend to occur in the same places, which by a process of positive feedback become modified to produce specific landforms.

The landforms produced by avalanching include both ero-sional and depositional forms. The erosional forms occur high up on the hillside where the snow accumulates and begins to move down, collecting its load that is deposited on the lower slopes. The main form of avalanche erosion is the avalanche chute or gully; these steep, straight gullies usually have a rounded cross profile and are very common on the slopes of the arctic regions of high relief, such as eastern Baffin Island. The gullies as they grow larger accumulate more snow and are thus self-enhancing. They also tend to carry melt water during the summer. The term "rasskars" has been used by Ahlmann (1919) to describe these straight, rounded gullies leading down the arctic hill slopes. Simi-lar avalanche chutes also occur in the Alps, where they tend to be rather fewer and larger, owing to the greater concentration of avalanche into favorable sites.

The depositional features characteristic of avalanche activity include the avalanche boulder tongues described in Paper 27. Rapp has also assessed quantitatively the amount of material produced by avalanche activity on the open artic slopes of Kärkevagge in northern Sweden. One major slush avalanche in this area carried 200 m³ of rock material over a distance up to 350 m down the steep slopes, extending onto the flatter ground of the valley floor. Very large boulders can be carried by avalanches; one such boulder measuring 5 by 3 by 2 m was carried 120 m down a 5° slope. Such large boulders can do considerable damage on impact as they move rapidly downslope.

Large boulders on the hillside or valley floor can also result in another minor deposition form of interest. This is the avalanche debris tail, which is a tapered deposit collected in the shelter of a large boulder and forming a small straight ridge that can attain a length of 5 to 10 m. The tails are all parallel and elongated down the main feature. The features described by Rapp are avalanche boulder tongues. He distinguishes two main types, the road-bank type and the fan type. The former has a lobate front and flat top, one example being 330 m long from the mouth of the chute and 70 to 80 m broad, with a thickness of 5 m. The long profile flattens from 28 to 15°. Because of the great mobility of avalanches, the tongues can extent right to the valley floor, and even a short distance up the opposite slope; they, therefore, sometimes dam small lakes in the valley bottom. The greater distance avalanche fans extend down the slope differentiates them from talus fans, a distinction that is made in Paper 28 in the discussion of the effect of slush avalanches.

25

Reprinted from *Geog. Jour.*, **94**, 153–161 (1939)

SNOW-PATCH EROSION IN ICELAND

W. V. LEWIS

SNOW-PATCH erosion seems to be very important under peri-glacial conditions, and yet, as Bowman [1] has noted, there has been little mention of it. The subject was dealt with briefly after the Cambridge Iceland Expedition, 1932,[2] and in order to amplify the field-work carried out on that occasion a second visit was arranged for the summer of 1937. The site chosen was Snaefell, a Late Glacial volcano [3] with subsidiary cones, a few miles distant from the northern edge of Vatnajökull. We hoped to see much of the summer melting, and although Snaefell was not reached until mid-July, we were fortunate in that the season was a late one. Countless snow-patches lay within easy reach of the base camp; many were permanent, but most melted away during our stay. Snaefell rose 3500 feet above us, so that by climbing we could examine snow-patches in which the cycle of seasonal melting had not proceeded so far as with the lower ones. The snow-patches were rather more frequent on slopes facing north and west, but protection from the sun's rays seemed to be only one of the factors tending to preserve them, the chief factor being their bulk.

The processes involved in snow-patch erosion fall into the larger category of nivation phenomena, and the hollow in which the snow lies will here be referred to as the "nivation hollow."

The snow-patches in the vicinity of Snaefell can be divided into three classes. First transverse snow-patches whose major axes lie transverse to the lines of drainage, secondly longitudinal snow-patches elongated downhill, and thirdly circular snow-patches which, though transitional between the other two in ground plan, are frequently of far greater thickness. This last type appears to represent an initial form of cirque glacier.

Transverse snow-patches. The transverse snow-patches varied in length from 100 yards to a mile or more, and in width from 20 to several hundred yards. Near our camp was one which, though small, was typical. It had just subdivided when we arrived on July 16, and the major portion was then about 120 by 40 yards. It occupied the angle at the foot of the eastern slopes of a subsidiary ridge of Snaefell. In Plate 1, taken on July 31, it is much reduced in size, and it finally disappeared on August 19. The back slope of the nivation hollow was steeper than that of the mountain above. A section drawn from a line of levels run across the left side of the snow-patch on July 18 is shown on p. 154. The nature of the surface along this line was representative. The low ground was covered by coarse grasses, frequently in hummocks, and interspersed with dwarf willow where the ground was slightly drier. Near the snow-patch the grass cover ended abruptly and the surface dropped a foot or more in level. Superficial digging here showed that the subsoil was frozen,

1 Bowman, Isaiah, 'The Andes of southern Peru,' New York, 1916, p. 311.
2 Lewis, W. V., "Nivation, river grading, and shoreline development in south-east Iceland," *Geogr. J.* 88 (1936) 431–47.
3 Ahlmann, H. W., "Vatnajökull; scientific results of the Swedish-Icelandic investigations, 1936–37, ' *Geogr. Ann. Stockh.*, No. 39, 1937, h. 3–4, p. 168.

and this, together with the gentle gradient, largely accounted for the sodden nature of the top-soil. The latter consisted of stones and small boulders in a matrix of fine silt and clay. Water followed ill-defined channels through this material and was deflected by the vegetation front until it formed a stream large enough to maintain a course through this cover. Above the snow-patch the slope increased to a maximum gradient of 30° and the ground was again bare, but here it was much drier and consisted of boulders and stones in clay. The sudden decrease in grade above this zone corresponded to the change from bare soil to vegetation. The whole of the unvegetated zone was probably covered by the snow-patch in the spring.

Melting was greatest when rain was falling, but even under normal melting conditions tiny runnels flowed from underneath the snow-patch and contributed to the sodden state of the ground below. These small runnels constantly moved material up to the grade of a medium sand, even when the slope

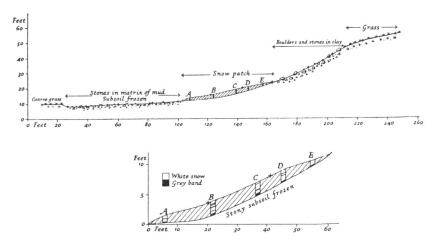

was gentle. During heavy rain, no runnels were visible on the ground free from snow-patch drainage.

On July 18 a series of pits were dug along the levelled line. The results are shown in the diagram above (lower portion). The vertical scale is double the horizontal to enable the thinner bands of grey snow to be represented. Heavy rain had fallen overnight which turned into intermittent mist and rain in the course of the day. Section E was dry to the bottom, the subsoil being frozen hard; it remained so for a quarter of an hour and then became damp. The ground below pit D was also dry and frozen hard when uncovered, but moisture began to form immediately. The ground below C was similarly frozen, but there was a lot of moisture passing over the surface, whereas below B the soil was covered by 3 inches of waterlogged grey snow. The bottom of the lowest pit A also had a layer of waterlogged snow, and like pit B had water passing over the surface of the ground when it was exposed. The pits were dug in the order A, B, C, so that when examined they were unaffected by any previous digging.

The steadily increasing dampness towards the foot of the snow-patch showed that the snow itself was the source of the moisture emerging from the foot of the patch. The waterlogged lower layer of snow resembled slush and the water both poured into the pits from this layer and seeped back into it on the lower side. The sections, when examined again two days later after more heavy rain had fallen, showed the pits greatly enlarged and the patch much reduced in size. The top pit E had now merged with the ground above and a tongue of clay and stones had sludged down into it. The next section D had a small runnel along the floor which was transporting clay and fine black sand, both of which were being carried under the snow on the lower side. Pit C contained a much smaller runnel, but this also removed both grades of material and several little pools had small debris-fans where runnels entered. Pit D contained a strong runnel entering a pool in which the coarse material formed a fan and the finer was deposited evenly over the floor. The bottom section, like the top one, had merged with the ground outside the snow-patch, and it contained a strong runnel. All grades of sand together with the clay were being carried by this little stream, and isolated pebbles of vesicular basalt and of tuff over a tenth of an inch in diameter were being moved intermittently. The presence of the pits did not greatly exaggerate the movement, for runnels as large as the one emerging from pit A, and similarly charged with material, occurred every 4 or 5 feet along the lower margin of the snow-patch. Sixty to eighty such runnels must have been moving a considerable amount of material from beneath the patch.

The mass of snow showed no signs of downhill movement, so that all the transport was done by melt-water. The runnels were clearly unable to erode or scour the rock surface; they merely transported material that had already been loosened and comminuted by other agencies. The very moist conditions of the lower pits could not have been permanent, otherwise at least the surface of the frozen subsoil would have melted, as it had beyond the snow-patch. Thus at night, during cold dry spells, the whole of the terrain below the snow probably froze hard. Such alternate melting and freezing would readily shatter the bedrock, the loosened fragments would themselves be broken up, and each shattering would yield a supply of fine material. The abundance of this fine material at the toes of snow-patches bore witness to the power of frost action to comminute material, a point stressed by Matthes.[1] Neither frost action nor the runnels of melt-water could alone excavate the nivation hollow. The rate of comminution of the underlying bedrock and the slow supply of material from above by downhill creep were evidently well able to keep pace with the removal by the runnels under the snow-patch, for the solid rock beneath the patch had an ample cover of predominantly fine material.

In this way nivation hollows probably recede a little farther into the hillside each season, increasing the height of the steep back slopes. Each winter a correspondingly greater thickness of snow accumulates as drifts banked up against the little cliffs, and so the excavation continues. The accumulation of snow in the form of drifts is important, for the processes of nivation can only last as long as the supply of moisture remains adequate, and this in turn depends on the volume of accumulated snow. It is clearly not a case of

[1] Matthes, F. E., "The glacial sculpture of the Bighorn Mountains, Wyoming," *Rep. U.S. geol. Surv.*, 1899–1900, Pt. II, p. 180.

unequally distributed snowfall, but merely of unequal capacity to collect drift-ing snow. Eventually the accumulation becomes sufficient for the snow-patch to last throughout the year. A second factor which tends to concentrate such action in patches is that the vegetation cover must be absent for active niva-tion. Small snow-patches aid the vegetation because they afford both a supply of moisture and insulation from excessive cold in the early spring, and the hollow gives some shelter from the wind.[1] But when the snow-patch is large anough to last into the summer it kills the vegetation and so accelerates nivation. At elevations greater than about 3000 feet the vegetation cover is absent, and so this factor becomes inoperative.

The snow-patch described above was one of the smaller ones on the eastern flanks of Snaefell. The lower slopes of the mountain within a few miles of the base camp faced east and rose at a gradient of about 15°. At intervals elongated nivation hollows containing snow-patches broke this uniform slope. Some were over half a mile in length and formed minor terraces on the hillside. The initial location of the snow-patches was probably determined by softer beds in the more or less horizontally bedded tuff which, with minor intercalations of vesicular basalt, formed the underlying rock. A typical patch occupying a nivation hollow of intermediate size, though still very immature in form, is shown in Plate 2, taken on August 17, when it was much reduced in size. Only three weeks earlier this patch was 600 yards long and reached to the top of the steep unvegetated slope. This terracing of a hillside, though seen here in what is clearly an initial stage, might assume considerable physiographical importance if it were carried many stages further.

The west-facing side of this ridge, which was of much less vertical extent, had only one major snow-patch, and this was continuous for nearly a mile. It was situated immediately below the crest of the ridge. Several other sub-sidiary peaks and ridges of Snaefell had similar snow-patches on their western sides. The reason for this location seems to be that the snow falling during an east wind is blown clear of the summit but comes to rest on the more sheltered slopes immediately below the crest of the ridge. This location for snow-drifts is well known to Alpine glaciologists.[2]

Longitudinal snow-patches. Most of the snow-patches in our area fell into this category. The upper slopes of Snaefell are furrowed on all sides by narrow gullies, each containing a snow-patch early in the season. Plate 3 shows the southern flanks of the mountain on August 14, when only the larger patches remained. As each gully contained a little stream it was not so easy to estimate the amount of snow-patch erosion. Later in the season the patches subdivided and the streams issuing from ice-caves exposed the ground beneath. These caves showed much the same features as those in the pits described above. The stream courses were boulder-strewn and the water carried some fine material. The contact zone between the snow or ice and solid ground on either side of the cave was generally damp, and muddy material frequently sludged downwards into the stream and was washed down by numerous runnels. The dampness increased rapidly towards the lower end

[1] Falk, P., "Plant ecology in the vicinity of Snaefell, E. Iceland." In course of pub-lication in the *Journal of Ecology*.

[2] Seligman, G., 'Snow structure and ski fields,' London, 1936, p. 210.

1. Subdivided transverse snow-patch, much reduced by melting, 31 July 1937

2. Transverse snow-patch in immature nivation hollow, 17 August 1937

3. Southern slopes of Snaefell with longitudinal snow-patches, 14 August 1937

318

4. Small circular snow-patch, 11 August 1937

5. Circular snow-patch and ice-cave, 11 August 1937

6. Steep slope at the head of a circular snow-patch, 20 August 1937

of each patch and more water dripped from the roof in the smaller than in the larger patches. Although dampness increased as the patches dwindled, the total amount of erosion accomplished each season must be greater for the larger patches, because the disintegrating processes of freezing and thawing and the washing out of the material continue for a longer period.

Several gorges, cut deeply into a mass of tuff on the eastern face of Snaefell, indicated that stream erosion can, under certain conditions, proceed more quickly than nivation. Some of these gorges, only a yard across, were more than 40 feet deep and half a mile in length. They were usually bridged by snow to within 5 or 10 feet of the top, but where the rock face was exposed above this level swirl marks and pot-holes, though much weathered, were still visible, showing that nivation had been negligible during the period of gorge cutting. These instances were however definitely exceptional.

An estimate of the efficacy of nivation in longitudinal snow-patches could best be made in examples which showed definite departures from normal stream gullies. These were usually of the type with terraced cross-sections opening wide at the top so that they collected much snow. Plate 3 shows, in the foreground, a well-developed minor cliff formed in the side of a little valley. The valley and the greater part of the snow-patch are hidden. The snow occupied a bench in the valley side and the cliff above the snow was steeper than the slopes bounding the stream itself. There can be little doubt that here the snow-patch was widening the valley by sapping the base of the cliff. The soft tuff was probably much subject to frost-shattering and the resulting material was removed by runnels into the stream below.

Before leaving the question of longitudinal snow-patches reference should be made to another way in which they contribute to the downhill movement of material. On the east face of Snaefell a series of steeply inclined snow-patches led up to the summit. At intermediate levels they carried long thin tongues of black debris. When followed upstream these tongues usually started at the sides or at junctions of patches, and were due to a form of solifluction phenomena which we witnessed on several occasions. During the height of summer melting, when the subsoil at the margins of snow-patches was still frozen, material sludged down on to the snow and continued over the surface until the moisture, which initiated the movement and then acted as a lubricant, melted its way down into the snow. Fuller reference has been made to this action elsewhere.[1]

Circular snow-patches. The circular snow-patch was not as frequent as the other two types, but it was often of much greater thickness. The small patch depicted in the left of Plate 4 was all that was left on August 11 of a patch which, in the middle of July, had filled the entire hollow and also covered most of the bare ground visible as a scar extending across to the right-hand side of the photograph. The rounded form of the hollow containing the snow was quite unrelated to stream action, which had formed only the two small gullies visible in the photograph. Matthes [2] attributes the rounding of niva- tion hollows to marginal melting. In Iceland any attenuated limb became

---

[1] Lewis, W. V., "Dirt-cones on the north margins of Vatnajökull, Iceland." In course of publication in the *Journal of Geomorphology*.

[2] Matthes, F. E., *op. cit.*, p. 181.

pinched off or melted away quicker than the snow of a margin with lesser curvature, for the surrounding rocks and bare drift, owing to their greater capacity for absorbing radiation, reach a far higher temperature than the snow surface. Dust also blew on to the edge of the patch, and this further accelerated melting by acting as a heat absorbent. Thus it might well be that, in the absence of any other factor tending to determine the shape of the patch, the circular form, having the least periphery per unit area, will result. In the patch shown in Plate 4 the bare rock is visible at two points on the steep back slope and in neither instance is the exposure due to stream action. At the rock exposures the speed of removal was evidently in excess of the rate of comminution by freeze-thaw processes.

The largest permanent snow-patch near the camp lay in a ravine at an elevation of 2500 feet. It was well suited for study because a cave extended beneath the snow-patch for about 50 yards. The patch is visible in the top left-hand corner of Plate 1, and Plate 5 shows it much nearer on August 11. At the beginning of our visit the snow encompassed the cliffs on the right and extended beyond the limits of the photograph on the left. It thus considerably overlapped the ravine. The main patch was about 350 yards by 200 yards, and judging from the inclination of the floor of the cave and the surface of the patch the maximum thickness must have been quite 50 feet.

Where the snow spread on to the hillside and drained away from the ravine, conditions approximated to those at the transverse snow-patch first described. The surface was saturated and a considerable amount of fine material was being carried away. Far the greater part of the melt-water however flowed into the main stream. This stream entered the snow-patch from above, and derived its water from a number of tributary longitudinal patches. Just inside the cave the stream had cut a little gorge below the general level of the base of the snow-patch; beyond, the cave widened to 30 feet. The gorge was cut in the harder bands of tuff and lava which formed the cliffs on either side of the entrance. Still farther in we were stopped by a rock barrier down which the stream cascaded. The temperatures of the air and water on July 18 were as follows:

Outside air temperature	56° F.
Air temperature in the entrance	48
Air temperature 50 yards from the entrance	48
Temperature of water dripping from the roof ..	34
Stream temperature in the cave	36

Water streamed from the hundreds of little cusps in the roof. The latter consisted of hard ice produced by the freezing of melt-water, for the thickness of the snow was not nearly sufficient for the transformation into ice by pressure. Near the gorge the ice rested on bare rock, but usually there was an intermediate weathered layer of coarse angular stones, silt, and clay, the stones being frozen into the finer matrix to a depth of 6 inches below the surface of contact with the ice. Below this level, where the material was exposed to the air, it was unfrozen.

By August 11 the much-diminished snow-patch seemed damper underneath and the sides of the cave were covered in places with a thin layer of liquid mud. Yet the total amount of water in the stream was less than previously. The basal layers of ice were still frozen to the loose material beneath in the drier places, though the moisture seemed more widely distributed than on earlier occasions when the only really wet parts were below the little circular melt-holes which pierced the roof. These varied from a quarter to more than an inch in diameter and usually carried a trickle of water. By August 20 melting had revealed the back wall to the extent shown in Plate 6. The abrupt fall of the ground at the beginning of the snow-patch is well shown. The stream itself is sculpturing only a narrow zone of this cliff and the latter is steeper where it is free from stream action. Thus the snow-patch seems to be actively sapping this headwall. Radiation from the headwall had caused the snow to recede and so to augment the supply of melt-water for frost action.

Thus wherever the contact between the snow-patch and the underlying material was examined there were signs of erosion, so that the patch appeared to be incising itself ever more deeply into the hillside. Although the adjacent slopes were steep and the lower layers of ice, no thrust planes or other signs of active downhill movement were seen. The basal layers were frozen hard to the ground except where water or radiation had caused melting along the surface of contact.

Melt-water in the cave increased as the patch became thinner near the entrance. This was due to some of the surface melt-water not reaching the ground below where the patch was thick, probably because it was held up on meeting the ice layer. Here any moisture that refreezes would help to thicken this layer and the remainder found its way to the bottom by percolation and through melt-holes. Consequently most moisture reached the floor near the edges of snow-patches and where the patches were thin. Most material is thus washed out from beneath the edges, particularly those on the lower side. Here also the severity of frost-action is probably greatest as the oscillations of the temperature across the freezing-point reach the underlying material, partly by conduction through the thin snow cover, and largely through the agency of the numerous melt-water passages. This zone of maximum destruction would recede with the receding margins as envisaged by Matthes.[1] The ice-cave allowed temperature changes to reach many parts of the floor which otherwise might suffer far less frequent frost-shattering. The severity of frost-shattering was probably greatest in spring when day and night temperatures regularly crossed the freezing-point, for in July and August the night temperatures rarely fell below freezing-point at the base camp 400 feet below. Higher up Snaefell the night temperatures fell below 0° C. both in July and August.

In a snow-patch of this type with a large stream running through it, full allowance must be made for the work done by the stream itself. It quickly removes the fine material carried into it by runnels of melt-water, and in addition does a considerable amount of erosion, such as gorge cutting, through the normal agencies of stream sculpture.

This snow-patch was present late in the summer of 1936 when one member of the 1937 party saw it, and in July 1935, as shown in a photograph taken by

[1] Matthes, F. E., *op. cit.*

S. Thorarinsson.[1] The limit of melting in 1936 is probably represented by the top dust-band a few feet above the heavy black line over the entrance of the cave shown in Plate 5. The heavy band seems to represent the concentration of dust resulting from several seasons' melting. The sagging of this band over the entrance suggests a certain amount of differential movement in the unsupported snow and ice. Were it not for this movement the size of the cave would increase by melting each summer until the roof would collapse.

On the east face of Snaefell were two steep valleys at about 4300 feet above sea-level. Both had well-marked steps formed by horizontal basalt flows below which were large snow-patches. They had already developed the half-funnel shape so characteristic of large nivation hollows, but were clearly in the transitional stage between nivation hollows and cirques, for the *névé* in one case had begun to move downhill forming a fine series of arcuate crevasses where the surface gradient was about 15°. At the lower limit of this snow-patch there was no sign of movement or of end-moraines; it fed the stream which had cut the deep narrow gorge mentioned above. This example shows that the transition from a large snow-patch to a small cirque glacier might be a very gradual one. The most satisfactory distinguishing feature is that of movement. The presence of ice in the lower layers can obviously not be taken as the criterion. Whether or not the crevasses were due entirely to differential movement within the mass or in part to movement of the mass over the ground beneath could not be determined. Their arcuate form, their parallelism, and their V-section suggested that they were due in the main to differential movement within the mass. When, as in the case of a glacier proper, a composite mass of snow and ice begins to move over the rock floor a new set of sculpturing agents is introduced which lies outside the scope of this paper.[2]

A fine series of funnel-shaped hollows or embryo cirques occurred high up on the southern flanks of Snaefell. A particularly massive series of horizontal basalt flows forms a terrace and scarp, seen on the sky-line in Plate 3, at an elevation of about 5000 feet. The largest nivation hollow is in the centre of the photograph and two minor ones are seen just to the left. The largest snow-patch seemed to be semi-permanent, judging from the exposure of dirty *névé* which rapidly increased in area towards the end of August. There was a marked rounding of the contour of this nivation hollow near the crest. The waterfalls pouring down the face of the cliff from the summit ice-cap have barely notched the cliff, so that it must recede by nivation as quickly as the stream channels do by waterfall recession, a clear indication of the potency of nivation processes at this altitude. The rounding of the contours of semi-permanent and seasonal snow-patches has been dealt with elsewhere.[3] Briefly, in a snow-patch the removal of material down the back slope takes place at an approximately uniform rate. This movement converges on the apex of the snow-patch, from which point it is removed by the stream. Hence the slopes

[1] Ahlmann, H. W., *op. cit.*, p. 166.

[2] See Lewis, W. V., "A melt-water hypothesis of cirque formation," *Geol. Mag.* 75 (1938) 249. Also Lewis, W. V., "The function of melt-water in cirque formation." In course of publication in the *Geographical Review*.

[3] Lewis, W. V., *op. cit.* (1936), p. 434.

converging on this point will tend to be uniform and so a rounded contour will result. This contrasts with a normal stream-head below the spring line, where the slope is least along the bed of the stream and more or less equivalent to the angle of rest on either side, thus giving the contours a V-pattern.

Thus in Iceland nivation hollows different in form result from the three classes of snow-patches. The transverse snow-patches, whose shape is generally due to horizontal outcrops of varying resistance or to the junction of the hill-slopes with the flood plain, give rise to elongated nivation hollows which terrace the hillsides. The longitudinal snow-patches follow drainage channels and sometimes modify the cross-section of the stream gullies. The circular snow-patches which form in slight concavities, or when horizontal outcrops cross stream channels, are perhaps of greatest physiographical importance because they give rise to the cirque which is one of the dominant forms of mountain glaciation.

Much of the field-work for this paper was carried out on the Cambridge Expedition to East Iceland, 1937. Grants towards the cost of this expedition were received from the Royal Society, the Royal Geographical Society, and the Worts Fund of Cambridge University. The members of the expedition wish to record their indebtedness for this financial assistance.

26

Reprinted from *Biul. Peryglac.*, No. 15, 85–98 (1966)

TWO NIVATION CIRQUES NEAR ABERYSTWYTH, WALES

Edward Watson *

Aberystwyth

[*Editor's Note:* In the original, material precedes this excerpt.]

CWM DU

Cwm Du has no moraine. Its basin is fronted by a drift scarp, 60 feet (18 metres) high in the centre, which resembles a moraine when viewed from below, but in reality is the front of a drift terrace. The main stream in the cirque has cut a deep gully parallel to this front but instrumental levelling in the area of the profile JK (fig. 5), shows that the top of this scarp is lower than the cirque floor. The gully shows up to 60 feet (18 metres) of head without the bedrock being exposed and the smooth slope rising up to the backwall suggests that this is a solifluction deposit, accumulated beneath a snow patch.

The long axis of the cirque basin is southwest to northeast and its floor falls in the same direction. The snow patch would have sloped from the southwest corner towards the present stream exit. The angle of elevation from the front of the drift platform at the stream exit to the top of tde backwall in its southwest corner is 13°. The upper limit of the snow patch would be lower than the top of the backwall so that the surface of the snow would

[*Editor's Note:* Certain figures have been omitted owing to limitations of space.]

---

* Department of Geography and Anthropology, University College of Wales, Aberystwyth.

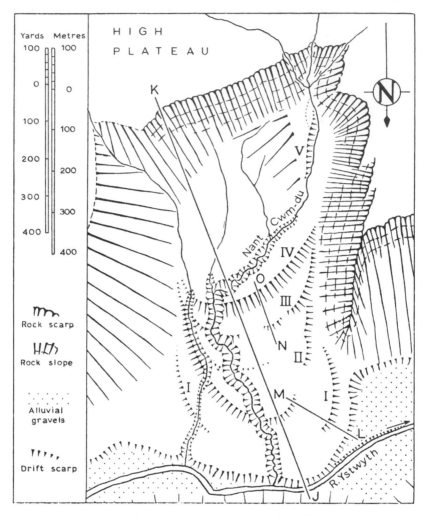

Fig. 5. Cwm Du and its associated fan

The scarps on the latter are numbered I, II and III. Between II and III are traces of another, (IIa on fig. 10) but it is less distinct than the others. IV is the scarp at the mouth of the cirque and V the protalus in its south-west corner. JK, LM and NO show the lines of the profiles on fig. 8. Rock outcrops in the bed of the Ystwyth at J and some 250 yards to east

be less steep than this angle but it is suggested that this reading gives a basis of comparison with other cirques. The absence of a protalus suggests that no superficial debris reached the foot of the gently-sloping snow surface in Cwm Du and that conditions may have resembled those shown in, Botch's block-diagram (1946, p. 221, fig. 7).

The smooth curve of the floor of the lower part of the cirque is replaced towards the head by a drift accumulation which fills the southwest corner of the cirque (see fig. 5 and V, fig. 6). The streams falling steeply into the cirque have filled in the area behind this accumulation at its southern end with bouldery alluvium but downstream (at profile RS, fig. 6) it is seen to be composed of head. This suggests that after the snow patch had disappeared from the cirque, it re-formed, filling only the southwest corner and built up this drift as a protalus. The angle of elevation from the top of this protalus ridge to the rim of the backwall on profile RS of figure 6 is

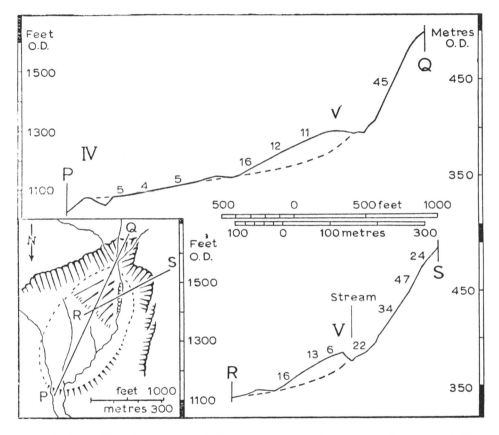

Fig. 6. Cwm Du. Profiles through Cwm Du (cirque only) along the lines PQ and RS on inset map

The Roman numerals IV and V correspond to those on fig. 5. The broken lines show a restoration of the cirque floor at stage IV. The vertical scale is twice the horizontal. The numbers along the profile indicate the slope in degrees

27°, indicating that this final snow patch belonged to the steeply-sloping class, like that in Cwm Tinwen where the comparable angle along profile AB of figure 3 is 25°.

The drifts in Cwm Du cirque as in its associated fan and in Cwm Tinwen, consist of two types of head. One is a tough bluish-grey silty deposit charged with angular and sub-angular rock fragments of all sizes from fine gravel to great boulders. The other is yellowish-grey, loose and charged with similar debris but with a smaller proportion of fines so that it may often be described as a muddy angular gravel. It often shows rusty mottling and in the more open beds manganese staining.

In Cwm Du cirque, exposures in Nant Cwm-du gully (50—60 feet; 15—18 metres deep where it leaves the cirque) suggest that these beds form the whole of the deposits, occurring in distinct layers, between 1 and

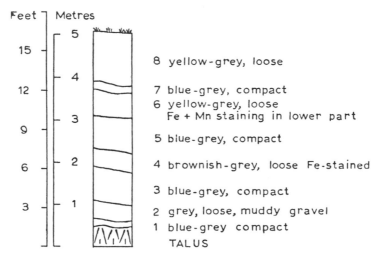

Fig. 7. The sequence of head deposits in Nant Cwm-du gully, 120 metres SW of the point where JK crosses it. The scale begins at stream level

3 feet (0.3 to 1 metre) thick. The bluish-grey type is typical of the soli-fluction deposits on the greywackes and mudstones of the region and the yellowish-grey type is probably basically the same except that it has suffered some degree of washing during deposition, the yellowish tinge being due to post-glacial weathering accompanying a more ready percolation of water (fig. 7).

The protalus, V, has many boulders scattered over its outer slope, and exposures show these two types of head; on the profile line RS of figure 6, 11 feet (3.3 metres) of the loose yellowish-grey type on 2 feet (0.6 metres) of the compact bluish-grey type, above 17 feet (5.2 metres) of talus.

The exposures of the cirque floor (stage IV) suggest that the drift may have been built up in layers as the theory of Botch states. It must be noted that these exposures occur on the outer edge of the cirque and may represent deposits laid down at the margin of the snow patch. This might account for the increased washing of some of them — in this connection, it may be recalled that Botch found "natural pavements" in the area uncovered by the snow in summer, due to the washing away of the fines by melt water (fig. 9).

The building up of such a platform of drift would be due to the fact that the ground below the snow patch is affected by summer thaw only to a very shallow depth and that as the debris derived from the weathering of the backwall moves as a solifluction layer beneath the snow patch it tends to thicken as the gradient lessens. In winter this muddy accretion is frozen to the permafrost below and never thaws out again fully so that summer after summer it is added to. The abrupt limit to the deposit, represented by the scarp 60 feet (18 metres) high, seems to imply that the actual front of the snow patch fluctuated relatively little, otherwise the thawing out of the uncovered ground must have been accompanied by soil flow.

Little data is available on the thickness at which névé becomes ice, but R. F. Flint believes it to be "30 metres or more" (1957, p. 19). J. Tricart accepts this figure — "30 or 40 metres" (1963, p. 157). For slow plastic deformation — the more significant change in this context — "the minimum thickness of ice and firn required is not known... but thought to be 30 or 60 metres in a temperate climate and greater in a polar climate" (Flint 1957). At stage V in Cwm Du, if the snow extended to the top of the backwall, its maximum thickness would be 80 to 100 feet (25 to 30 metres), so that it probably was entirely a mass of snow. The same applies to the younger (protalus) stage in Cwm Tinwen where snow to the top of the backwall would give a maximum thickness of 125 feet (40 metres). The névé mass extending to the outer edge of the cirque in Cwm Du (IV on figure 6) would, if the snow extended to the top of the backwall have a maximum thickness of 230 feet (70 metres), while if it extended half-way up, the figure would be 150 feet (45 metres). In this case, the lower layers of this extensive "snow patch" (1,800 feet or 550 metres in length), may have been ice.

CWM DU FAN

The drift scarp enclosing the cirque basin (IV, fig. 5), appears to be the highest and most continuous member of a series which is developed across the surface of the drift fan below it. This series, marked I, II and III on figure 8 gives the fan a stepped profile. These steps or scarps are not continuous across the fan so that profile JK does not give the complete series; profiles LM and NO restore the missing steps in the general picture. The upper step of the series, III, is parallel to the terrace front of the cirque, IV, and like it falls in elevation towards the east. Unlike IV, it appears to have been worn down at two points as if erosion had lowered the top of the scarp. Scarp II extends as a continuous scarp to the east of the Nant Cwm-du gully except where it is cut by the unnamed stream. West of this stream it appears to have been eroded as in the case of scarp III. Scarp I exists only on the flanks of the fan; in the central area the fan gives the appearance of having been built up so that there is a continuous slope from scarp II to the limiting bluffs of the fan, which are due to erosion by the River Ystwyth.

S. G. Botch in his block-diagram of the features associated with a typical snow patch (1946, p. 221, fig. 7) and in his diagram showing the evolution of a nivation niche (1946, p. 219, fig. 6) shows solifluction debris accumulating underneath the lower part of the snow patch and ·moving on sub-aerially down the slope as a series of solifluction terraces. The similarity between these two diagrams and the situation at Cwm Du suggests that the steps of the fan may be solifluction terraces (*cf.* fig. 9), but the author has come to believe that these steps — I, II and III — are similar in origin to scarp IV at the mouth of the cirque and that they mark earlier snow patch limits.

One argument against the view that the steps are solifluction terraces is their arrangement in plan. They do not form a series concentric with scarp IV but appear to enclose a series of lobes which have a progressively changing axis. The lobe enclosed by step II has an axis running west of north (about 350°); the axis of the higher lobes swings clockwise until in the cirque (the area enclosed by step IV) it runs north-northeast (about 25°). This would be consistent with an area of snow emerging from a gully which was being extended towards the south and west as freeze—thaw made its maximum attack on the backwall on these sides (fig. 10).

The position of the stream, Nant Cwm-du, on the centre line of a convex fan — not of its own construction but predominantly of head — also favours the view that the steps on the fan mark snow patch limits. Inside the cirque, Nant Cwm-du and the stream to the east occupy positions which would have been sub-marginal to the snow patch of stage IV, and leave it at its lowest point, where one would expect melt water to escape. The present stream appears to be a direct descendant of such a melt water stream, having become entrenched in later times. Downstream of step IV, the stream passes through the lowest point of each crescentic step so that

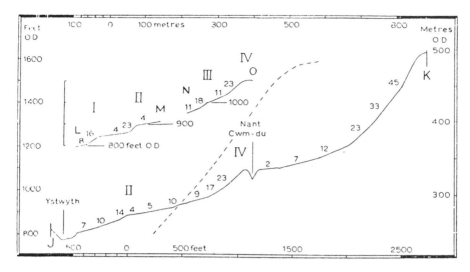

Fig. 8. Profile through Cwm Du and its fan along the line JK on fig. 5 with supplementary profiles along LM and NO

The Roman numerals I to IV correspond to the scarps on fig. 5. The broken line shows the profile of the Ystwyth valley side, obtained by continuing the contours of the valley side across the mouth of the cwm. The vertical scale is twice the horizontal

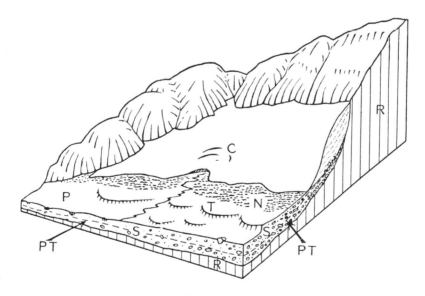

Fig. 9. Block diagram of a snow patch during the melt season, after S. G. Botch (1946, fig. 7)

C — crevasses in the névé above a melt water cave; N — natural pavement; T — solifluction terraces; P — patterned ground (formed by autumn frosts); S — solifluction debris; R — bed-rock; PT — permafrost table, rising quickly at the base of the névé

the stream course cuts across the fan to a central position where it joins the Ystwyth (fig. 10).

These arguments are strengthened by a consideration of the tributary valley on the south-facing side of the Ystwyth valley northwest of Cwm Tinwen (see fig. 1). This is drained by two streams and the area of their

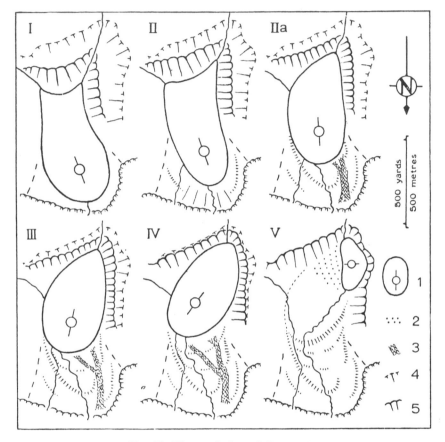

Fig. 10. The evolution of Cwm Du

1. the snow patch showing the axis of the lower part; 2. the protalus of stage V; 3. melt water courses on the fan; 4. existing backwall; 5. the backwall at successive stages. Step IIa (between II and III on fig. 5), is shown here as marking a snow limit; if these steps do represent climatic pauses, research elsewhere in Wales may help to decide the status of IIa

basin would be comparable with that of Cwm Du. Snow must have collected in it during the winter when Cwm Tinwen had a permanent snow patch, but it appears to have melted during the summer as there is no rocky backwall and no clearly defined floor comparable, with Cwm Du. The basin at Blaen-y-cwm is filled with solifluction debris having a surface slope of

7°—9° (fig. 12). This slope is interrupted by low terraces which have lobate fronts and are nowhere so imposing as the steps on Cwm Du fan. Furthermore, although this infilling has not a convex surface, the two streams are on its outer edges flowing between rock slopes and drift scarps 30—40 feet (10—12 metres) high (fig. 11). Both this basin and Cwm Du are

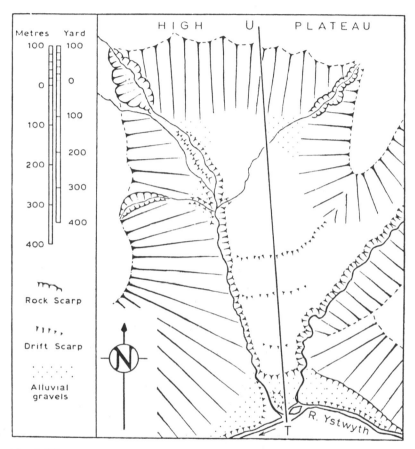

Fig. 11. Solifluction deposits in the valley north-north-west of Blaen-y-cwm. These form a wedge-shaped terraced slope between the two streams

developed on the same rock type; the difference lies in their orientation. It is suggested that at Blaen-y-cwm we have a slope developed by sub-aerial solifluction without a permanent snow patch.

Figure 13 is an attempt to reconstruct the evolution of Cwm Du and its fan by advancing the backwall of the cirque to compensate for the building of the fan, along the profile JK of figure 8. One of the difficulties here is the fact that the cirque was not extending along the axis of the fan. Only

a fraction of the material laid down at stage IV came from the backwall
on the profile JK; the bulk of it came from the southwest corner of the
cirque. This is less true of the earlier stages, but the reconstruction in
any case is only very approximate. The "initial" profile shows a nick-point
just above 900 feet, as does the Ystwyth and several of its tributaries in
the area (Brown 1952). Though no rock is seen in the stream bed, the
Nant Cwm-du gully is shallowest at this point, only 25 feet (8 metres)

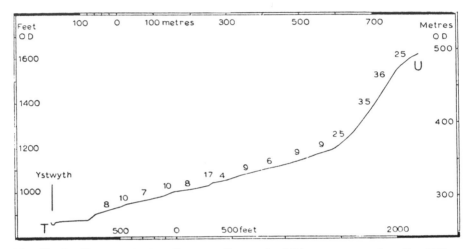

Fig. 12. Profile through the tributary valley north of Blaen-y-cwm, along line TU on
fig. 11. The numbers along the profile show the slope in degrees

deep compared with more than 50 feet (16 metres) upstream and downstream of this, and the steepest stretch of the present stream profile in the
fan occurs just below this.

From the elevation of the High Plateau here and the position of scarps I,
II and III, it seems impossible that a snow patch extending to them, could
fall into the steeply-sloping class. On the reconstruction shown in figure 13,
the angles from the rim of the backwall are, for scarps I, II and III similar
to that for scarp IV, which is 18°. This is in harmony with the fact that
each suggested snow limit is marked by a terrace front and not a protalus
rampart.

When one tries to estimate the thickness of snow at these stages on the
basis of figure 13, the maximum thickness, if the cirque were filled to the
top of the backwall, would be 250 feet (82 metres), 250 feet, and 220 feet
(72 metres) respectively for stages I, II and III (cf. 230 feet, 75 metres,
for stage IV). In this respect, the problem of stages I, II and III is the same
as that of stage IV. One returns to the question of the thickness reached
by snow and ice before it begins to behave as a cirque glacier.

The main exposures of the deposits making up Cwm Du fan are shown on figure 14. The interbedding of tough bluish-grey head and loose yellowish-grey head seen in the floor of the cirque also occurs in the fan just behind scarp II at exposures 2 and 3. At other points, 36 feet (11 metres),

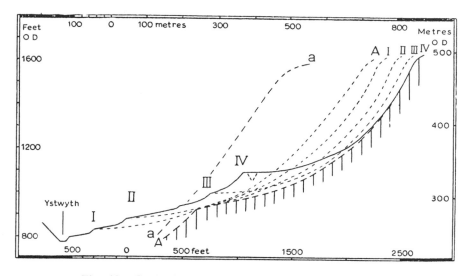

Fig. 13. Stages in the development of Cwm Du profile

The continuous line shows a composite profile along JK of fig. 5. The vertical lines show the bed-rock, in part hypothetical. A — the initial profile of the gully; I—IV — the successive portions of the backwall corresponding to the steps on the fan; a — a — the profile of the main valley side

upstream of exposure 3, the west side of the gully shows only the blue-grey head and at exposure 5 only the blue-grey head is seen for 23 feet (7 metres) above the stream.

In front of scarp II, small calibre waterlaid gravels (less than 2 inches or 5 cms long) interbedded with thin layers of grey silt at the top, from the lower 9 feet (3 metres) of exposure I. Again, in front of scarp III, exposure 4 shows blue-grey head overlain by 7 feet (2 metres) of similar small gravels capped by 1.5 feet (0.5 metres) of sand and silt, on top of which is blue-grey head. In both cases, these water-laid beds might be older than the step behind, having been laid down during a recession phase when summer melting was more pronounced, and then been overwhelmed by the solifluction deposits of the snow patch of the succeeding rigorous phase. They might also represent melt-water deposits laid down while the step behind was being built up. The former interpretation is favoured by the sequence in front of scarp IV, where exposure 5 shows 2 feet (0.7 metres) of waterlaid sands, gravels and silt resting on the blue-grey head of step III.

These are overlain by a stony bouldery yellow-grey head (in places a muddy gravel) which thickens when followed upstream (fig. 14), and would appear to be material that has been partly washed by snow melt as it sludged down the face of scarp IV. At exposure 4, the top of the

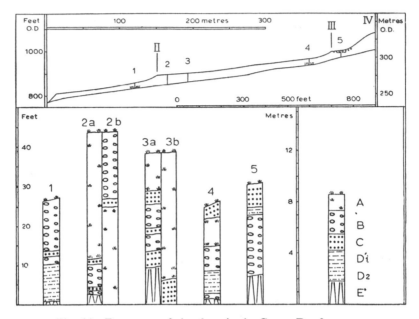

Fig. 14. Exposures of the deposits in Cwm Du fan

A — vegetation-covered slope; B — the blue-grey head; C — the yellow-grey head; D_1 — waterlaid silts and sand; D_2 — waterlaid gravels; E — talus. The sites of the numbered exposures are shown by vertical lines on the profile which represents the west bank of the gully projected on to profile line JK (fig. 5). The profile also shows the position of the waterlaid beds of exposures 1, 4 and 5, those at 5 being overlain by poorly washed yellow-grey head

sequence also shows a gravelly head, roughly stratified parallel to the present surface, showing similar conditions at stage III to those of stage IV.

With this escape of melt water from the snow limit of stage IV, may be associated the fluting of the face of scarp IV suggesting wide shallow gutters leading down to step III. As shown on figure 10, this drainage probably escaped by a shallow channel which breaches scarp II. It may be pointed out that this drainage and the deformation of the scarps is developed on the western, "warmer", side of the fan; the protalus structures of Cwm Tinwen are similarly best preserved on the eastern side, while the western shows considerable deformation.

The exposures in Nant Cwm-du gully show that the material in the fan has been laid down in several stages and that the steps are not a series of terraces formed contemporaneously with the build-up of the platform of head at stage IV. The water-laid gravels and silts seen in three places each one a short distance from a scarp suggests that the building of the steps may have followed on milder climatic interludes.

CONCLUSIONS

S. G. Botch recognized two types of snow patch. The first, the steeply-sloping has been described in North America and Europe, but the second type, the gently-sloping, has not received such widespread recognition. In many parts of upland Wales, some hundreds of feet below the elevation of cirques showing the features of glacial erosion, the protalus ramparts of the first are found, close to steep headwalls. At similar elevations more open cirque-like basins with their terraced superficial deposits are believed by the author to be due to the second.

Cwm Tinwen represents the first type. Its very narrow basin helps to distinguish it from the typical glacial cirque (*cf.* Sharp 1942, p. 496). The estimated thickness of névé, allowing for the fact that it would be considerably less than full to the rim, is less than that generally believed necessary to form a cirque glacier.

Cwm Du, which falls into the second class, has the basin form and steep headwall of a glacial cirque but the existence of a platform of head, 20 metres high on the cirque floor with a surface slope similar to that of solifluction slopes on the same rocks in the Aberystwyth region seems to indicate the operation of a solifluction process beneath an inert mass of névé as suggested by Botch. But the accumulation of head in Cwm Du appears to have been on a much greater scale than that indicated by Botch. Two factors may have operated here. One is the rock type. In this area, the bed rock is the Aberystwyth Grits, a series of greywackes of Silurian age. Under frost weathering, the grits form joint blocks but the well-cleaved mudstones are rapidly broken down to silt so that the whole is very susceptible to solifluction. This undoubtedly influenced the rate of accumulation in Cwm Du and also accounts for the composition of the protalus in Cwm Tinwen which has a greater proportion of silt, grit and small gravel than is often the case. The second factor is the size of Cwm Du; indeed, it may be too large for the accumulation of snow alone. Any reasonable estimate of the thickness of snow at stage IV would be more than the generally accepted depth at which névé becomes ice, but it may be less than that required for plastic deformation and flow. The author submits that this is the critical distinction — that between moving ice which erodes the floor and an inert mass beneath which a soliflual movement of debris takes place. More research on the larger contemporary snow patches is required to test the validity of this.

References

Behre, C. H. 1933 — Talus behaviour above timber, in the Rocky Mountains. *Jour. Geol.*, vol. 41; p. 622—635.

Botch, S. G. 1946 — Snejniki i snejnaya eroziya v severnykh tchastyakh Urala. *Bull. Geog. Soc. USSR*, t. 78; p. 207—222. (Translation by C. E. D. P., Paris: Les névés et l'érosion par la neige dans la partie Nord de l'Oural).

Botch, S. G. 1948 — Eshtche neskolko zametchanii o prirode snegovoy erozii. *Bull. Geog. Soc. USSR*, t. 80; p. 609—611. (Translation by C. E. D. P., Paris: Encore quelques remarques sur la nature de l'érosion par la neige).

Boyé, M. 1952 — Névés et érosion glaciaire. *Rev. Géomorph. Dyn.*, année 3; p. 20—36.

Brown, E. H. 1952 — The River Ystwyth, Cardiganshire; a geomorphological analysis. *Proc. Geol. Assoc.*, vol. 63; p. 244—269.

Bryan, K. 1934 — Geomorphic processes at high altitudes. *Geog. Review*, vol. 24; p. 655—656.

Flint, R. F. 1957 — Glacial and Pleistocene geology. New York.

Lewis, W. V. 1939 — Snow patch erosion in Iceland. *Geog. Jour.*, vol. 94; p. 153—161.

McCabe, L. H. 1939 — Nivation and corrie erosion in West Spitzbergen. *Geog. Jour.*, vol. 94; p. 447—465.

Matthès, F. E. 1900 — Glacial sculpture of the Bighorn Mountains, Wyoming. *Twenty-first Annual Report U. S. Geol. Survey*, 1899—1900; p. 167—190.

Russell, R. J. 1933 — Alpine landforms of the western United States. *Bull. Geol. Soc. America*, vol. 44; p. 927—949.

Sharp, R. P. 1942 — Multiple Pleistocene glaciation on San Francisco Mountain, Arizona. *Jour Geol.*, vol. 50; p. 481—503.

Tricart, J. 1963 — Géomorphologie des régions froides. Paris, C. D. U.

[*Editor's Note:* The discussion has been omitted.]

Reprinted from *Geog. Ann.*, **41**(1), 34–48 (1959)

AVALANCHE BOULDER TONGUES IN LAPPLAND

Descriptions of little-known forms of periglacial debris accumulations

BY ANDERS RAPP

Geographical Institute, University of Uppsala

"Avalanche boulder tongues" is the author's term for certain accumulations of rock debris caused by snow avalanches. This article is primarily a short description of such features from two localities above the timber line in the northern mountains of Swedish Lappland. The tongues can give valuable and interesting information concerning the localisation, frequency and eroding capacity of snow avalanches. The formations in question have probably not been thoroughly described earlier, possibly because they have not been distinguished from the other, well-known types of debris accumulations ("ordinary" talus slopes, alluvial cones or rock-slide tongues).

Introduction

Snow avalanches probably cause great denudation in many high mountain areas. It is quite evident that they do so in the highest massifs within the amphibolite mountains of Lappland, where many slopes are characterized by the transport and accumulation of rock debris by avalanches.[1] Although avalanche erosion probably is an important morphological process, very little is known about it and in many cases it is not mentioned at all in the textbooks.

Because of the great damage done to communication lines, houses, forest etc. in many mountainous countries, avalanches have necessitated relatively comprehensive research, with the purpose of organizing protection against them. The leading research body is the Swiss Institute for Snow and Avalanche Research in Davos (see "References"), but also in Austria, Norway and other countries similar practical work is carried out. But this is not directly concerned with the problem of the morphological effect and importance of avalanches.

Among the few scientists who have treated this problem is Allix (1924), who has described many examples of very great avalanche erosion in the French Alps; he mentions for instance, one single avalanche which carried roughly 2,000 cubic metres of soil and rock debris. In the last few years avalanche erosion has also been reported on or mentioned by the following authors: Tricart-Cailleux 1953 (Eastern and Western Alps), Matznetter 1956 (Eastern Alps), Peev, 1957 (Bulgarian mountains), Pillewizer 1957 (Karakorum), Jäckli 1957 (Swiss Alps).

Following field studies on Spitsbergen, in the Scandinavian mountains and in the Alps, the present author dealt briefly with the question of avalanche denudation in two previous articles (Rapp 1957 and 1958). On his initiative a number of students at the Geographical Institute in Uppsala have performed field studies on this subject at different localities in Lappland (see "References").

One of the main reasons why avalanche boulder tongues have not been described and reported on is probably that they have not been distinguished from other types of debris accumulations. From this point of view it is convenient to start with a short comparative description of some related types of debris accumulations.

Debris accumulations formed by rock-falls, mud-flows, snow avalanches and rock-slides

Fig. 1 shows some of the accumulations which are formed on mountain slopes by various types of mass movement.

1: 1. Talus cone. An accumulation of rock debris, formed close to a mountain wall, mainly through many small debris falls from the wall. The slope profile of the talus is straight or very slightly curved and the inclination is 30—40°. The material is assorted, angular rock debris,

---

[1] "Avalanche" in this article always means snow avalanche. They can be either "white" or „dirty". Some avalanches erode and carry rock debris, etc, mixed with the snow. These are termed "dirty avalanches".

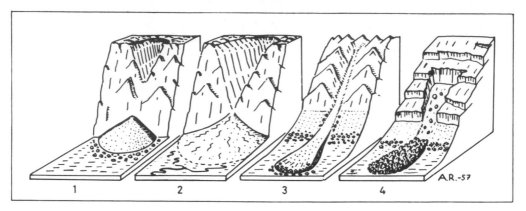

Fig. 1. Sketch of four types of debris accumulations. 1:1 Talus cone, 1:2 Alluvial cone, 1:3 Avalanche boulder tongue, 1:4 Rockslide tongue. For further description, see the text pages.

with the big boulders at the talus base (fall-sorting), where they form a so-called base fringe. The latter can reach out a little on ground that is less steep than about 30°, but it is of importance to note, that if the finer material of the talus (gravel, cobbles, pebbles) covers the slope down to about 25° or a lower inclination, this must depend upon some other type of mass movement than small rock-falls.

If the rock-wall is weathering uniformly over its whole free surface, a simple talus slope is formed instead of a series of cones. This simple talus slope has straight contours but in other respects it has the same characteristics as the cone.

1:2. Alluvial cone. An accumulation of rock debris formed by mudflows or torrents. The profile line is generally concave from top to base. The inclination at the top can be as high as the angle of repose (30—40°) but is generally lower. At the base the inclination may be as low as about 3—10° (HEIM 1932, p. 7) with a continuous transition to the valley bottom. Compare, for instance, the small alluvial cone on the map (Fig. 9) situated between the contours of 20 and 60 m.

The surface of alluvial cones is uneven with curving torrent or mud-flow gullies and levées (SHARP 1942, p. 222). The material is generally edge-rounded. Alluvial cones can be formed through the transformation of talus cones by mud-flows and water transport. Because of this, the sorting of the debris is the reverse of that of a talus cone, i.e. there is coarse debris

on top and fine debris at the base of an alluvial cone.

1:3. Avalanche boulder tongue. (More detailed descriptions in later sections of this article.)

Accumulation of rock debris, formed by erosion and deposition by snow avalanches. The form is often very regular. The slope profile is markedly concave. The distal part of the tongue may reach far out on flat ground in the valley bottom and sometimes it may continue upwards a little on the opposite valley side. The surface of the accumulation has not semicircular contours like an ordinary talus or alluvial cone, but is flattened out, as if a large bull-dozer had gone down the slope, pushing all debris either down to the valley bottom or to both sides of its track. The broad, straight, avalanche track has a smooth surface sometimes vegetation covered and often with very characteristic detailed forms, inter alia, the so-called "avalanche debris tails" (RAPP 1957, p. 183). See Fig. 6 and the following text.

In this article two types of avalanche boulder tongues are distinguished. One is preliminarily called "Road-bank tongues" because they are upraised and flatridged like road-banks. The other type is preliminarily called "Fan tongues". They generally consist of a thin cover of debris and they go farther out over the valley bottom than the first mentioned tongues do.

Many avalanche boulder tongues have an asymmetrical transverse profile with a much more distinct limit at one side than at the other. See Fig. 7 and the following text.

Fig. 2. Tarfala Lake and Mt. Kaskasatjåkko from the south. The avalanche boulder tongues discussed in the text are marked by numbers 1 to 6. A = Kebnepakte Glacier, B = a large lateral moraine. Compare the same locality seen from the air in Fig. 3. Photo A. R. 13/8 1956.

The rock debris in the tongues is generally angular and is not strictly sorted according to size. However, there is often a tendency towards a greater percentage of the biggest boulders along both sides and at the front of the tongues. On slopes, poor in rock debris, the avalanche tracks are often marked by a fringe of scattered boulders, bordering a tongue of pure grass-covered ground.

It is of importance to emphasize that the boulder tongues are only one of the characteristic forms caused by snow avalanches and that they seem to be common and distinct only on slopes which are rich in rock debris and situated above the timber line. Below this limit broad lanes of destroyed forest are one of the most common and clear signs of the occurrence of avalanches.

1: 4. Rock-slide tongue.

A tongue-like accumulation of rock debris

generally with concave profile and small inclination in the distal part. In this they resemble the avalanche boulder tongues, but are mostly easy to distinguish because of their very rough and uneven surface, which consists of very large, angular boulders without any sorting or detailed forms (see HEIM 1932, p. 99; SHARPE 1938, Pl. IB and VIII B).

This type of accumulation has not been formed by more or less continuous debris supply in portions as have the three earlier types, but has been formed in one single catastrophic movement.

Of these four types, Nos. 1, 2 and 4 are well known and have been well described. This is, however, not the case with No. 3. The present author has not found any detailed description of avalanche boulder tongues, but even if there should be such descriptions, there is good reason

För publicering godkänd av Försvarsstaben

Fig. 3. The S. slope of Mt. Kaskasatjåkko with the avalanche boulder tongues Nos. 1—6 as seen from the air. The chutes above tongues Nos. 1—2 in realy are straight and not curved not curved as un this picture, which is the upper, right-hanp part of a larger photrography. A = Kebnepakte Glacier, B = a large lateral moraine. Compare Fig. 2. Air photo: Royal Swedish Air Force 1946.

for saying that they are so little known that we are well justified in taking up the subject for a closer examination here.

Avalanche boulder tongues at Tarfala, Kebnekaise (The "Road-bank type")

Position: 67 55′ N., 18° 35′ E. Map. "Kebnekaise" 1 : 50,000; also in AHLMANN 1951, Pl. I.

Kebnekaise is the highest mountain massif of Swedish Lappland. The timber line in this district is at about 700 m. There are several glaciers which go down to about 1,200 m. The ground above 1,200 m outside the glaciers is mainly naked block-fields or rock walls, so the locality is situated within the "frost-shatter zone" (Frostschuttzone, BÜDEL 1948, p. 31). The

estimated yearly precipitation is about 1,300 mm (map by WALLÉN 1951).

The formations described in the following are located on the S-facing slope of Mount Kaskasatjåkko. They reach down to the N. shore of the small Tarfala Lake (1,190 m). The bedrock consists of amphibolite. (See further JOHANSSON 1951.)

Fig. 2 shows Mount Kaskasatjåkko (2 076 m) photographed from the S. shore of Tarfala Lake. On the S.-facing wall there are four parallel chutes[1] running downwards and going over into three debris tongues which reach down to the shore of the lake. These last-mentioned features are

[1] Gullies cut in the rock walls. The term after Matthes 1938 (see further page 47).

Fig. 4. Debris tongue No. 2, Tarfala, from the steep side. The tongue is about 6—8 m thick. Note the angular debris and the tendency to fall-sorting on the steep side. Some "fresh" cobbles are visible on the snow patch behind the lady in the foreground. Photo A. R. 23/8 1955.

the avalanche boulder tongues. Fig. 3, which is an aerial photograph, shows the same locality from another perspective. The chutes in the wall can be seen as bright bands, partly with narrow stripes of snow. The avalanche boulder tongues are sharply outlined as straight, bright bands over the lower parts of the slopes. The dark-coloured parts between the avalanche chutes and the tongues are caused by older, weathered and lichen-covered surfaces of bedrock and debris, respectively.

The avalanche boulder tongues have been marked by numbers on Fig. 3. Nos. 1 to 3 are situated below the chutes in the mountain wall. It is these tongues that represent what in the following sections will be called boulder tongues of Road-bank type. No. 4 starts from a valley-side depression of other kind and the tongue has another shape. For instance, it is not built up so high and distinct over the substratum as tongues Nos. 1—2 are. No. 4 thus belongs to the "Fan

type". Nos. 5 and 6 have not been examined in the field by the author, but to judge from the photograph they look rather like Nos. 1—3.

Description of the tongues

We choose boulder tongue No. 1 as an example for closer examination.

It is about 330 m long from the mouth of the chute down to the lake and all the way it is about 70—80 m broad. Its thickness is greatest about 150 m from the lake where the eastern edge of the tongue rises abruptly to a height of 5 m over the substratum. Tongue No. 2, which is fed from two converging chutes in the rock-wall, is larger: 100—120 m broad and as a maximum about 8 m thick.

The material in the tongues consists mostly of angular debris from gravel to boulders of 2 m in size. There are also some edgerounded stones mixed in. As a whole, it is not sorted, but there are two exceptions from this rule. One is that

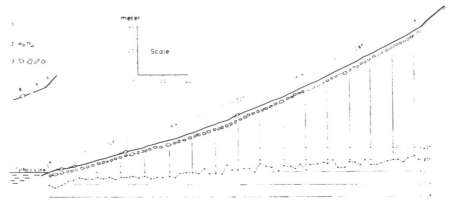

Fig. 5. Slope profile and inclination diagram of avalanche boulder tongue No. 1, Tarfala. The diagram shows the inclination at every 5th meter on the slope by means of thick, short lines, placed right under where they were measured in the profile. The thin lines on the diagram only connect the measured points. 1 = fine debris (sand gravel), 2 = cobbles, pebbles, 3 = boulders, 4 = bedrock wall, 5 = debris slope, 6 = debris tail.

the biggest boulders seem to be deposited along the two sides and in the distal part of the tongues. The other is that the debris in the steep and very unstable eastern side of the tongues has clearly been fall-sorted (see Fig. 4). One gets the impression that this steep side was formed when the eroding avalanches from time to time pushed a portion of debris from the ridge of the tongue obliquely downwards and outwards, leaving it to tumble down when the snow melted. But why is not the western margin of the tongues as steep and distinct as the eastern?

The down-slope profile of the tongue is shown in Fig. 5. It is markedly concave. Only the proximal part of the tongue, close to the rock wall, has an inclination as great as the angle of repose (30—35°). The distal part of the tongue goes down to about 10° at its end in the lake. Tongue No. 4 gives a still clearer proof of the strength and range of avalanche erosion, for it continues upwards over a broad lateral moraine, smoothing out and cutting off its ridge.

From the inclination diagram in Fig. 5 it is further evident that the flat ridge of the tongue is very smooth, especially in its lower part. It consists of gravel and cobbles with scattered boulders. Many of the cobbles and boulders on the surface are orientated with their longaxes in a downslope direction (cf. Fig. 9). The smooth ridge surface is practically free from vegetation except for some few tufts of *Ranunculus glacialis*, *Salix herbacea*, lichens, etc. The cross-profile of

the tongue is illustrated in Fig. 7 and discussed later on.

In spite of its smoothness there are some distinct detailed forms on the ridge of the tongue. Many small torrent gullies and levées (or mud-flow levées, Sharp 1942) occur in the proximal part of the tongue. The chute in the rock wall sometimes carries running water and a water-eroded gully goes from the chute down on the debris tongue, where it disappears about 50 m from the rock wall. These forms are common on similar tongues. They show that mudflows and running water often play a certain role in the supply of debris to the proximal part of the tongues.

Avalanche debris tails

Among the detailed forms which occur on the boulder tongues, the "tails" mentioned above are of special interest, because they seem to be very characteristic of avalanche activity. An avalanche debris tail is a small, straight ridge of debris located close to and on the distal side of a large, fixed boulder in an avalanche track (Fig. 6). The tail is usually 5—10 m long. Close to the boulder it is almost as high and broad as the latter (often about 0.5—1 m), then it gets smaller and smaller in a distal direction. The debris in the tail is very loosely heaped together and orientated like roofingtiles. The tails run strictly parallel to each other in the direction of the tongue or are slightly divergent

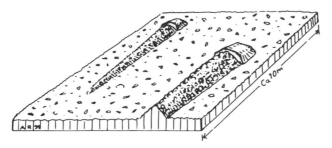

Fig. 6. Sketch of avalanche debris tails. They consist of a fixed boulder with a 5—10 m long tail of loose, unstable debris on its distal side, and they occur upon the surface of many avalanche tracks.

towards the sides of this (cf. Fig. 9). Sometimes there is a little accumulation of debris also on the proximal side of the boulder, but this proximal tail is much smaller than the distal one. Most of the tails seem to be located in the accumulation zones of the avalanche tracks, but they are also found within the erosion zone.

Generally the tails described above seem to be accumulation ridges, comparable with similar but greater features formed by flowing masses of other kinds than snow avalanche (compare, for instance, crag-and-tail moraines formed at the distal side of protruding rocks under a moving glacier). In some cases a double tail with a depression in between has been observed on the distal side of a big boulder in an avalanche track, thus giving a clear proof that the feature must be an accumulation and not an erosion form. However, there have also been observed flatridged avalanche debris tails of compact soil covered with old vegetation and formed through an erosion notch on each side, starting at the fixed boulder. These tails are not so common.

It is possible that the tails are generally developed in one of the following ways or as a combination of both:

(a) The fixed boulder acts like a stopper, halting the stones that collide against it when they are moving with the avalanche. The stones that hit the boulder roll over it and are dropped on its distal side.

(b) The boulder acts like a breaker on the avalanche snow and thus protects from removal all the loose debris which has been deposited on its distal side by all the previous dirty avalanches. After some tens or hundreds of ava-

lanches have passed the same boulder, there will be a tail of debris formed in this way.

One could perhaps illustrate these ideas by saying that in first case (a) the boulder is acting both in an "active" way (stopping, collecting the stones) and in a "passive" way (protecting against further erosion). In case (b) the boulder is throughout acting only in a "passive" way. Anyhow, the debris tails can surely not be formed by one single dirty avalanche.

Good examples of debris tails indicating the strong morphological effect of avalanches have been observed by the author at many places in the high mountains of Lappland (for instance Tarfala, Siellavagge, Singivagge in Kebnekaise; Pallenvagge, Nissonvagge at Abisko) and also in some cases in lower mountains (for instance, on Nipfjället in N. Dalecarlia at a height of 900 m). In 1957 the author visited some localities in the Alps but could not find such well-developed "tails" in the avalanche tracks there, except in one place: Pfeis near Innsbruck, Austria, on slopes probably very much exposed to winds near some cols at 2,000 m.

The asymmetrical cross-profile

Some of the avalanche boulder tongues at Tarfala (primarily Nos. 1, 2 and 3) have an asymmetrical cross-profile, mentioned earlier in this article and also shown in Fig. 3. In this picture it is evident from the contours of tongues Nos. 1 and 2 at the shore of the lake that the eastern side of the tongues is steep and straight, the western side not so distinctly marked. The same asymmetrical cross-profile (Fig. 7, C—D) characterizes many (but not all) boulder tongues of

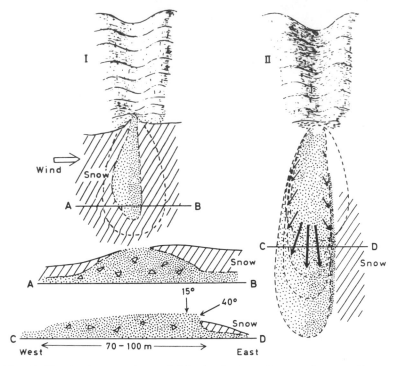

Fig. 7. Sketch showing the probable development from a talus cone (I) to an asymmetrical "Road-bank tongue" (II) through continuous eroding avalanches. The large diagrams show a wall with a rockfall chute and the accompanying debris accumulation as seen from the front. The arrows show strength and direction of debris removal by avalanches as indicated by the debris tails. I = winter, II = summer season. The small diagrams A-B and C-D show vertical sections through the above.

the Road-bank type, in many other localities in the Kebnekaise, Abisko and Sarek mountains.

Close to most of these Road-bank tongues, there is a snow patch in the same position as the snow to the east of tongues, Nos. 2 and 3 in Fig. 3. The existence of these snow patches, which remain until the end of the summer, has given rise to the following hypothesis as an explanation of the asymmetry of the tongues.

It is probable that the snow patch found on the steeper side of the tongues in summer has been caused by the prevailing wind direction and very strong snow-drifting during the winter. At Tarfala Lake the prevailing and snowtransporting wind would thus come from the west and would almost clear the ridge and the western side of the tongues of snow cover and accumulate it on the eastern (lee) side. This was observed to be the case in March, 1959 (T. STENBORG, oral communication). Prevailing winds from the west

are typical of large parts of the Scandinavian mountains. In this locality the westerly winds should be especially prevalent and strong, because there is a col down to 1 500 m just W. of the boulder tongues.

It is these circumstances that, in the author's opinion, cause the asymmetry of the tongues. The development would be as follows:

(a) Accumulation of a small talus cone or alluvial cone under a rock-fall chute in the wall.

(b) Because of the prevailing wind direction during the winter the cone will be more or less swept free of snow on the windward side and covered by a great snowdrift on the lee side. For the same reason there will be a great accumulation of snow in the chute over the cone, which will mean a great likelihood of avalanche erosion (snow accumulation above, bare ground below in the avalanche track). When an avalanche is released, it runs down the chute and out over the cone, where it will *cause the greatest erosion on the bare ridge and the windward side of the cone.* Consequently the greatest deposition of debris will be in front of the bare parts of the cone roughly in proportion to the length of the bare surface. At the same time, the avalanche will diverge and turn over a little bit more to the bare part of the debris cone, because its lee side is built up somewhat higher by the snow. Perhaps the greater friction between the bare ground and the avalanche may also cause it to turn more to the windward side. That the avalanches really are diverging some what is shown by the direction of the debris tails (see Fig. 7).

In this way the original talus cone will continue its development into a more and more pronounced, asymmetrical, "Road-bank tongue" as time goes by and it is fed with debris through rock-falls and mud-flows from above and now and then run over by an avalanche (see Fig. 7).

As a complement to this the following may be added. A study of particularly broad and open chutes ("funnels") in the E. wall of Mt. Pallentjåkko and the N. wall of Mt. Siellatjåkko, Abisko, shows that asymmetry of the tongues can arise without the "help" of a talus or an alluvial cone as an initial form. This means that in some cases the avalanches may themselves accumulate a boulder tongue, which acts as a snow screen and will therefore be asymmetrical by and by.

The rotation of the earth may conceivably play some part in turning the avalanches to the right in the cases described. The fact that tongues on north-exposed slopes often "turn" to the left, however, gives further support to the wind-drift theory, outlined above. Fine examples of this have been reported from Pastavagge, Sarek, by Hammar and Svensson. Pastavagge is a trough valley running in W-E direction. Its length is about 12 kms and according to a map by Svensson, there are in all 37 avalanche boulder tongues. 30 of these are asymmetrical (29 with their steep side towards the E., only one towards the W.)

Very likely the snowpatch on the lee side of the tongue contributes to some extent towards keeping the side of the tongue steep by a certain nivation effect. But nivation is not the main reason for the asymmetry, as is clear for at least two reasons:
(1) there is no noticeable accumulation of debris in front of the snow patches. If much material

had been removed from the tongue and transported over the snow, there ought to be such a debris accumulation.
(2) the side of the tongue is straight all the way. If there was a strong nivation effect from the snow, there should be some form of nivation hollow moving backwards in a bow and eating into the steep side of the tongue. This is not the case.

Avalanche boulder tongues at Pallenvagge, Abisko (the "Fan type")[1]

Pallenvagge is a trough valley situated about 15 km south of Abisko, Lappland. (Position: 68° 15′ N, 18° 45′ E. Map: "Abisko" 1 : 100,000) The valley bottom has a height of 1,000 m. The surrounding mountains Tjåmohas and Pallentjåkko consist of amphibolite and their tops reach 1,745 m. The height of the timber line north of the valley is 600—700 m and the lower limit of the "frost-shatter zone" marked by the block fields runs about 1,000—1,200 m above sea level. A glacier in the innermost part of the valley has its front at 1 250 m.

The yearly precipitation at this locality is probably somewhat lower than in the Kebnekaise mountains. Exact records are not available, so it is only possible to estimate it roughly at around 800 mm per year (map by WALLÉN 1951).

From the eastern side of Mt. Tjåmohas some avalanche boulder tongues go down to the bottom of Pallenvagge. Two of these are especially large and reach far out on flat ground in the valley. The large southern tongue (see Fig. 8) will be described in more detail here as an example of a type of tongue that is called preliminarily the *Fan type*. The name comes from the low, flat, sometimes fan-like shape of these tongues.

Description (Fig. 8 and Fig. 9.)

The avalanche track starts just under the top of Mt. Tjåmohas in a 200 m wide but shallow depression which acts as an accumulation hollow for snow on the lee side of the top. This depression is much wider and not so distinctly marked as the rather narrow, distinctly cut, avalanche chutes which are often combined with tongues of the Road-bank type mentioned above. The total height of the avalanche track above the debris tongue is about 650 m, its inclination is around 30° in the higher part and

Fig. 8. Mt. Tjåmohas (1745 m) with large debris tongue of the "Fan type" in the valley bottom (1,000 m). Compare Figs. 9 and 11. Note the snow accumulation in the broad depression under the top of the mountain and the distinct boulder-free part of the avalanche track to the right of the alluvial cone from where the debris tongue continues down-wards. Compare Fig. 9. Photo A. R. 11/8 1958.

35° in the lower part of the mountain-side. This means that the mountain-side is not steeper than the angle of repose for rock debris and it is also to a large extent covered by a mantle of boulders except for the avalanche tracks, where the bed-rock is swept clean of debris cover.

The sketch map (Fig. 9) is partly based on contours and other details measured in the field by Flink and Söderlund. The figures give relative heights of contours (0 m = about 1,000 m above sea level). The details of the map are generalized.

Fig. 9 shows the debris tongue and the lowest part of the erosion track on the mountain-side. Three main units can be seen on the map:

(a) An alluvial cone between the contours of 60 and 20 m.

(b) A broad erosion track on both sides of this alluvial cone, indicating that the avalanches descend on a far wider track than do the streams

that formed the cone. The width of the avalanche track is most evident on the N. side where the protruding rocks as well as the vegetation slope is swept free from boulders.

(c) The "real" avalanche boulder tongue reaching far out on the slightly sloping valley bottom. This tongue thus *reaches no less than 300 m beyond the range of the alluvial cone* on a substratum with an inclination of about 5° to 0.5°. A comparison between Fig. 8 and Fig. 9 will show, that the base of the talus slope N. of the tongue is situated roughly at the 20 m contour of Fig. 9. This is an example of the limited range of small rock falls in comparison with avalanches.

The S. border of the tongue is quite distinct but does not rise more than about 0.5 m or so over the surrounding grass-covered ground. The N. border shows a continuous transition from debris cover to scattered boulders over the grassy

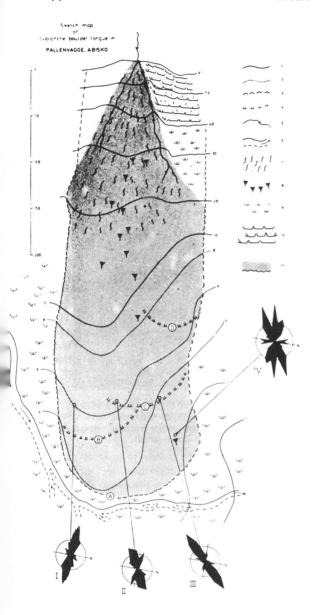

Fig. 9. Sketch map of an avalanche boulder tongue of the "Fan type" in Pallenvagge. Compare Fig. 8. For description, see the text pages. Fieldwork by P. Flink, M. Söderlund and A. Rapp.

Legend: 1 = contour, 10 m interval, 2 = contour, 2 m interval, 3 = limit of avalanche track, 4 = indistinct boulder front line, 5 and 6 = streams, 7 = alluvial cone with mud-flow levées, 8 = avalanche debris tails, 9 = vegetation covered surface, 10 = bare bedrock slope, 11 = rock debris (alluvial cone and debris tongue).

ground. Probably it is the N.-sloping substratum that has caused this "asymmetry" of the sides, and also caused the tongue to curve slightly to the N.

The characteristic features listed above are typical of tongues of the Fan type and they indicate that the tongues are formed by very large, broad and far-reaching avalanches in combination with transport by running water. Sometimes the alluvial cone is very small or totally absent.

The alluvial cone has a concave longitudinal profile with an inclination of 21° at the top and about 5° at the base. Its surface is rather uneven and covered by small gullies and levées. The material at the base is of a size from gravel to cobbles with scattered boulders up to 1 m in size lying on the surface.

The large debris is angular, the small is edge-rounded. As can be seen on the map, the first avalanche debris tails occur already on the alluvial cone, and the whole cone surface bears witness to the smoothing action of avalanches; many levées and gullies are nearly obliterated. At the top of the cone there is also in summer a small stream which parts into two main branches running one on each side of the cone and creating two open gullies.

The transition from the alluvial cone to the debris tongue is, of course, not direct and distinct, although it is well marked by the zone where the mud-flow levées cease. The debris within the real tongue consists of mixed sizes of angular pebbles and boulders of amphibolite. Generally speaking the average size of the boulders is somewhat larger (about 0.2—1 m) in the distal part of the tongue than in the proximal part of it. The maximum height of the tongue over its substratum seems to be about 2 m to judge from a simple extrapolation on the map.

Upon the tongue there are some indistinct boulder lines, forming curved fronts. They have been marked out on the sketch map at the points B, C and D. They indicate that the tongue has been accumulated by avalanches of various ranges and ages. Most of the dirty avalanches nowadays probably reach as far as the D-front or a little bit further on the S. side of the tongue, where the debris is very fresh and unstable, but no front is to be seen. A few recent dirty avalanches reach to the B- and C-fronts, while the A-front seems to have been formed by very large avalanches that either occur very

För publicering godkänd av Försvarsstaben

Fig. 10. The amphibolite mountains of Vistas N. of Kebnekaise (Stuor Reitavagge in the foreground) with their typical walls, very regularly dissected by parallel chutes leading down to debris tongues, a landscape characterized to a high degree by avalanche denudation. The photo was taken so early in the year that the wall sculpture is clearly marked out by the still snow-filled chutes and the naked ridges in between them. Summit heights about 1,700—2,100 m, valley bottoms about 1,000 m. Air photo B. Lindskog 22/6 1950.

seldom or belong to some bygone climatic phase. Between A and B the debris is to a great extent covered by large lichens (*Rhizocarpon* and others). It could also be supposed that the part of the tongue lying in front of B and C was created by creep movements in the wet soil near the stream. The existence of a well-developed, old avalanche debris tail near the A-front (at the 2-m contour on the map) indicates, however, that large avalanches could at one time reach so far. So does the orientation of the boulders at this place (Fig. 9; Rose diagram IV). If the boulders had been moved by creep, they would point at right angles to the 2-m contour. It may be added that on many localities where the tongues end higher up on the valley sides their fronts have moved a little down-slope by creep. This is indicated by the pushed and piled soil in front of the tongues.

The different lichen zones mentioned above brings us to the interesting problem of the correlation between climate and size and frequency of dirty avalanches. On many tongues of the Fan type there are indications of a zoning like the one described, which gives rise to the question: do the old, lichen-covered, distal parts of many debris tongues of the Fan type indicate a bygone climatic period with greater size and frequency of dirty avalanches than now? It would, however, lead too far to try to enter upon this question here.

One thing more is worth pointing out on the sketch map. The tongue has forced the small stream to curve round it, but there is no direct connection between boulder tongue and stream. On the contrary, there is a 5-m-broad band of grass-covered ground without traces of river erosion between the tongue and the stream, indicating that there has been no noteworthy removal of debris from the tongue by the stream at least during the last couple of years.

Fig. 11. The S. slope of Mt. Tjämohas. Abisko. The wall is dissected by steep, narrow and deeply incised chutes, very regularly formed. Below the chutes are debris accumulations showing various stages of avalanche transition. Some of them are "Road-bank tongues" of the asymmetrical type with their steep side towards the E. (right-handside).
Summit 1,745 m, valley bottom 1,000 m above sea level. Bedrock : amphibolite.
Compare with Fig. 8 and note the striking contrast between the E. and S. side of the mountain. Photo A. R. 11/8 1959.

Some remarks on dissected mountain walls

Tongues of the Road-bank type are often combined with mountain walls dissected by steep and distinct rock-fall chutes, either narrow and parallel or converging to funnel-like forms with narrow outlets. Fig. 10 gives an idea of how the high amphibolite mountains are characterized by these regularly dissected walls. This picture may also indirectly give an idea of how common the avalanche boulder tongues are, as most of the chutes have an accumulation of this kind at their mouths. Naturally this is not the case with the chutes that lead down to glaciers, because the ice movement will transport the debris away and prevent its accumulation.

It is evident that the chutes are now and then swept free from part of the debris in their bottoms by avalanches, as has been reported, for instance, from the Sarek Mts by JOSEFSSON 1957. Thus the avalanches are most probably responsible for very important denudation effects in this type of mountain, partly by sweeping the bedrock bare of debris and leaving it open for continued weathering (Erneuerung der Exposition, Penck 1924, p. 52), partly by scratching the bedrock and partly by transporting the debris downslope in a more effective way than any other transport process has until now succeeded in doing since the deglaciation. It is possible that this typical chute sculpture has arisen and been developed mainly by rock-falls and avalanches in combination.[1] We can only touch upon the very interesting question of the age and origin of the chute sculpture.[2] But so far it is quite clear that the chutes are no glacial forms and that they are quite foreign to simple trough-valley walls which have been severely eroded by the glacier ice. For the moment, it will only be pointed out that there seem to be three possible alternatives as regards the time of origin and first development of the chutes;

(a) before the last glaciation or

(b) in a long nunatak stage during the glaciation or

(c) after the glaciation.

The last alternative does not seem to be so likely at least in some localities, where the debris accumulations under the chutes are much too small to "fill out" their erosion chutes in the wall and this discrepancy cannot be explained by post-glacial removal of debris from the accumulations (Fig. 11). Matthes (p. 637) has made the same observations in the Sierra Nevada Mts.

Regularly dissected walls of a similar type seem to exist in many northern mountain districts. Compare, for instance, pictures in the following articles from Alaska (DETTERMAN 1958, Fig. 9), Labrador (FORBES, 1938. Fig. 125 a.o.) and Spitsbergen (BÜDEL 1948, Fig. 7).

Summary and conclusions

Two types of avalanche boulder tongues have been described in this article. They are preliminarily called (a) the Road-bank type (or "Road-bank tongues") and (b) the Fan type (or "Fan tongues"). Both types have been formed by avalanche transport, but very often they also seem to be influenced by mud-flows and running water. The Road-bank tongues can be characterized briefly as slightly transformed talus or alluvial cones. In the Fan tongues this transformation by avalanches has gone very far. *Transitional forms certainly exist between the two types, so it is not always possible to draw a definite distinction between them.*

The tongues of the Road-bank type are rather thick and raised up over their substrata. They have a distinctly concave longitudinal profile but do not reach far out on flat ground. Many of them (but not all) have an asymmetrical cross-profile with one side steep and straight, rising abruptly some meters above the ground level, the other side not so high and well-marked. Most probably these tongues are developed from talus cones or alluvial cones which have been eroded and transformed by dirty avalanches with a limited range, coming from rather narrow chutes or rock-fall funnels. The asymmetry is very likely caused by strong snow-drifting from one side. The steep side of these tongues is thus situated in a lee position as regards the prevailing winds.

Tongues of the Fan type may also be developed from talus or alluvial cones, but the transformation has gone on longer than in the previous case. They are only a little raised up over their substrata. They often reach far out on flat valley bottoms and they can also diverge to the sides to form a fan-shaped tongue. They are most probably formed by great avalanches with a very strong eroding and transporting capacity, usually under small hanging valleys with steep waterfalls or on valley sides with large and wide-open depressions where large masses of snow can accumulate.

To judge from preliminary surveys and aerial photographs, Road-bank tongues are common within the Sarek, Kebnekaise and Vistas mountains. The Fan tongues are not so common. Both types seem to occur mostly upon the eastern, northern and southern sides of the mountains probably because the accumulation of snow in the upper parts of the walls is greater there than on the western, windward slopes.

Avalanche boulder tongues are of great morphological interest from the following points of view:

(1) They give proof of avalanche erosion as a separate and in certain climates probably very important type of slope denudation. In favourable

---

[1] Matthes 1938, has described similair chutes in the Sierra Nevada Mts. of California, which he considers to be formed almost exclusively by avalanches (p. 636). To judge from his photograph, Fig. 2, also pure rock-falls are effektive in »his» cases too, because of the steep and talus-like cone-form of the debris accumulations under the chutes.

[2] The chute sculpture of rock-walls will be treated by the author in a thesis which is under preparation.

cases it should be possible to make a quantitative calculation of how large a breaking-down during postglacial times they represent.

(2) They are most probably periglacial forms (recent or fossil respectively) that are only(?) developed above the timber line. They can give information on the climate (precipitation, wind direction) that has existed during the time of their development. Some of the Road-bank tongues, namely the asymmetrical ones, are especially suitable as wind-direction indicators.

(3) They are of both theoretical and practical interest because they show the localisation and range of at least some types of avalanches.

It is therefore the author's hope that morphologists interested in periglacial or slope morphology will take up the following questions for study at different localities:

(a) In which mountain districts do the avalanche boulder tongues occur? Are the tongues recent or fossil?

(b) What role does the avalanche denudation play as compared with other denudation processes in various climates and mountain relief?

Acknowledgements

Financial contributions for the field work have been received from the following institutions and funds to which the author wishes to express his thanks: Jernkontorets Lundbomsfond, Svenska Turistföreningen and Matematisk-Naturvet. Fakultetens fältforskningsanslag, Uppsala. Mr N. Tomkinson, B. A. has revised the English text. The author also wishes to thank colleagues from Stockholm and Uppsala for valuable discussions.

Literature

Published references.

AHLMANN, H. W., 1951: Scientific investigations in the Kebnekajse massif, Swedish Lappland, I. Geograf. Annaler, h. 1—2 Stockholm.

ALLIX, A., 1924: Avalanches. The Geograph. Review, Oct.

BÜDEL, J., 1948: Die klimamorphologischen Zonen der Polarländer. Erdkunde Bd II, Lfg 1—3. Bonn.

DETTERMAN, R., BOWSHER, A. and DUTRO, T., 1958: Glaciation on the Arctic slope of the Brooks range, N. Alaska. Arctic, Vol. 11, No. 1. Montreal.

FORBES, A., 1938: Northernmost Labrador mapped from the air. Am. Geogr. Soc. Spec. publ. nr 22. New York.

HEIM, A., 1932: Bergsturz und Menschenleben. Vierteljahrsschr. d. Nat. forsch. Gesellschaft Zürich, Vol. 77.

JOHANSSON, H., 1951: Scinetific investigations in the Kebnekajse massif, Swedish Lappland, II. Geograf. Annaler, h. 1—2. Stockholm.

JÄCKLI, H., 1957: Gegenwartsgeologie des bündnerischen Rheingebietes. Beitr. zur Geol. d. Schweiz, Geotechn. Serie, Lief. 36. Bern.

MATTHES, F. E., 1938: Avalanche sculpture in the Sierra Nevada of California. Union int. de géodesie et de géophysique, Ass. intern. d'hydrologie. Bull. 23 Riga.

MATZNETTER, J., 1956: Der Vorgang der Massenbewegungen an Beispielen des Klostertales in Vorarlberg. Geogr. Jahresber. a. Österreich. Bd XXVI. Wien.

PEEV, C. D., 1957: Laviny kak sovremennyj denudacionnyj faktor ("Avalanches as a contemporary denudation factor"). Izvestija vsesojuznogo geograficeskogo obscestva, Nr 89. Moscow.

PILLEWIZER, W., 1957: Bild und Bau des NW-Karakorum, Teil II. Photographie und Forschung, Bd 7, Hft 2. Stuttgart.

RAPP, A., 1957: Studien über Schutthalden in Lappland und auf Spitzbergen. Zeitschr. f. Geomorphologie, N. F., Bd 1, Hft 2. Berlin-Nikolassee.

— 1958: Om bergras och laviner i Alperna. Ymer, Hft 2. Stockholm.

Schnee und Lawinen in den Schweizeralpen. Winterbericht des Eidg. Inst. för Schnee- und Lawinenforschung. Davos.

SHARP, R. P., 1942: Mudflow levées. Journ. of geomorphol., Vol. 5, No. 3. New York.

SHARPE, C. F. S., 1938: Landslides and related phenomena. New York.

TRICART, J., and CAILLEUX, A., 1953: Cours de geomorphologie, II Fasc. I, 2: Le modele glaciaire et nival. Paris.

WALLÉN, C. C., 1951: Nederbörden i Sverige. Medelvärden 1901—1930. Meddel. fr. SMHI Ser. A, Nr 4. Stockholm.

Unpublished references.

FLINK, P. and SÖDERLUND, M., 1958: Sluttningsstudier i Pallentjåkkomassivet, Abisko. Manuscript Geogr. Inst., Uppsala.

HAMMAR, B. and SVENSSON, P., 1959: Sluttningsstudier i Pastavagge och Rapadalen, Sarek. Manuscript. Geogr. Inst. Uppsala.

JOSEFSSON, B. 1958: Studier över temperaturgången i fast berg och lavinaktivitet vid Mikkaglaciären sommaren 1957. Manuscript. Geogr. Inst. Uppsala.

28

Reprinted from *Jour. Geol.*, 77, 92–97, 100 (1969)

A MODEL FOR ALPINE TALUS SLOPE DEVELOPMENT
BY SLUSH AVALANCHING[1]

N. CAINE

Institute of Arctic and Alpine Research, University of Colorado, Boulder, Colorado 80302

ABSTRACT

Two separate processes are shown to be involved in accumulation on talus. Direct free fall from the cliff gives accumulation amounts which decrease with distance downslope from the cliff foot. Accumulation and redistribution by slush avalanching then reverse this trend to give a depositional layer, the thickness of which increases directly with downslope distance. These two processes can be shown, by means of a simulation model, to be capable of accounting for the form of the talus. Increasing the relative importance of slush avalanche redistribution in this model leads to an increase in the length of the basal concave segment on the slope.

The development of near-vertical cliff faces has been simulated by symbolic or geometric models on many occasions (e.g., Penck, 1924; Lehmann, 1933; Bakker & Le Heux, 1952; Scheidegger, 1961). Most of this work has, however, been concerned mainly with the form of the "denudation slope" derived from retreat of the cliff. Where a talus deposit has been introduced to the model, it has been assumed to have a rectilinear slope of constant angle and uniform accumulation rate. In studying the development of the denudation slope such an assumption is probably essential, but as a model for the talus slope it may not be satisfactory.

Although many surveyed slope profiles on taluses demonstrate rectilinearity, others tend to be concave, in whole or in part. A basal concavity seems particularly frequent and is probably the reason for early suggestions that a talus profile approximates a logarithmic curve (e.g., Marr, 1909, p. 96). More recently, the work of King (1956), Young (1956), Rapp (1960), and Andrews (1961) proves the existence of this concavity. In other cases, however, a convex talus profile has been reported (e.g., Rapp, 1960).

Even where the slope is approximately rectilinear, local variations in angle of up to 10° may occur, as Hamelin (1958), Andrews (1961), Rapp (1960), and Caine (1967) show.

Such a variety of forms suggests that the process of talus formation is not adequately described as one of constant, uniform accumulation over the whole slope. An attempt has, therefore, been made to define more closely the processes operating on some talus slopes and to test their effect in a geometric model. This paper comprises two parts: one in which a hypothesis is developed from measurements of accumulation on active taluses, the other in which that hypothesis is tested in a simulation model.

FIELD RESULTS

The taluses studied.—This work forms part of a wider inquiry into rock glacier development in the Two Thumbs Range, Southern Alps, New Zealand, and, for that reason, only the two active taluses which feed waste onto a major rock glacier below the southeastern face of Mount Chevalier (2,410 m.) have been studied (fig. 1). Although these slopes appear typical of many taluses in the alpine zone of New Zealand (where active accumulation is indicated by a lack of weathered fragments and of vegetation and lichens), the conclusions reached

[1] Manuscript received December 1, 1967; revised June 17, 1968.

[JOURNAL OF GEOLOGY, 1969, Vol. 77, p. 92–100]

[*Editor's Note:* Certain figures have been omitted owing to limitations of space.]

here must be considered tentative, since they are based on scanty data.

The talus slopes studied are both concave in profile (fig. 2) and, since they consist of laterally coalescent cones, are of the compound type defined by Rapp (1960, p. 5). Both the talus material and the cliffs from which it is derived are composed of poorly sorted, highly indurated sandstone of Jurassic age which is dark gray when freshly exposed but soon weathers to a light brown-pink color. The talus material itself has not been subject to detailed particle-size analysis, but it generally falls into the 5–10 cm.

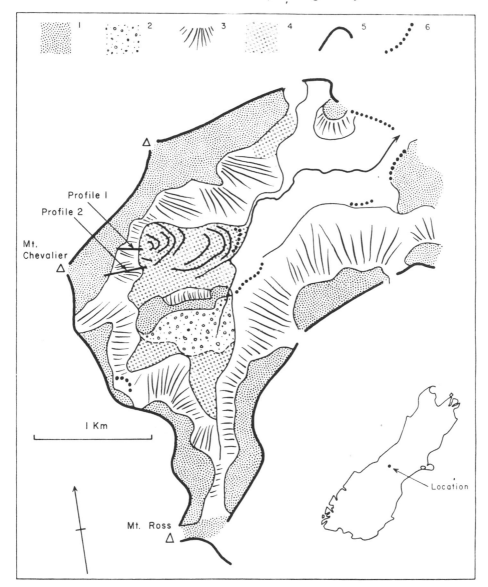

FIG. 1.—The field area. 1, Exposed bedrock in the form of steep crags and free faces; 2, glacially molded bedrock; 3, talus; 4, rock glacier and low angle block covered areas (showing arcuate rock glacier ridges); 5, ridge crests; and 6, morainic ridges.

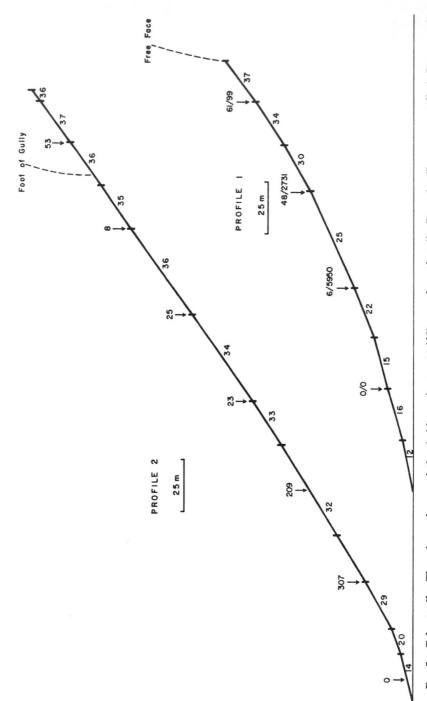

FIG. 2.—*Talus profiles.* The volume of accumulation (cubic centimeters × 100) on each quadrat (for December/January on profile 1; January only on profile 2) is indicated. Figures below the profile lines are slope angles (degrees).

range, with an almost complete lack of fines less than 2 mm. in size. Variations in size are, however, wide and do not seem to conform to a simple pattern.

Methods.—Accumulation rates were measured by methods similar to those used in Spitsbergen and northern Sweden (Rapp, 1960, 1961). Meter squares of plastic sheeting intended to trap accumulating materials were not, however, successful in the present inquiry, probably because of their small size and the short period of measurement. Field data were, therefore, obtained by making counts of the stones which had accumulated on and within the winter snow cover of the talus. They were derived from the collection of the accumulated fragments on 10 m. square quadrats spaced at 50 m. intervals along a line normal to the contours (the profile survey lines of fig. 2). The results approximate the volume of fragments accumulated during the period when the slope was snow covered or since the quadrat was last cleared. Under some circumstances the removal of waste in this fashion may influence further accumulation on the snow, but since the stones never covered more than 1 per cent of the area of any quadrat on the Mount Chevalier taluses, this will be assumed to be negligible.

After collection, the three axes of each particle were measured and their product taken as its volume. This is somewhat larger than the true volume of the particle, but since the overestimation involved is probably consistent, it is possible to compare amounts between quadrats.

Results.—The rock fragments on the snow cover of the four quadrats of profile 1 were first collected on December 12, 1966, when the whole talus was still covered with winter snow to a depth of more than 50 cm. Accumulation amounts vary inversely with distance from the head of the slope and diminish to nothing on the 25° segment at its foot. The correlation coefficient (r) between accumulation and distance is -0.9604, which is significant at the 98 per cent level on t-test.

The relationship is best summarized by a linear regression equation:

$$y = 7372 - 44.99x, \qquad (1)$$

in which y = accumulation in cubic centimeters per 100 sq. m.; x = distance from the head of the slope in meters, and with a standard error of the estimate (Sy) = 728. The mean particle size for each quadrat does not vary consistently.

Since the snow cover of the talus was continuous and undisturbed, it is unlikely to have been subject to slush avalanching prior to the date of collection. This accumulation is, therefore, thought to be the result of falls from the free face which collected on the slope from the time it was snow covered, early in the winter. The inverse relationship with distance downslope agrees with the observations of Rapp (1961, p. 102) but may be rather different when the talus is not snow covered. A continuous soft snow cover would probably minimize the amount of bounding and rolling of particles after their initial fall onto the slope. Thus, the relationship between direct accumulation and downslope distance may be different on talus slopes which are not snow covered for long periods.

The four quadrats of profile 1 were again examined on January 10, following a period of thawing, when a number of small avalanches of slush and wet snow had disturbed both the snow cover and, especially on the higher parts of the slope, the talus beneath it. Above the basal segment, where there was again no deposition, the waste collected was related directly to downslope distance. This situation is also summarized by a linear regression:

$$y = 5833.8x - 124300 \qquad (2)$$

(Sy = 5560 and r = 0.9973 with p = 95 per cent on t-test). On this occasion, the mean size of the particles in each quadrat was found to increase with downslope distance such that:

$$\log_{10}y = 1.27167 + 0.02918x \qquad (3)$$

($Sy = 0.0686$ and $r = 0.9983$ with $p = 95$ per cent on t-test), where $y =$ the mean volume of the particles on each quadrat (in cubic centimeters). Accumulation of this kind is thought to be due to slush avalanches which carry material down from gullies in the free face. On the other hand, the geometric increase in block size with downslope distance suggests that the rolling of individual particles may have some influence on the result.

On profile 2, measurements were made only on December 12, when the slope, although entirely snow covered, had been subject to slush avalanching from the gully in the cliffs above, that is, under conditions similar to those prevailing on profile 1 when the second set of measurements was made. Considering the six quadrats on the talus below the free face, accumulation on the slope conforms to the pattern found in January on profile 1. The volume of accumulation once more increases in linear fashion downslope such that:

$$y = 156.6x - 8125 \qquad (4)$$

($Sy = 5012$ and $r = 0.911$ with p over 99 per cent on t-test) and again falls abruptly to zero when the slope angle becomes less than 25°. The difference between the regression coefficient obtained on this occasion and that derived from profile 1 in January is, perhaps, explained by the longer slope of profile 2. This would lead to a wider spreading of slush avalanche material, both laterally and longitudinally.

A TALUS ACCUMULATION MODEL

These field data suggest that two processes are responsible for accumulation on the alpine talus slopes of New Zealand: first, the falling of waste directly onto the slopes; second, slush avalanching over the talus. The former leads to deposition of new material, predominantly at the head of the talus, while the latter is responsible for both the introduction of new material and the redistribution of waste already on the slope. It gives greatest accumulation near the talus

foot. In what follows, these will be referred to as primary and secondary deposition, respectively, and their effects through time will be examined in a simulation model of a talus.

The model.—The model to be discussed is considered in profile only and its development is based on a number of assumptions concerning the initial slope and the accumulation processes affecting it. The initial slope consists of four rectilinear facets: upper and lower horizontal ones with, between them, a talus angle tan 0.8 (about 38°) and a free face of tan 10.0 (about 85°) (tangent values, used in computation, will be retained because the conversion to degrees is only approximate). Only the lower half of the slope is shown in figure 3. The development of the slope is governed by the following assumptions:

1. that the bedrock is subject to a weathering rate (R) which is constant in time and space;
2. that the development of the free face occurs by parallel retreat (this assumption is implicit in assumption 1; it is made solely to facilitate the computation of the volume of material released by weathering);
3. that the volume of talus due to free face retreat ($Vol._1$) equals the volume of bedrock from which it was derived (i.e., using the terminology of Bakker and Le Heux [1952], $c = 0$);
4. that, in the initial case, the talus and the free face constitute a closed system so that no material is lost from the foot of the slope (after case 1, this assumption is relaxed to allow the introduction of material during secondary deposition but the basal restraint is maintained throughout). This assumption may be made to appear more realistic if it is considered, with assumption 3, to suggest that the net volume of accumulation equals the volume of bedrock removed from the free face;
5. that the angle of rest of the talus remains constant at tan 0.8;
6. that the thickness of primary accumulation on the talus is an inverse linear function of distance from the foot of the free face;
7. that movement and a redistribution occur wherever the talus develops angles greater than tan 0.8;

8. that the thickness of secondary accumulation due to redistribution is a direct linear function of distance from the foot of the free face;

9. that neither primary nor secondary deposition occurs on or below a slope of less than tan 0.45 (about 25°).

The first four of these assumptions are made to simplify the model and are essentially identical to those made by Bakker and Le Heux (1946). The last five, which tend to define the developing slope, are based on the field evidence discussed above.

In relaxing assumption 4, the volume of secondary accumulation has been increased, since field evidence (equations 1 and 2 above) suggests that it greatly exceeds $Vol._1$ In the initial case, $Vol._2$ consists of the volume involved in redistribution on the slope as defined by assumption 7. In the other cases, this volume has been multiplied by a constant (A) to give $Vol._2$ as used in redistribution. In practical terms, this requires the introduction of new material during the slush avalanching process. Thus, A represents the importance of this, relative to the waste derived directly by falling from the free face.

The development of the model from the initial slope has been achieved by iterative procedures similar to those used by Young (1963) and Ahnert (1966). Each iteration is a sequence of four steps:

1. free face retreat which gives Vol_1;
2. accumulation on the talus in inverse relation to distance downslope from the free face; total accumulation equals Vol_1;
3. definition of the material on the talus which causes angles in excess of tan 0.8. This is achieved by taking a slope of tan 0.8 from the base of the steeper segments. The area between this and the actual slope profile then represents the volume to be removed. This volume is multiplied by A to give Vol_2;
4. further accumulation on the talus—on this occasion involving Vol_2 which is distributed in direct relation to downslope distance.

Following step 4, a return is made to step 1 and the iteration commenced on the slope derived from the preceding iteration. The sequence of iterations ceases with the disappearance of the free face. The repetitive computations involved in this kind of work are most easily done by electronic computer, an IBM 1620 being used in the present inquiry.

Results.—The products of this analysis are series of sequential profiles for values of A up to 8 (fig. 3).

For the main part of the slope, the development is similar in each of the five cases examined and consists of three stages. In the first iteration, a massive readjustment of the profile occurs because the initial slope becomes entirely unstable as soon as deposition occurs on it. As a result of this, the angle of much of the talus is significantly reduced (to about tan 0.73) and its surface becomes uneven and irregular. For about forty iterations (at $R = 0.1$) following the first one, the angle of the talus slope gradually increases to tan 0.79 and the profile becomes rectilinear, except for a basal concavity. Once rectilinearity at tan 0.79 is established, the slope grows by parallel increments, maintaining the basal concavity throughout.

Variations in the value of A visibly affect only the lower, concave element of the talus slope. Although the short concavity is developed in case 1, amounting to about 8 per cent of the length of the slope, the concave element is significantly lengthened as the value of A is increased (fig. 3). At the higher values examined, it occupies almost 25 per cent of the slope.

The form of the bedrock denudation slope developed below talus cover is defined by the successive positions of the foot of the free face and is identical to that due to the parallel rectilinear retreat model of Bakker and Le Heux (1952). Increasing the value of A in the present model has the effect that values of c greater than zero had in the work of Bakker and Le Heux.

CONCLUSION

The profiles of figure 3 are similar to the many surveyed talus profiles which include concave elements at their foot. Differences between these profiles and those surveyed in the field may be explained by the initial assumptions of the model, especially those concerning the nature of the material, the topography below the talus, and the processes acting upon it. Such a coincidence between the model and surveyed slopes does not necessarily imply that all taluses with basal concavities have developed in response to accumulation processes akin to those on which the model is based. The case of the Two Thumbs taluses, however, where free fall and slush avalanching processes have been demonstrated empirically, strongly supports the suggestion that these are sufficient to account for the present form of the slopes. Testing this possibility was the main reason for initially setting up the model.

It is also reasonable to extrapolate from this and suggest that the proposed model may summarize the development of other active alpine talus slopes, particularly where the slopes are supplied with material from the two sources on which the model is based. To extrapolate further and account for taluses at lower altitudes (where seasonal slush avalanche activity is insignificant) in the same terms is probably not justified. A redistribution agent which acts in a manner analogous to that of slush avalanches has yet to be demonstrated on talus slopes below the alpine zone.

ACKNOWLEDGMENTS.—I should like to thank J. T. Andrews, D. E. Greenland, J. D. Ives, and S. E. White for critically reading the manuscript and T. Baumann for drafting the diagrams. The use of facilities in the Computer Centre, University of Canterbury, New Zealand, is also acknowledged.

REFERENCES CITED

AHJERT, FRANK, 1966, Zur Rolle der elektronischen Rechenmaschine und des mathematischen Modells in der Geomorphologie: Geog. Zeitschr., v. 54, no. 2, p. 118–133.

ANDREWS, J. T., 1961, The development of scree slopes in the English Lake District and central Quebec–Labrador: Cahiers de géographie de Québec, v. 10, p. 219–230.

BAKKER, J. P., and LeHEUX, J. W. N., 1946 Projective geometric treatment of O. Lehmann's theory of the transformation of steep mountain slopes: Koninkl. Nederlandse Akad. van Wetenschappen Proc., v. 29, no. 5, p. 533–547.

————— ——— 1952, A remarkable new geomorphological law: Konikl. Nederlandse Akad. van Weten schappen Proc., v. 55, no. 4, p. 399–410, p. 544–570.

CAINE, N., 1967, The texture of talus in Tasmania: Jour. Sed. Petrology, v. 37, no. 3 p. 796–803.

HAMELIN, L. E., 1958, Le talus oriental d'éboulis de l'Aiguille rousse: Rev. Géographie alpine, v. 46, p. 429–439.

KING, C. A. M., 1956, Scree profiles in Iceland: Premier Rapport, Commission pour l'Étude des Versants: Amsterdam, International Geographical Union, p. 124–125.

LEHMANN, OTTO, 1933, Morphologische Theorie der Verwitterung von Steinschlagwanden: Vierteljahrsheft Naturf. Gesell. Zurich, v. 78, p. 83–126.

MARR, J. E., 1909, The scientific study of scenery (3d ed.): London, Methuen, 372 p.

PENCK, WALTER, 1924, Die morphologische Analyse (translated as Morphological analysis of landforms, by HELLA CZECH and K. C. BOSWELL: London, Macmillan, 1953, 429 p.).

RAPP, ANDERS, 1960, Talus slopes and mountain walls at Tempelfjorden, Spitsbergen: Norsk Polarinst. Skr., 119, 96 p.

————— 1961, Recent development of mountain slopes in Karkevagge and surroundings, northern Scandinavia: Geogr. Annaler, v. 42, p. 71–200.

SCHEIDEGGER, A. E., 1961, Theoretical geomorphology: Berlin, Springer-Verlag, 333 p.

YOUNG, ANTHONY, 1956, Scree profiles in west Norway: Premier Rapport, Commission pour l'Étude des Versants: Amsterdam, International Geographical Union, p. 125.

————— 1963, Deductive models of slope evolution: Akad. Wiss. Göttingen Nachr. 11, Math-Naturw. K1, no. 5, p. 45–66.

Part IV
WIND ACTION

Editor's Comments
on Papers 29 Through 32

Wind, like snow, can produce erosional and depositional features in a periglacial environment. Some of its erosional effects,
in connection with the blowing of hard snow crystals, have already been mentioned. This effect reaches extreme proportions
in such very cold and windy environments as the part of the
Antarctic that is not covered by permanent ice. The periglacial
environment is particularly susceptible to the action of wind for
several reasons. The winds in the Arctic and Antarctic tend to be
very strong, and they are rendered more effective because of low
precipitation and the paucity of vegetation, both factors that tend
to cause exposure of bare soil or rock. Another fruitful source of
windblown sediment is found in the great spreads of bare outwash material left by the rapid retreat of the Pleistocene ice
sheets, which formed the large sandar such as now occur
around the retreating glaciers of southeast Iceland. These
sparsely vegetated tracts of bare sand, gravel, and silt are of low
relief and therefore provide little shelter against the wind.

In this part some examples of wind erosion will be cited. The
erosional effects of periglacial wind processes in Arctic deserts
are the subject of Paper 29, by B. Fistrup, in the first section. The
second section is devoted to periglacial loess, about which a great
deal has been written. Papers 30 and 31 are examples of this

extensive literature. Paper 30, by W. H. Hobbs covers the loess associated with the Greenland ice sheet. Loess is of great significance in central Europe, and Paper 31, by M. Pécsi, concerns the loess of Hungary and provides valuable data. The coarser material, covered in the third section, includes sand that forms into dunes. Paper 32, by M. Seppälä, is devoted to the periglacial dunes of northern Sweden. The selection of papers is designed to cover as wide as possible a range of the world in which periglacial wind action has been important, from active forms in the present Arctic deserts to the fossil forms of central Europe and north Europe. Periglacial wind action is very widespread and is particularly associated with glacial processes in the proglacial zone of retreating ice sheets.

WIND EROSION

Erosion by wind produces characteristic forms that include wind-faceted pebbles, called "ventifacts." These etched and polished stones are sometimes sufficiently numerous to form a desert pavement. An example occurs beneath the loess of northern Germany, where it is called the *Steinsohle*. The erosion of solid rock by wind-carried grains of sediment or snow crystals produces corrasion forms referred to as "tafoni," which can cause considerable hollows and flutings in even hard rocks such as granite. The cavernous weathering produced by wind erosion in the periglacial regions is rather similar to that produced in desert areas in other latitudes.

The pattern of periglacial wind erosion can often be used to assess the direction of the prevailing wind in those areas which are no longer in a periglacial region. Thus wind-eroded landforms provide useful palaeoclimatic evidence concerning former wind direction and, by inference, former pressure patterns.

LOESS

Loess is by far the most important periglacial deposit because of its very widespread occurrence over much of central Europe, very large tracts of Asia and the Far East, and large areas of North America. Loess covers one tenth of all the land surface of the earth, the major areas excluded being the tropics and the areas covered by the ice sheets during the last glacial advance. Loess varies in thickness from 1 to 100 m and occurs in a variety of relief forms. It is most important in the flatter areas, such as the North American prairies, the basins of the Mississippi, and the Hwang Ho, La Plata, and Danube valleys, but it also occurs in more hilly areas. One reason for the great importance of loess is that it provides one of the most fertile soils for agriculture; on the other hand, its mechanical properties provide problems for engineering operations.

Loess is defined as a homogeneous, largely unstratified silt. It is usually permeable, porous, and unconsolidated, but it can stand in vertical bluffs. It is often yellow or buff as it contains limonite, but can be gray. The term "loess" comes from the Rhine Valley, where it was used in 1821, with the meaning "loose." Lyell used the term in 1834. The material is of aeolian origin and is associated with the arid conditions of the proglacial zone. In many areas it is now known that there are several layers of loess, which are separated by palaeosols and which relate to different glacial events; no preglacial loess is known. Loess consists largely of silt-sized particles with diameters of 0.01 to 0.05 mm. The origin of this particular grain size has been attributed to sorting by the depositing medium, the coagulation of clay, or to the comminution of the rock debris to this particular size.

Loess consists largely of quartz grains derived by frost comminution, which produced large clouds of dust in the cold arid climate zones as well as the warmer ones. The dust has been transported for considerable distances. Loess is fairly uniform over the earth, with the main constituents being 40 to 80 percent quartz, with a mean of about 65 percent; feldspar accounts for 10 to 20 percent, and calcium and magnesium carbonates from 0 to 35 percent. Clay particles can form authigenically in loess either as it collects or afterwards. Loess also contains heavy minerals that may reflect the origin of the material and are therefore useful diagnostically. The heavy minerals generally occur in the silt fraction. Lime in loess is of significance, and its migration subsequent to deposition may lead to secondary effects such as the accumulation of caliche. The porosity of loess is probably due to a number of factors, of which frost action is likely to be one of the most important. Others include electrical charges of the grains during deposition, the activity of plants and animals, and the redistribution of carbonates.

Various classifications of loess have been suggested based on the type of material, the relief position, and the time of formation. Loess has probably not all formed by the same method, and

50 theories have been advanced to account for it in various areas. This indicates the diverse nature of loess formation and the processes that give rise to it. The most favored theory is the aeolian theory, supported by the majority of workers, who consider that loess is windblown in origin. The association of loess with ventifacts supports the aeolian theory.

Other sedimentary processes have also played a part and some loess may have been redeposited, such as the brick earths of southern England, which were redeposited by water. The terms "niveolian" and "pluvionivational" have been suggested to describe loess redeposited under different conditions, including solifluction or downslope creep, which could give rise to the deposit called "colluvium." Lyell suggested a fluvioglacial origin for loess; others have stressed the pedogenic processes that may modify the loess once it has been deposited. The polygenetic theory provides for a variety of modes of formation, with a combination of processes operating.

The nature of the included fauna and flora provide useful information concerning the conditions under which the loess was deposited. The Great Plains loess probably accumulated under climatic conditions more favorable to plant and animal life than those of the present day, conditions that existed over a wide area and for a long time. The source of loess appears to have been the outwash from glacier ice of the main ice sheet as well as glaciers in the Rocky Mountains. It seems loess collected in glacial rather than interglacial periods in a generally cool environment. The dust probably collected mainly in the vegetated zones beyond the border of the barren tundra zone around the margins of the ice masses, the vegetation helping to trap the dust as it settled.

SAND DUNES

Where the material blowing in the periglacial environment consisted of sand-sized particles, the forms that accumulated were typical of sand dunes. The early work of Högbom in 1923 first drew attention to the pattern of Pleistocene dune building in Europe. Most of the dunes of periglacial origin date from the last ice advance or the postglacial period; older ones can no longer be recognized owing to burial under loess, erosion, or destruction by subsequent ice advance. The dunes derived their sand from the same source as the silt that formed the loess, but the dunes occur closer to the outwash spreads, many of them being found in the *urstromtäler*. Others occur in Poland along the middle Vistula Valley. The glacially lowered sea levels also provided spreads of sand that could form dunes, and the varying extent of the large water bodies, such as Lake Agassiz, provided exposed sand during low-water-level periods.

In form the dunes cover most of those now found in desert areas with the exception of the barchan, which requires a complete absence of vegetation and a unidirectional wind pattern. Vegetation played a part in the dune pattern in periglacial areas, as it does in coastal dunes now. The same types of U-dune occur in both environments, as these dunes are affected by vegetation. This type of dune has points facing upwind as a result of deflation removing the central part of the dune, while the lower edges are held more effectively by the vegetation. Longitudinal dunes are also common. Most of the dunes of Europe are the transverse or U-dune type, which are also referred to as parabolic dunes. These dunes and the longitudinal dunes provide an estimate of the dominant dune-building wind direction, which may differ from the prevailing wind in some instances. Paper 32, on the dune-building process, indicates a clear preferred orientation of dune pattern. In places the dunes do not have an opportunity to form, and instead the windblown sand forms an amorphous sheet, which is described as cover sand. These bedded sands have been described in the Vendée Massif and in the Trent Valley of the English Midlands; they also occur in Belgium and the Netherlands.

29

Reprinted from *Geog. Tidsskr.*, **52**, 56–60, 64–65 (1953)

THE WINDEROSION IN ARCTIC DESERTS

B. Fistrup

In the arctic deserts the windpolishing is very pronounced. From Søndre Strømfjord J. A. D. Jensen (1889) has already given a very vivid description of stones that are smoothly polished to a plate supported only by a thin stem, that the whole stone resembling a table. Jensen describes rifled or grooved stones, where the hard layers protude as sharp edges or the softer layers have completely vanished, so that the stone is riddled, while again other stones, where the disintegration is not yet so advanced, look like a sponge. In Peary Land the hard dolerite is completely polished, while the softer sandstone is pitted and eroded into the most bizarre forms. There are numerous observations of windpolishing from the arctic regions made under such conditions that the polishing is due only to drifting snow and not to drifting sand. Some examples have been given by Carl Samuelsson (1926) and since then numerous others have been added. The hardness of the ice stated in the geological handbooks is 1,5 to 2,0 and it has therefore been denied that drifting snow should be able to polish the hard dolerites. By experiments K. J. V. Steenstrup (1893) tried to prove that the hardness of the ice grew by declining temperature and he found that the ice at 70° reached a hardness of about 3, and thought this fact a proof of the inability of the ice-needles to polish harder materials. Koch und Wegener (1930) found, that the hardness of drifting ice was exactly 4 at —44° and 3—4 at —15°. In a paper from 1939 Curt Teichert however, from observations in the Alps and in Green-

[*Editor's Note:* In the original, material precedes this excerpt.]

land showed that the hardness of the ice had to be considerably greater and found it to be about 4 at $-44°$ and about 6 at $-50°$. From later, especially Russian observations it appears that the physical qualities of the ice are conditioned by the temperature and the pressure, thus the coefficient of viscosity of ice according to Tsytovich and Sumgin (1937) (Muller, 1945) will raise violently by falling temperature. At the request of Eliot Blackwalder (1940) a

Fot. B. Fristrup.

Ventefacts at the coast of Melville Land, Peary Land.

number of laboratory experiments on the hardness of ice were carried out by examining blocks of ice, frozen to $-78,5°$ by aid of solid carbon dioxide, and by this temperature the ice was found to have a hardness of 6 (on the Mohr's scale) matching orthoclase feldspar. Within the arctic areas the temperature of the iceneedles will be considerably lower than $-40° -50°$ found by the macromatic observations and according to the experiments referred to it is thus to be regarded as having been proved that the iceneedles in drifting snow often will be of a hardness of about 6 or perhaps even more.

In Peary Land the strongest winderosion occurs during the winter. The country itself at that time is more or less free from snow, and the fallen snow will drift over the surface until it settles in a drift sheltered by a slope or behind a big stone. The snow surface is pressed hard with a strong sastrugi. The winderosion is strongest in the lawlands because the foehn winds are much more stronger here than on the plateaus. During the winter the interior tablelands will be covered by a layer of snow and only the larger stones submerged. The upper parts of these stones as a rule are windpolished, while the lower parts, protected by the snow, will not get polished.

Fot. B. Fristrup.

Dried clay with saltcrustations, Heilprin Land, Peary Land.

The windpolishing seems to take place very quickly. Investigations upon measuring the winderosion in Peary Land have been done by J. Troelsen (1952). A standard for the windpolishing may be obtained by observing the polishing of the stones in Esquimaux tentrings. Eigil Knuth (1951, 1952 a) has described a number of these tentrings and judging by the archæological findings he has been able to define them as belonging to the Dorset-culture. These stones are polished by the wind and often greatly diminished since they have been raised in their present position by the Esquimaux. A precise dating has not yet taken place with carbon 14. At present time Peary Land is not inhabited, and it is probable, that the age of the Esquimaux tentplaces is within a thousand years, and during that time some of the stones have been diminued to nearly the half seize.

In many valleys the windpolishing at the stones are rather insignificant. This is owing to the formation of an icelayer that protects the stones. The ice is formed during the autumn and the beginning of the winter. The frost sets in quite suddenly, and the upper surface of the earth therefore freezes, and on many places the water

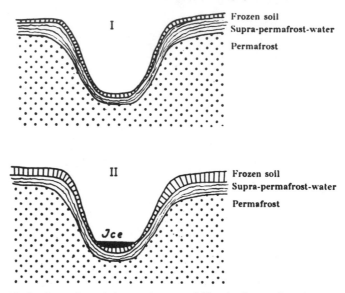

Diagram of the formation of ice at the ground, for further explanation see the text pag. 59.

in the active layer between the permafrost-layer and the frozen surface will be exposed to pressure, and at the weak places, especially in the floor of the valleys and in dried-up rivers the water will be forced ahead, overflow the frozen earth and there freeze. The icelayers thus formed will vary from one year to another, but they can obtain quite significant dimensions.

At all times of the year drifting sand may occur in Peary Land. On account of the dry earth even in the middle of the winter the earth is at many places not firmly frozen and can be kicked loose. After long periods where the snow had been swept from the country, drifting sand was not rare to a smaller extent, and during the winter small heaps of sand were deposited in the snow. In spring

370

the sand and dust will be gathered in pits and melt down in the
ice on account of their dark colour and consequently great ability
of absorbing the heat. When the sea ice melted, a number of holes
in the ice have been formed similar to the cryoconite holes in the
glacier-ice. In most arctic regions loess-deposits are playing an important part, but sand deposits and especially sand dunes are rare.
In Peary Land dunes existed in very few places along the river-
beds in Etukussuks Dal and in Børglum Elvens Dal. The dunes are
very small, but protected by slopes along the rivers and at the
seacoast smaller drifts can be formed reaching a height of about
two meters or more. The sand of the dunes is quite dry in summer,
and the permafrostlayer is deep under the sand. On the area around
Søndre Strømfjord the dunes are better develloped than in Peary
Land and may reach much bigger dimensions here. Landscapes may
be formed, which are similar to the Danish "Indsande" in physiographical respect with smaller dunes, partley covered by vegetation
and great open sandplanes with drifting sand. Outside of Greenland
dunes in arctic regions have been studied in Alaska by Robert F.
Black (1951). The dunes are especially found along the north-coast
and in Central Alaska, and the dunes here have much larger dimensions than those in Greenland.

[*Editor's Note:* Material has been omitted at this point.]

REFERENCES

Black, Robert F., 1951: Eolian Deposits of Alaska, Arctic. Vol. 4.

Blackwalder, Eliot, 1940: The Hardness of Ice. Amer. Journ. of Science.
    Vol. 238.

Jensen, J. A. D., 1889: Undersøgelse af Grønlands Vestkyst fra 64° til
    67° N. B. 1884 og 1885. Medd. om Grønland. 8.

Knuth, Eigil, 1951: Pearylands Dorsetkultur og de første Skrællinger.
    Grønl. Selskabs Årsskrift 1951.

Knuth, Eigil, 1952: The Danish Expedition to Peary Land, 1947—50.
    Geogr. Journ. CXVIII.

Koch, J. P., und *Wegener, A.,* 1930: Wissenschaftliche Ergebnisse der dä-
    nischen Expedition nach Dronning Luise-Land und quer über das
    Indlandeis von Nordgrønland 1912—13. Medd. om Grønland. 75.

Muller, Slemon, Wm., 1945: Permafrost or permanently frozen Ground
    and related engineering problems. Washington.

Samuelsson, Carl, 1926: Studien über die Wirkungen des Windes in den
    kalten und gemässigten Erdteilen. Bull. of the Geol. Instit. of Upsala.
    Vol. XX.

Steenstrup, K. J. V., 1893: Bliver Isen saa haard som Staal ved høje
    Kuldegrader. G. F. F. Bd. 15.

Teichert, Curt, 1939: Corrasion by Wind-blown Snow in Polar Regions.
    Amer. Journ. of Science, Vol. 237.

Troelsen, J. C., 1952: An Experiment on the Nature of Wind Erosion,
    Conducted in Peary Land, North Greenland. Medd. fra Dansk Geol.
    Foren. bd. 12.

30

Reprinted from *Jour. Geol.,* **39,** 381–385 (1931)

LOESS, PEBBLE BANDS, AND BOULDERS FROM GLACIAL OUTWASH OF THE GREENLAND CONTINENTAL GLACIER

WILLIAM H. HOBBS
University of Michigan

ABSTRACT

This paper discusses the action of the glacial anticyclone of Greenland upon the outwash deposit at its front, and the manner of deposition of loess, pebble layers, and stranded boulders etched by drifting sand which is now collected in dunes. Of much significance is the fact that the braided streams on the outwash plain flow only for a few months of the year.

The station of the University of Michigan expeditions at Mount Evans, Greenland, supplied perhaps the best opportunity yet made available to study existing climatic and resulting geological conditions near the border of an ice sheet of the kind that lay over Northern North America. This is true because within the antarctic region comparatively small land areas lie exposed, and, further, because all expedition bases there, save only that of Sir Douglas Mawson in Adelie Land (1911–14), were located far from the inland ice border. So far as Greenland is concerned, most former expedition bases have been placed close to the sea, and in some cases far from the inland ice border where domination of the wind by the glacial anticyclone is less complete.

The Mount Evans base was purposely selected where the land area marginal to the inland ice has its greatest width. For a distance of about 175 miles to the north and south of the station, the ice margin has an enclosing strip of land of an average width of nearly 100 miles. At the station itself, which is about 25 miles distant from the present ice border, throughout the two-year period during which the station was in operation all strong surface winds blew outward off the inland ice. These included in each year many violent storms, some of which attained hurricane velocities; and, though more common in winter time, in at least one instance a storm of hurricane force arrived in midsummer. That the force of these surface winds falls off somewhat rapidly after passing the ice margin was clearly indicated in several ways, and especially by their effect upon the vegetation. The willows and birches, which are generally low shrubs from one to two feet in height, in some exposed places near the ice margin were found growing flat on the ground with their branches spread out to leeward like the sticks of a Spanish fan.

Wherever a tongue of ice pushes out from the parent mass of the inland ice—and in Greenland this is generally at the head of a fjord or of a fjordlike valley—a valley flat of outwash material has been built up, and through this during the summer season the thaw water from the ice flows in braided streams (Fig. 1).

FIG. 1.—A valley flat of outwash at the head of a fjord valley below an outlet glacier from the inland ice of Greenland.

At the front of the glacier the ice cliff is, during the summer months, in a weakened state due to thawing temperatures. At frequent intervals blocks of ice fall from the cliff, and fragments which may include rock boulders are in part carried out in the braided streams which course over the outwash plain (Fig. 2). These streams are large or small according as the weather is relatively warm or cool, but with a lag of the stream behind the weather which is measured in hours only. After a warm spell which extends over several days, the outwash flat may be found in large part submerged; but after a corresponding cool period, the flowing water is reduced to a few easily fordable streams. In the early fall when thawing of the ice comes to an end, the flow of streams ceases entirely, and the plain is everywhere dried out. It is this seasonal life of the streams, and their extinction during the season when the storms off the ice have greater frequency and increased violence, which plays so important a rôle in the behavior of the materials composing the outwash plain.

Dried out at the surface, the finer materials of the outwash become the easy prey of the wind and cause sand storms which are comparable to those of deserts. Even in the summertime, sand storms which are extremely difficult to face are not unusual. In

FIG. 2.—Boulders scattered by floating ice over an outwash plain now uplifted and forming a terrace near the head of the Söndre Strömfjord. In the foreground a boulder strongly etched by the sand blast.

winter these storms are so much the more violent, and the sand is carried far and laid down over snow to produce a "sandpaper" surface particularly serious for sledging parties.

From the mixture of rock-flour with coarser materials which compose the outwash, the wind carries away the dust and finer sand, to leave behind a surface layer which near the ice front is a pebble band and further away a gritty sand—in either case a layer composed of the coarsest material which had been present in the outwash. Earlier flats of outwash materials have in the Greenland fjords been subsequently uplifted, and after dissection by the streams they now stand out as terraces of variable width lining the valley walls.

During storms the air is so filled with dust that traveling is difficult when directed toward the ice front. For days during the characteristic foehn storms the air about the Mount Evans Observatory, distant 25 miles from the ice front and nearly 1,300 feet up from the fjord, is so filled with dust that visibility is greatly reduced. This dust is caught in the tundra-covered district so that some miles distant from the ice border walking in the tundra raises a cloud of

FIG. 3.—Small sand dunes formed about vegetation along the sides of the valley flats of the Söndre Strömfjord, Southwest Greenland.

dust and the clothing becomes covered with it. That the surface of the ground is in some places materially raised as a result of this loess deposit is indicated by caribou antlers, which are sometimes buried in the tundra to as much as a third of their height.

The sand removed from the valley flats is collected in great dunes within sheltered places along the sides of the valley, and in smaller accumulations it is collected about vegetation covering much larger areas along the sides of the flats (Fig. 3). This sand, while in transit, has attacked the boulders so that examples of "stone lattice" are found in great numbers (Fig. 2).

All the characteristic effects of the wind within the better-known desert areas of lower latitudes are here found reproduced on or about the outwash plains of the continental glacier of Greenland—there are the Armored floor (pebble bed), the loess deposits within a zone marginal to the glacier, scattered boulders profoundly etched by sand-blast erosion—and all explained by the wind of the glacial anticyclone generated by the glacier itself and operating upon its own outwash deposit.

There is, of course, one notable difference between the area about the present continental glacier of Greenland and that about the continental glaciers of the Pleistocene in North America and Europe. The Greenland ice sheet is surrounded by a rugged plateau deeply dissected and representing differences of level of several thousand feet; whereas the continental glaciers of North America lay over much flatter country. Within the middle Mississippi Valley the zone of outwash, instead of being restricted to the heads of fjord valleys, as it now is in Greenland, must have formed an almost continuous zone marginal to the ice sheet. Under such conditions one may picture a zone of loess marginal to the ice, deposited in places upon till, and exhibiting thin pebble beds and scattered boulders which are etched by the sand blast as these have been found in Iowa particularly. Deposits of till would also to some extent supply materials for wind transport. In Southeast Greenland, however, there is found but little till to be affected.

31

Reprinted from *Acad. Geol. Hungary*, **9**, 65–67, 70–76, 83–84 (1965)

GENETIC CLASSIFICATION OF THE DEPOSITS CONSTITUTING THE LOESS PROFILES OF HUNGARY

By

M. Pécsi

INSTITUTE OF GEOGRAPHY OF THE HUNGARIAN ACADEMY OF SCIENCES, BUDAPEST

Observations concerning the packs of loess and fossil soils occurring in several upper Pleistocene profiles indicate that the formation of the individual packs of deposits took a time considerably shorter than hitherto believed. The evolution of a fossil soil or even of an entire fossil soil complex may not be attributed unconditionally to the full duration of an interglacial or interstadial phase. Furthermore, only a part of the loess packs occurring in the individual loess profiles is of primary aeolian origin. In the hills and mountains deluvial — stratified — varieties of loess prevailed and on the low-lying alluvial fans — the ancient floodplains of the rivers — fluvial deposition was dominant.

In the loess profiles of the Carpathian Basin, the polygenetic accumulation of a vertical sequence of various types of loesses can be proved. The aeolian deposition *in situ* is not as ubiquitous as was heretofore believed and the deluvial, proluvial, fluvial and eluvial deposits were at least equally widespread. A detailed genetic classification of these types is given in Table 4.

The genetic classification of the continental deposits and soils constituting the loess profiles of the Pleistocene, the establishing of their sequence of deposition and the determination of their probable conditions of origin is indispensable for any further and more comprehensive study into Pleistocene geology or geochronology, whether it is of palaeontologic, palaeobotanic, archaeologic or palaeopedologic nature in its principal orientation. In the past the microstratigraphic and palaeopedologic study of the Hungarian loess profiles was restricted almost entirely to the Paks profile (ÁDÁM—MAROSI—SZILÁRD 1954, P. KRIVÁN 1955, STEFANOVITS—RÓZSAVÖLGYI 1962, etc.).

Of late, the present author has made an attempt at the reconstruction of the depositional rhythms of the slope loesses on the basis of terrace and slope morphology, periglacial ground frost phenomena, loess genesis and the texture and structure of the deposits (PÉCSI 1961, 1962). This analysis was directed in the first place towards establishing the sequence of deposits occurring between two zones of fossil soil (Fig. 1). In case there occur several such fossil soil zones in the profile, the sequence of the interbedded deposits between the individual soil horizons may vary even within the same profile. On the other hand, soils of different origin and various types of loess might be observed to substitute each other in the sequence.

[*Editor's Note:* Certain figures and tables have been omitted owing to limitations of space.]

After having examined every loess profile of some importance within the Carpathian Basin, the author has distinguished the following phenomena and types of deposits.

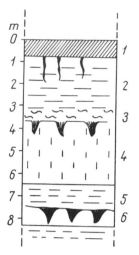

Fig. 1. Generalized profile of the phases of slope deposit formation between two fossil soil horizons of the last glaciation (PÉCSI, 1962). 1 — fossil steppe soil — chernosem — from whose lower part a polygonal pattern of lime-filled cracks extends down to a depth of 1.5 to 2 m. 2 — rhythmically stratified slope loess or sandy loess. Sediment deposited out of the slopewash. Little precipitation in early summer, considerable amount of snow, periodically frozen subsoil. 3 — finely stratified slope loess, with a slightly cryoturbated structure. Deposited by solifluction in the course of a wet cold atlantic spell of climate. 4 — unstratified, typical loess. Deposited out of the air in the driest, coldest spell of the continental stadial climate. 5 — stratified slope loess, substituted locally by likewise stratified loessy slopewash soil (loessy semipedolite). The deposition occurred by pluvionivational slopewash at the beginning phase of the stadial. 6 — fossil forest soil (or wooded steppe soil). Developed under a warm humid forest climate. The forest was exterminated by the gradual cooling of the climate. In the upper part of the forest soil traces of cryoturbation are locally observed.

1. Unstratified loess packs* (aeolian accumulation)

a) The deposits mentioned in literature under the heads "typical loess", "true loess", "aeolian loess", meaning unstratified packs of loess, do not constitute in general more than one third of the loess profile (Figs. 2, 3, 4, 5). Similar profiles have lately been described by KUKLA (1961), LOŽEK (1964), STEFANOVITS—RÓZSAVÖLGYI (1962), REMY (1960) and by GÜNTHER (1961).

b) The packs of unstratified loess are frequently substituted by unstratified sandy loess, differing only as to its grain size distribution from the above type.

The maximum thickness of the unstratified loess packs is generally less than 5 metres, being mostly about 1.5 to 4.0 m. The thicker packs of 4 to

* The term "pack" is intended to signify a number of adjacent layers, of more or less homogeneous lithology, deposited on dry land.

5 metres are bound to occur near the top of the profile (maximum of the Würm). Their physico-chemical properties, mineralogical composition, etc., are listed in Tables 1, 2 and 3.

These deposits presumably consist largely of aeolian dust. They were observed to contain in some profiles local intercalations of tuffite (KRIVÁN— RÓZSAVÖLGYI 1962, 1964). In this type of loess the percentage of biotite among the heavy minerals also may be rather high locally (30 to 50 percent). This type of deposit has been a subject of a great deal of controversy as mere lack of stratification is not an irrefutable proof for pure aeolian origin.*

In the upper Pleistocene loess profiles the number of unstratified loess packs varies from 2 to 5.

2. Stratified slope loesses and loesslike slope deposists (Deluvial accumulation)

a) The homogeneous type of stratified loess forms thin layers of 1 to 15 mm thickness readily visible even to the naked eye. Their dip (3 to 15°) is generally the same as the slope of the Recent relief. Sometimes, however, they follow the contours of valleys and dells filled up since. Although the deposits of stratified slope loess locally present a highly homogeneous aspect, the thin layers that can be taken apart in the original humid state of the substance are sometimes of markedly different grain size distribution. The aeolian dust fraction (0.02 to 0.05 mm) seldom exceeds 45 percent, and the sandy or clayey fraction or both may attain considerable importance (Tables 1, 2, 3). On the periphery of the basin where precipitation was more abundant, particularly in the western part of the Transdanubian hummocks up to the eastern foothills of the Alps, the clayey fraction is gradually enriched: stratified loess thus grades into a kind of adobe (stratified clayey loess). In the case of these marginal facies the difference in grain size distribution of the fine sandy and clayey strata becomes more pronounced.

This relatively homogeneous variety of stratified slope loess is called hummock loess.

b) As opposed to the above subtype, the piedmont and foothill slopes of the Hungarian Mountains exhibit a heterogeneous type of stratified slope loess. The packs consisting of thin layers of loess and loess-like deposits are rhytmically intercalated by layers of rock detritus. The detritus gradually grows finer with increasing distance from the mountains and the number

---

* In the grain size distribution of this type of loess the 0.02 to 0.05 mm fraction represents 45 to 60 percent by weight. The grains are encrusted by lime. The carbonate content is 20 to 25 percent, which includes some $MgCO_3$. The CO_2 content is 15 to 20 percent. Porosity is 50 to 55 percent, against that of clayey loamy loess (35 to 45 percent) and of sandy loess (60 percent) (I. LÁNYI-MIHÁLYI 1953, L. MOLDVAY 1962, I. MIHÁLTZ 1953, A. RÓNAI 1961, T. UNGÁR 1964).

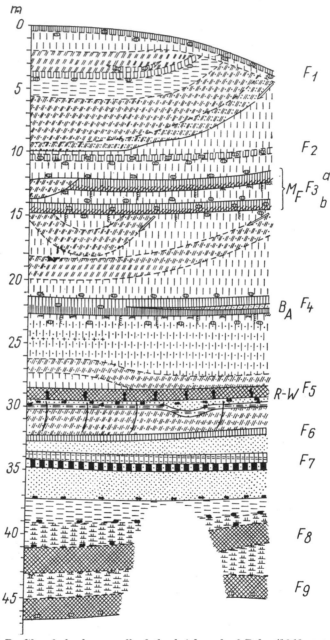

Fig. 2. Profile of the loess wall of the brickyard of Paks (1962 state)

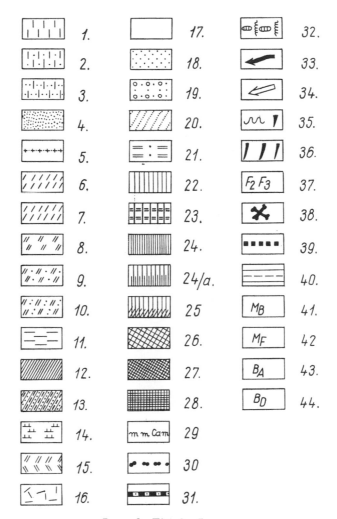

Legend: *Fig. 2—7*

I. Predominantly aeolian deposits: 1. — unstratified true loess; 2. — sandy loess; 3. — loessy sand; 4. — wind-borne sand, fine sand; 5. — volcanic tuffite

II. Predominantly deluvial-eluvial deposits: 6. — slope loess; 7. — sandy slope loess; 8. — stratified slope loess; 9. — stratified sandy slope loess; 10. — stratified loessy sand; 11. — stratified clayey loess; 12. — loessy slopewash soil (loessy semipedolite); 13. — sandy, clayey slopewash soil

III. Altered loesses. 14. — clayey loess (adobe, loess loam, ancient loess); 15. — decalcified clayey loess; 16. — gleyified loess; 17. — loess pack in general, genesis not specified

IV. Fluvial-proluvial deposits: 18. — fluvial sand; 19. — fluvial sand and gravel; 20. — proluvial sand; 21. — sandy clay, clayey sand (mud, silt)

V. Recent and fossil soils, illuvial horizons of soils: 22. — slightly humified horizons; 23. — swamp and marsh soil; 24. — chernosem-type soils; 24/a. — chestnut soil (kastannosem) (?); 25. — chernosem-brown forest soil; 26. — brown earth; 27. — brown earth with clayey illuvium ("Parabraunerde"); 28. — red (clayey) soils; 29. — horizon of lime accumulation ("Kalkilluvialhorizont"); 30. — lime dolls; 31. — limy sandstone bank, layer of concretions; 32. animal burrows, wasp holes

VI. Miscellaneous: 33. — erosional hiatus; 34. — phase of derasion (dell scooping); 35. — features of cryoturbation, solifluction; 36. — desiccation cracks; 37. — fossil soil horizons and accumulation of humus; 38. — finds of vertebrate fossils; 39. —charred wood fragments; 40. — limits of the formations: *a*) well-defined, *b*) vague, *c*) unconformity; 41. — "basal soil complex of Mende"; 42. — "upper soil complex of Mende"; 43. — "lower soil complex of Basaharc"; 44. — "upper soil complex of Basaharc"

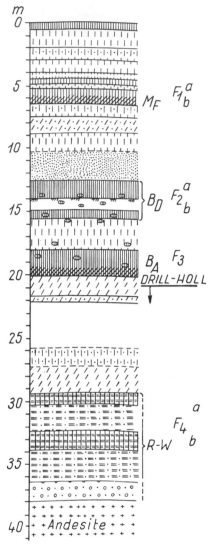

Fig. 3. Profile of the loess wall of the brickyard of Basaharc. The loessy series overlies the second terrace above the floodplain of the Danube

of detrital layers also diminishes. To the diverse regional types of stratified heterogeneous loess the author has given the comprehensive name of "mountain loess" (stratified slope loess with detritus). In the loess profiles of the mountains the packs of unstratified ("typical") loess play but a subordinate part.

The alternation and the proportions of stratified slope loess and un-stratified "typical" loess within a given profile are determined by local as well as regional factors. Thus e.g. within a given area the slopes of southern exposure may be covered exclusively with stratified slope loess.

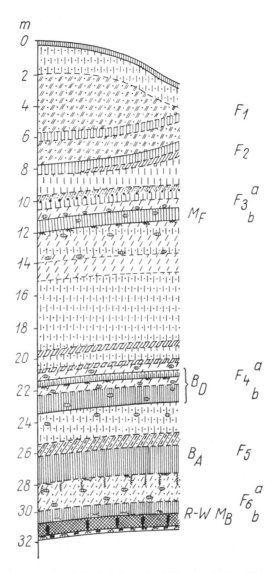

Fig. 4. Profile of the loess wall of the brickyard of Mende, 40 km SE of Budapest

c) The hummock and mountain subtypes of stratified slope loess occur in a number of varieties hard to tell apart. Some of them cannot even be classified as loesses any more, in which case they are to be termed loess-like slope deposits. This term comprises the loamy and clayey varieties on the one hand and the sandy varieties on the other.

A widespread type of stratified loess-like deposit is the loessy slope-wash soil (loessy soil sediment).

Fig. 5. Sketch map of the loess profiles. 1 — Basaharc, 3 — Mende, 4 — Paks, 5 — Erdut,
7 — Stari Slankamen, 9 — Szombathely, 12 — Békéscsaba

The abundance of reworked soil or humus in the slope deposits can be as high as to render the layers of loess or sandy loess wholly inconspicuous between the fine layers of slopewash soil. In this case we are dealing with a stratified slope sediment soil ("semipedolite" PÉCSI 1961, "Lehmbröckelsand" KUKLA 1961, LOŽEK 1964).

The geomorphologic, microstratigraphic and lithologic study of the stratified slope loesses and loess-like deposits reveals a deposition by melt-water and precipitation running in the form of slopewash over permanently or periodically frozen ground. Most of the accumulation must have taken place in the stadials as the stratified slope-loess packs frequently exhibit the traces of syngenetic cryoturbation. However, the cryoturbated forms do not warrant the conclusion that the climate was a uniformly humid and cold one. It seems that the transportation and deposition of the material was rather sluggish and intermittent, several interruptions being indicated by embryonal soils and phases of humification (Fig. 4).

3. Unstratified slope loesses and loess-like deposits, glacial loams

The subtypes of this group and their areal distributions recall those of the above group. A number of transitional types are known, from slope loesses that appear practically homogeneous, to glacial loams that had lost most of the features they once had in common with loess.

a) *Unstratified slope loess* cannot in general be distinguished from unstratified aeolian loess except if thoroughly examined in the field and in the laboratory. The best indicator is the presence of diverse scattered inclusions in the loess pack, rock detritus of various size, pebbles, minute clods or bands of fossil soil, clayey, loamy or sandy lenticles of vague outlines offering irrefutable proof of a deluvial redeposition. The considerable scatter of the grain size distribution, the rather high percentage of humus and of organic matter at large, the compacted structure and the microscopic aspect of the sample serve as clues in the laboratory.

In certain instances it is apparent that slope loess, homogeneous at present, was stratified to start with and subsequently lost its stratification in the course of its diagenesis — a process whose intimate relationship with the overlying fossil or Recent soil horizon can readily be demonstrated locally.

b) The subgroup of *unstratified loess-like slope deposits* is very large. Also the genesis of the deposits belonging to this subgroup has been interpreted in a number of different ways. Some authors do not consider it as a loess type at all; others classify but a few of its varieties under the head of loess; others consider it as a reworked loess or, in the case of certain subtypes, as the mountain facies of loess.*

If the loess-like slope deposit contains irregularly distributed gravel or detritus of varied size, it is relegated to the subtype of slope loess with detritus (1). On the piedmont slopes of southerly exposure of the Hungarian Mountains there frequently occur slope loesses containing detritus mixed with fossil soil (2). On the more humid western periphery of the basin a loamy slope loess (3) occurs in vast patches.

In distinguishing the loesses and loess-like deposits of diverse derivation the physical and chemical laboratory methods have not proved satisfactory. In our opinion the decisive clues are to be expected from the three-dimensional microstratigraphic and structural analysis of the individual layers constituting the profile.

c) A rather widespread variety of *loamy slope loess* is *a glacial slope loam,* showing but very vague, if any, loess-like features. This variety too carries some detritus.**

---

* Based on his own observations the present author calls "loess-like slope deposits" such deposits which contain a dust fraction not as high as that found in typical loess but still high enough (30 to 40 percent), a certain amount of $CaCO_3$, and feature a mineralogical composition resembling that of loess, a high porosity and forms of relief that could have developed on top of loess as well.

** This formation is a type reworked by slopewash of the Pleistocene deposit known by the names listed below: German, Staublehm (FINK), Russian, suglinok (MOSKVITIN, GERASSIMOV), Hungarian, nyirok (KEREKES), brown earth (SÜMEGHY) or glacial loam (BULLA).

d) The accumulation of the slope deposits in question in their present position is due — besides slopewash by precipitation and meltwater — in the first place to *periglacial slope solifluction* (Pécsi 1961, 1962, 1964). Even in the past the clayey types of this group have been attributed in this country to solifluction and slopy-tundra phenomena (Szádeczky-Kardoss 1963, Kerekes 1941, Bulla 1939, 1960, Peja 1959, Székely 1964 and many others). These deposits or packs of layers frequently exhibit traces of freeze-and-thaw solifluction, "Streifenböden", involutions and cryoturbated forms at large.

[*Editor's Note:* Material has been omitted at this point.]

REFERENCES

ÁDÁM, L.—MAROSI, S.—SZILÁRD, J.: Földrajzi Közlemények **2**, (78), 239—254, 1954. — ÁDÁM, L.—MAROSI, S.—SZILÁRD, J.: A Mezőföld természeti földrajza. (Abstracts: Die Geomorphologie des Mezőföld, Fizhicheskaya geografiya Mezefelda.) Földrajzi Monográfiák II, Akadémiai Kiadó, Budapest, 1959. — BORSY, Z.: A Nyírség természeti földrajza. (Physical geography of the Nyírség region.) Földrajzi Monográfiák V, Akadémiai Kiadó, Budapest, 1961. — BERG, L. S.: Klimaty zhizn. Moscow 1947. — BULLA, B.: Földtani Közlöny. **67**, 196—215, and 289—309, **68**, 33—58, 1937 and 1938. — BULLA, B.: Földrajzi Közlemények **67**, 268—281, 1939. — BULLA, B.: Quelques problèmes géomorphologiques interglaciaires de la zone périglaciaire du Pléistocène. Studies in Hungarian Geographical Sciences. Akadémiai Kiadó, Budapest, 1960. — FINK, J.: Mitteilungen d. Geol. Gesellsch. Wien **53**, 249—266, 1960. — FINK, J.: Mitteilungen der Geol. Gesellsch. Wien **54**, 1—25, 1961. — FÖLDVÁRI, A.: Földtani Közlöny **86**, 351—356, 1956. — FRANYÓ, F.: Magyar Állami Földtani Intézet Évi Jelentése 1961-ről, 31—46, 1961. — GERASIMOV, I. P.: Petermanns Geogr. Mitt. **105**, 234—248, 1960. — GERASIMOV, I. P.: Lessoobrazovaniye i pochvoobrazovaniye. (Loess and soil formation.) INQUA VI. Congress. Abstracts of Papers. Lodž 1961. — GÜNTHER, W.: Sedimentpetrographische Untersuchung von Lössen. Köln 1961. — HAASE, G.: Geographische Berichte. **8**, 97—129, 1963. — HORVÁTH, A.: Állattani Közl. **44**, 171—184, 1954. — INKEY, B.: Földtani Közlöny. **8**, 15—26, 1878. — KÁDÁR, L.: Közlemények a KLTE Földrajzi Intézetéből. Debrecen, **19**, 1954. — KÁDÁR, L.: Földrajzi Közlemények **4**, (80), 143—163, 1956. — KÁDÁR, L.: Features of the loess plains in the regions of alluvial fans. INQUA VI. Congress. Abstracts of Papers. Lodž 1961. — KEREKES, J.: Beszámoló a Földtani Intézet Vitaüléseinek munkálatairól, 97—149, 1941. — KRETZOI, M.: Acta Geologica **2**, 67—79, 1963. — KRIVÁN, P.: Magyar Állami Földtani Intézet Évkönyve. **43**, 364—512, 1955. — KRIVÁN, P.—Rózsavölgyi, J.: Földtani Közlöny **92**, 330—333, 1962. — KRIVÁN, P.—RÓZSAVÖLGYI, J.: Földtani Közlöny **94**, 257—268, 1964. — KUKLA, J.: Vestnik Ustr. Ustavu Geologického **36**, 369—372, 1961. — KUKLA, J.—LOŽEK, V.—ZABURA, A.: Quartärband **13**, 1—29, 1961. — LÓCZY, L.: A Balaton környékének geológiája és morfológiája. I. rész. A Balaton környékének geológiai képződményei és ezeknek vidékek szerinti telepedése. I. szakasz. (Geology and relief of the Lake Balaton region. Part I. Geological formations of the region Lake Balaton and their areal distribution. Section I.) A Balaton Tud. Tanulm. Eredményei 1913. — LOŽEK, V.: Quartärmollusken der Tschechoslowakei. Prague, 1964. — MARKOVIĆ—MARJANOVIĆ, E.: Less v Jugoslavii. (Loess in Jugoslavia.) INQUA VI. Congress. Abstracts of Papers. Lodž, 158—159, 1961. — MIHÁLTZ, I.: Magyar Állami Földtani Intézet Évi Jelentése 1950-ről, 113—138, 1953. — MIHÁLTZ, I.: Acta Geologica **2**, 109—121, 1953. — MIHÁLYI—LÁNYI, I.: Alföldi Kongreszszus. Az Alföld földtani felépítése. Akadémiai Kiadó, Budapest, 1953. — MOLDVAY, L.: Acta Univ. Szegediensis, Ser. Min.-Petr. **14**, 75—109, 1961. — MOLNÁR, B.: Földtani Közlöny **91**, 300—315, 1961. — MOSKVITIN, A. I.: Byuleten Kom. po izucheniyu chetvertichnogo perioda **28**, 33—55, 1963. — PEJA, GY.: Borsodi Földrajzi Évkönyv **2**, 5—23,1 959. —PÉCSI M.: Inst. Geol. Prace **34**, 187—311, 1961. — PÉCSI, M.: Földrajzi Értesítő **11**, 19—35, 1962. — PÉCSI, M.: Petermanns Geographische Mitt. **107**, 161—182, 1963. — PÉCSI, M.: Biuletyn peryglacjalny. **14**, 279—293, 1964. — PÉCSI, M.: Ten years of physico-geographic research in Hungary. Studies in Geography. Akadémiai Kiadó, Budapest, 1964. — PINCZÉS, Z.: Földrajzi Értesítő **3**, 575—584, 1954. — RÓNAI, A.: Die Bedeutung der Quartärforschung in Ungarn. Institut Geologiczny Prace, INQUA VI. Congress. Lodž, 241—245, 1961. — SCHERF, E.: Versuch einer Einteilung des Ungarischen Pleistozäns auf moderner polyglazialistischer Grundlage. Verhandl. d. III. Internat. Quart. Konf. Wien, 1938. — SOKOLOVSKY, I. L.: Regionalniye i geneticheskiye tipy lessovych porod. (Regional and genetic types of loess formations.) INQUA VI. Congress. Abstracts of Papers. Lodž, 164—165, 1961. — REMY, H.: Eiszeitalter und Gegenwart **11**, 107—120, 1960. — STEFANOVITS, P.—RÓZSAVÖLGYI, J.: Agrokémia és talajtan **11**, 143—160, 1962. — STEFANOVITS, P.: Magyarország talajai. (The soils of Hungary.) Akadémiai Kiadó, Budapest, 1963. — SÜMEGHY, J.: Magyar Állami Földtani Intézet Évi Jelentése 1953-ról, 394—404, 1955. — SZÁDECZKY-KARDOSS, E.: Földtani Közlöny **66**, 213—228, 1936. — SZÉKELY, A.: Földrajzi Közlemények **12**, (87), 199—218, 1964. — TREITZ, P.: Földrajzi Közlemének, 225—277, 1913. — UNGÁR, T.: Hidrológiai Közlöny **44**, 537—545, 1964.

32

Reprinted from *Geog. Ann.*, **54**(2), Ser. A, 85–88, 94–96, 102–104 (1972)

LOCATION, MORPHOLOGY AND ORIENTATION OF INLAND DUNES IN NORTHERN SWEDEN

BY MATTI SEPPÄLÄ

Department of Geography, University of Turku and
Department of Physical Geography, University of Uppsala

ABSTRACT. By interpreting aerial photographs periglacial inland dunes in the northernmost parts of Sweden were studied. These dunes are now mostly stable. Aspects studied were their position, morphology, orientation together with recent deflation and the direction of the ancient winds that formed the dunes.

The most common type of dune is the parabolic dune. The dunes are situated on and in the vicinity of eskers, glaciofluvial deltas, outwash plains, valley trains and glacial drainage channels. Seven detailed geomorphological maps were made of different areas where the dunes are encountered using information from the aerial photographs.

There are both palsa and string bogs in the dune areas.

In many of the deflation basins of parabolic dunes there are deflation lakes which have formed as a result of a rise in the level of the ground water after the formation of the dunes.

In all dunes were encountered at 94 localities and the direction of the winds forming the dunes was determined. The directions of the winds varied between N 15° W and N 80° W. The direction of the effective winds coincides fairly closely with that of the edge of the melting ice sheet. There was no evidence at all to show that the shape of the dunes might have been influenced by fall winds blowing off the ice sheet.

Present-day eolian action is principally of a deflationary nature even though small new parabolic dunes are forming in several areas. One particularly dynamic dune area was noted in the north of the area studied. There the parabolic dunes are still moving.

Introduction

New methods of research have stimulated studies of dunes in northern regions where the formation of dunes is closely linked with glaciofluvial sediments devoid of vegetation and which were deposited by the continental ice sheet. New methods of dating together with sedimentological research and the interpretation of aerial photographs have been the most notable steps forward which have affected research on dunes.

When attempting to find examples of studies of dunes resembling those of the area of Scandinavian glaciation, it may be noted that it is perhaps the inland dunes of Poland which have been most intensively studied since the Second World War (e. g. Galon 1959, Stankowski 1963, Wasylikowa 1964, Pernarowski 1966, Dylikowa 1969, Prusinkiewicz 1969, Kobendzina & Urbaniak 1969).

In Finland, too, scholars have increasingly turned their attentions to inland dunes which have become anchored by vegetation (Ohlson 1957, Aartolahti 1967, Seppälä 1969, 1971, Jauhiainen 1970, Lindroos 1972). Some of the studies mentioned have been carried out in nothern regions once covered by the Scandinavian ice sheet. In these regions the margin of the ice sheet retreated in the course of deglaciation to the S and SW.

Eolian forms have also been the subject of research in Norway, in particular using aerial photographs (Klemsdal 1969).

The inland dunes of the northern parts of Sweden, on the other hand, have not yet been examined thoroughly with the aid of new methods of research. It is true, however, that there are in existence a number of old studies (e.g. Högbom 1923, pp. 157—170; Lundqvist 1943, pp. 135—143). The area covered by Fromm (1965) does not extend farther north than 67°15', so that it covers only the southern part of the area dealt with in this paper. Hoppe (1959) has taken dunes into account to some extent in his study of the soils to be found in the river valleys of northern Sweden: this study was made with the help of aerial photographs.

In this paper an inventory of the dune areas of northern Sweden has been drawn up by first locating the most important glaciofluvial deposits on geological maps of northernmost Sweden. Dunes were located on the aerial photographs of these areas (Fig. 1) on

[*Editor's Note:* Certain figures and tables are omitted owing to limitations of space.]

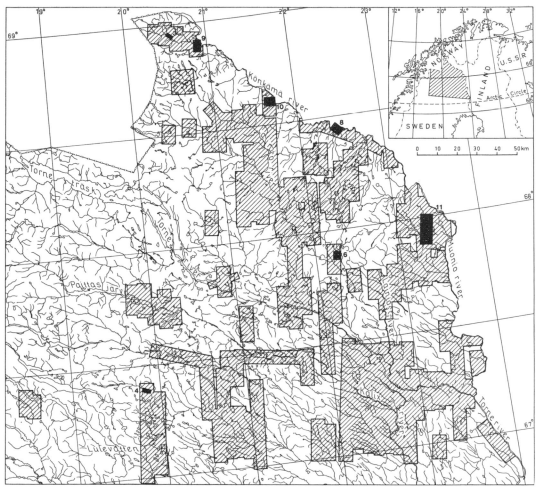

Fig. 1. Investigated area. Areas of which aerial photographs were studied are hatched. Black rectangles indicate locations of which detailed geomorphological maps were made. Numbers correspond to the numbers of figures.

a scale of approximately 1 : 30,000 (photographed at an altitude of approx. 4 600 m). Altogether 680 photographs were examined. The whole of the nothern Sweden was not, therefore, covered and the inventory does not provide an exhaustive picture of the occurrence of dunes in the area. However, compared with earlier studies (e.g. Högbom 1923, Fig. 6; Lundqvist 1943, Fig. 39; Fromm 1965, Fig. 94) numerous new dune areas were found. A number of small low dunes in areas covered by the aerial photographs may have passed unnoticed in the southern parts of the area because of a dense covering of vegetation.

The problems dealt with in this study are: 1) the position of the dunes and their relation to other geomorphological forms, 2) the shape of the dunes and their orientation together with the conlusions that may be drawn on the basis of these two factors as to the direction of the winds that caused the dunes to be formed, and 3) observations of present-day deflation from the aerial photographs and the regional distribution of deflation.

Location

Scholars who have studied dunes have noted on many occasions earlier that the inland dunes of northern Fennoscandia are often closely linked

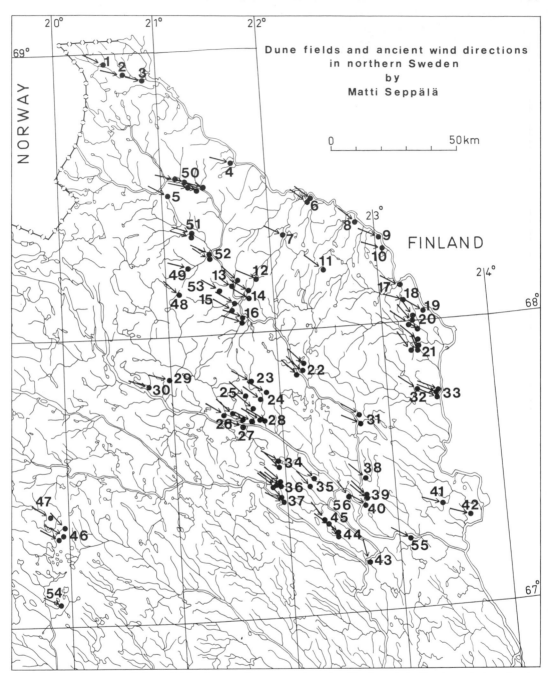

Fig. 2. Location map of dune fields. Arrows show approximately effective wind directions during the dune formation period at each point where dunes have been observed. Numbers refer to Table 1. One number may refer to several points (see Table 1).

with deposits of glaciofluvial material. One precondition for the formation of the dunes was that there was available material of suitable grain size which had not been anchored by a covering of vegetation.

Since the majority of glaciofluvial sediments

have been deposited at low altitudes, most of the dunes are encountered in valleys. The position of dune areas is shown on the map (Fig. 2) and in Table 1 the dune areas have been given names according to the nearest named localities. On the topographical maps are given the approximate altitudes of the dune areas, which vary from 155 to 620 m a.s.l. (Table 1). The highest areas are to be found in the N and SW parts of the investigated area (Nos. 1—5 and 46—51, Fig. 2, Table 1). Dunes found on the lowest altitudes (below 170 m a.s.l.) are situated in the SE parts of the investigated area (Nos. 41—43, Fig. 2, Table 1). The lowest dune areas are found below the highest ancient shore line (Lundqvist & Nilsson 1957).

Most of the dunes are situated in the immediate vicinity of eskers (see Table 1). Depending on the direction in which the eskers run the dunes are found either to the E, S or N of the eskers. Where the esker runs in a more or less NW—SE direction, then small dunes have formed on both sides of the ridge (Fig. 3). In such cases the material of the dunes has not necessarily originated from the esker itself but from glaciofluvial deposits on either side of the esker. In cases of this kind the direction of the wind was approximately the same as that of the esker. In the upper course of the river Kummajoki (No. 1, Fig. 2) there are small transverse and parabolic dunes only to the N of the esker. Since the esker, which have run in the same direction as the wind, is rather wide and is made up of sufficiently fine material a dune field has formed partly at the end of the esker, partly on both sides of it and partly as a continuation of the esker (Fig. 4 and 5, No. 47, Fig. 2).

Fig. 7 shows a case where dunes have formed on glaciofluvial deposits that form a continuation of an esker: the dunes lie to the SE of the esker indicating that the wind must have blown in more or less the same direction as the esker.

Many of the eskers of northern Sweden lie in a SW—NE direction. In such cases the dunes have formed on the eastern and southeastern sides of the esker (Fig. 6, No. 22, Fig. 2) (cf. Högbom 1923, pp. 164—166). Their position shows that the wind blew across the eskers.

In the neighbourhood of the eskers there are to be found many different kinds of glaciofluvial deposits which may have been de-formed by earlier or by recent river action. The Karesuando dunes (No. 6, Fig. 2) lie close to flat glaciofluvial deposits and eskers(Fig. 8) but it is possible that part of the material of which they are formed may have originated from the alluvial deposits of Muonio river. These latter deposits now probably lie in part beneath bogs. There is one special feature of this dune field which warrants mention and that is that the dunes have drifted in part over and above drumlins. It is unlikely that the material of the drumlins could have provided the material for eolian deposits.

Both the dunes of Vittangi (No. 27, Fig. 2, Table 1) and of Junosuando (No. 56, Fig. 2, Table 1) lie right on the banks of the Torne river and parts of them at least may be river-bank dunes originally of alluvial material (cf. Högbom 1923, p. 167).

The old channel of the Merasjoki river has cut through the dunes lying north of the river at a number of points (Fig. 11) but elsewhere river action has not greatly affected dunes in the investigated area.

One particularly notable geomorphological form connected with dunes are glaciofluvial valley trains which either in part or completely fill old river valleys. Dune fields of this type are exemplified by Nos. 13, 32, 33, 37, 49, 50, 51 and 52 (Fig. 2 and Table 1). Furthermore, many other dune fields are situated in broad river valleys in which are found glaciofluvial deposits with no particular characteristic morphological features.

Dune fields Nos. 3, 18, 21, 41 and 43 (Table 1 and Fig. 2) have formed either on the surface of or in the vicinity of glaciofluvial deltas. The last two of these deltas formed as marginal deltas below the highest ancient sea level. The others are ice-dammed lake deltas: as an example of these a geomorphological map was drawn showing in detail the delta lying near the Kummajoki river (Fig. 9). Here a large group of parabolic dunes has formed at the distal end of the actual delta in an area of glaciofluvial deposits where there is a low esker, too. At the western edge of the delta there can be seen four shorelines of a former ice lake in the form of erosion terraces (e. g. Tanner 1915, Fig. 94, p. 509). On both sides of the overflow channel there are two small residual delta plains higher than the large one (Fig. 9).

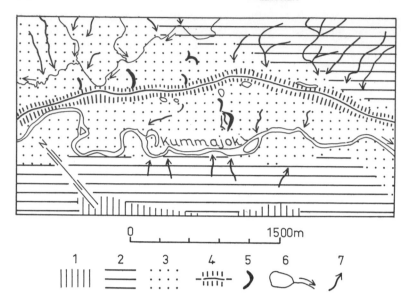

Fig. 3. Detailed geomorphological map of dune field near Kummajoki (see Fig. 1; Point No. 2 in Fig. 2 and Table 1). Key to symbols: 1, Bare rock. 2, Till-covered slope. 3, Glaciofluvial deposits with no special morphological features. 4, Esker. 5, Dune. 6, Lake and river. 7, Snow melt-water gully.

Only a few typical large outwash plains have been found in the northernmost parts of Sweden. With the help of aerial photographs a large sandur-like field near Siikavuopio was charted and a map drawn to illustrate these outwash plains. The surface is covered with numerous glacial drainage channels and a parabolic dune has formed on the outwash plain (Fig. 10). At the time the outwash plain formed there was still ice in the river valley to the east of it so that a steep edge was formed to the outwash plain at the point of contact with the ice and the channels at the eastern edge of the plain terminate in thin air. Meltwater and a tribuatry of the Könkämä river have later eroded a deep valley along the northern

edge of the outwash plain. The winding eastern edge of the plain and ravines caused by meltwater action do not support the view that this glaciofluvial form might have formed as an ice lake delta (cf. Tanner 1915, pp. 511—513).

Many of the dune fields are characterised by glaciofluvial drainage channels which have now dried; many of them become filled with bogs (See Table 1). As examples of this phenomenon mention may be made of the dunes near the Kiimajoki eskers (No. 20, Fig. 2 and 11) and the Jukkasjoki dunes (No. 44, Fig. 7). Glaciofluvial drainage channels are also encountered in the vicinity of dune fields 13, 21, 23, 25, 27, 31, 35, 37, 38, 39, 40, 42, 55 and 56 (Table 1). The dunes have in part drifted into the chan-

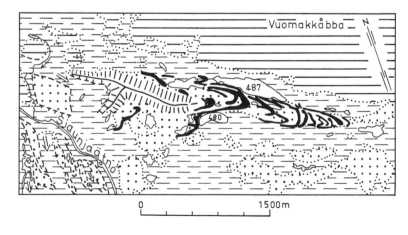

Fig. 4. Detailed geomorphological map of dune field near Vuottasjaure (Fig. 1; No. 47 in Fig. 2 and Table 1). Key to symbols in Fig. 11. See Fig. 5.

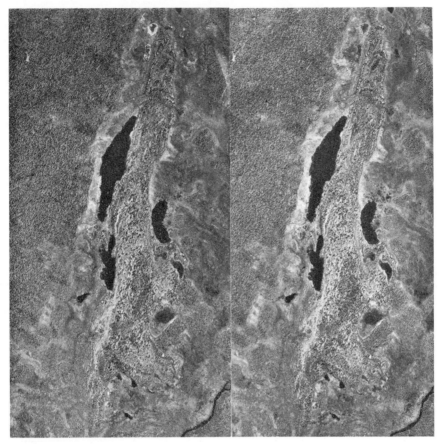

Fig. 5. Stereogram of the area in Fig. 4. Aerial photograph by Rikets Allmänna Kartverk (64 Je 274 22 and 23).

ncls and across them.

In several areas there are small pools which have no outlet near the dunes. These pools are to be found in the deflation basins of parabolic dunes and partly surrounded by the dunes. Deflation lakes of this type were observed in the case of dune fields Nos. 6, 9, 13, 16, 20, 21, 22, 28, 43, 44, 50 (Figs. 2, 6, 7, 8 and 11). The author has described similar deflation lakes in Finnish Lapland in detail earlier (Seppälä 1971, pp. 18—23). In the same paper (pp. 69—70 and 78) the author came to the conclusion that the formation of deflation lakes took place during the Atlantic period when the level of the ground water rose. On the basis of numerous observations made in the course of the present study it may be said that level of the ground water rose in general after the time of the

formation of dunes in Fennoscandia. The deflation lakes encoutered in the Kaamasjoki-Kiellajoki river basin (Seppälä 1971) cannot therefore have been the result of purely local factors but the climate in the area must obviously become more humid after the dunes had been formed.

Very many of the dunes are bounded at the distal end by bogs (Figs. 4—9, 11). Both palsa and string bogs are to be found in the area investigated (Table 1). These were easy to identify on the aerial photographs (cf. Hoppe & Blake 1963). There are large palsa bogs near dune fields Nos. 1—6 and 50—52 (Figs. 8—10). The southern limit of the region in which palsa bogs occur in the investigated area runs a little to the south of the 68° N latitude (Lundqvist 1943, Fig. 45; Hoppe & Blake 1963, Fig. 1; Rapp & Clark 1971, Fig.

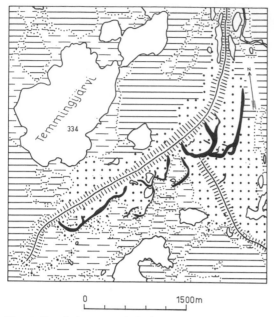

Fig. 6. Detailed geomorphological map of dune field near Temmingijärvi (Fig. 1; No. 22 in Fig. 2 and Table 1). Key to symbols in Fig. 11.

1). Palsa bogs are therefore to be found in the vicinity of other dune areas than those referred to above even though in many cases string bogs predominate in the terrain. Palsa and string bogs are found, at least partly, in the same areas; at the very north of the area studied palsa bogs are in the majority while in the south it is string bogs which are dominant.

Morphology of dunes

The most common type of dune encountered in the area is the parabolic dune (Figs. 3—11, Table 1): this is the most characteristic type of all dunes in periglacial areas (e.g. Högbom 1923, Hörner 1927, Louis 1929, Kádár 1938, Black 1951, Galon 1959, Bird 1967). Since the formation of parabolic dunes is brought about by vegetation (e.g. Kádár 1938, p. 169), the sand of the dune is partly bound by vegetation at the time the sand is deposited. Parabolic dunes are often to be found in clusters which also take the form of a parabola (Figs. 4, 9 and 11) (cf. Solger 1910, Högbom 1923, Louis 1929, Galon 1959, Kobendza 1958, Seppälä 1971). The noses of

the parabolae point E or SE. The longest arms of parabolic dunes measure as much as 2 km approximately (Fig. 11). The height of the dunes varies between 1 and 15 m. The highest point is at the nose of the dune. The gradient of the slopes of the dunes is difficult to determine using aerial photographs but, in general, it may be said that the arms of the dunes are almost equilaterally triangular in cross-section whereas nearer the nose the lee side becomes progressively steeper than the windward side (cf. Lundqvist 1943, Fig. 40; Seppälä 1971, p. 16).

In addition to parabolic dunes a number of transverse dunes were found (Table 1); these often had slightly curved ends to the west. Their existence reflects wind action that was both weak and of short duration causing the sand to drift but not to be carried very far. The development of dunes of this type is to be seen as a stage in the forming of a parabolic shape by way of a transverse form and, under ideal conditions, would lead to two longitudinal dunes (e.g. Kádár 1938, Fig. 2; Galon 1959, Fig. 6; Lindroos 1972, Fig. 42).

Only a few examples of clearly defined longitudinal dunes were observed (e.g. No. 27 and 47, Table 1, Figs. 4 and 5). Dune No. 27, which lies on the south side of the Torne river, is clearly a riverbank dune as was noted earlier (p. 88). Longitudinal dunes following the direction of an esker (No. 47) and large parabolic dunes are evidence of strong wind action in the places where they are encountered.

[*Editor's Note:* Material has been omitted at this point.]

Conclusions

Aerial photography has proved to be an extremely practical method for studying dune areas, deflation and other geomorphological features connected with dunes covering wide areas in periglacial regions with little vegetation. From the aerial photographs it is possible to make topographical maps of dune areas but detailed geomorphological maps have proved to be more useful for determining the origin of the dunes and the directions of ancient winds which caused the dunes to form.

The dunes are found together with deposits of glaciofluvial material. The position of eskers, deltas, outwash plains and valley trains in relation to the dunes together with the form of the dunes and the gradient of their slopes provide support for the view that the winds that brought about the formation of the dunes blew more or less from the NW, from between N 15° W and N 80° W.

The winds blew along the edge of the ice sheet during the deglaciation so that there can be no question of winds blowing off the ice sheet. Quite clearly there was high pressure above the ice sheet and this caused the normally prevailing westerly winds to change direction. Because of differences in direction of the winds which could be determined from different dune areas it is possible to conclude that the dunes were in all probability not formed at the same time. Rather those dunes in the SW of the area studied are younger than those found in the N and E. It is clear that soon after the melting of the continental ice sheet areas of sand became covered with vegetation judging by the general absence of deformation in the dunes and the incomplete parabolic shape of some dunes. Only in rare cases have the dunes wandered more than 2 km from the place from which the material of the dunes is assumed to have been transported (cf. Fig. 11). Most of them are situated less than a kilometre from a source of sand, for example, an esker.

In the deflation basins of parabolic dunes there are in many places deflation lakes which are hollows in the deflation basins that have filled with water after a rise in the level of the ground water. This indicates that the climate has become more humid after the time at which the dunes ceased to move any more.

The present-day periglacial nature of the northern part of the investigated area is evidenced by palsa bogs and strong deflation.

Nowadays wind action in the area tends on the whole to destroy the forms. Small recent parabolic dunes are to be found in several of the dune areas, however; these have resulted from deflation caused by forest fires. The areas of strongest wind action today is in the upper course of the Lainiojoki river, where a number of large parabolic dunes are still at a dynamic stage of their development (No. 50, Fig. 2). The area is more than 500 m a.s.l. and obviously has a particularly dry and cold climate as there are large palsa bogs in the vicinity.

The strength of deflation lessens gradually the farther south the dunes are situated and some of the dunes found in the lowest river valleys are completely covered and bound by vegetation.

Acknowledgements

Professors Åke Sundborg and John O. Norrman of the Department of Physical Geography, University of Uppsala have been of the greatest help in encouraging me to write this paper by the way they have given me advice and assistance. The department placed the necessary aerial photographs at my disposal and also provided me with all facilities. Mr. Rolf Å. Larsson, Uppsala, gave me considerable practical help. In particular I should like to mention the staff of the photograph archives of Rikets Allmänna Kartverk, Stockholm, whose experience and advice contributed in no small measure to bring this work to its conclusion. Mrs. Leena Kiiskilä, Turku, has made the final drawings for the figures and Mr. Christopher Grapes, Turku, has translated the paper into English.

The work was made possible by financial support from the Finnish-Swedish Cultural Foundation and the National Research Council for Natural Sciences, Finland.

To all these people and institutions I should like to express my deep-felt gratitude.

Dr. Matti Seppälä, Department of Geography, University of Turku, SF-20500 Turku 50, Finland.

Flygfotograferingen utförd av RAK 1964 och 1968. Godkänd för reproduktion och spridning av RAK. Kartorna samtidigt godkända från sekretessynpunkt för spridning.

References

Aartolahti, T., 1967: Über die Dünen von Urjala. *C. R. Soc. géol. Finlande 39:* 105—121.

Ångström, A., 1958: *Sveriges klimat.* Stockholm.

Bird, J. B., 1967: The Physiography of Arctic Canada, with special reference to the area south of Parry Channel. Baltimore.

Black, R. F., 1951: Eolian deposits of Alaska. Arctic 4: 89—111.

Dolgin, I. M., 1970: Subarctic meteorology. UNESCO Symposium on the Ecology of the Subarctic Regions, Proceedings of the Helsinki Symposium: 41—61.

Dylikowa, A., 1969: Le problème des dunes intérieures en Pologne à la lumière des études de structure. Biul. Perygl. 20: 45—80.

Fromm, E., 1953: Glaciation and changes of level in Quaternary Age. Atlas över Sverige: 19—20.

— 1965: Beskrivning till jordartskarta över Norrbottens län nedanför lappmarksgränsen. Summary: Quaternary deposits of the southern part of the Norrbotten county. Sveriges geol. unders. Ser Ca. Nr 39: 1—236.

Galon, R., 1959: New investigations of inland dunes in Poland. Przegląd Geogr. 31: 93—110.

Hoppe, G., 1959: Om flygbildstolkning vid jordartskartering jämte några erfarenheter från jordartskartering i norra Lappland. Geol. Fören. Stockholm Förh. 81: 307—315.

— 1967: Case studies of deglaciation patterns. Geogr. Ann. 49 A: 204—212.

Hoppe, G. and Blake, I. O., 1963: Palsmyrar och flygbilder. Summary: Palsa bogs and air photographs. Ymer 83: 165—168.

Högbom, I., 1923: Ancient inland dunes of northern and middle Europe. Geogr. Ann. 5: 113—243.

Hörner, N. G., 1927: Brattforsheden. Ett värmländskt randdeltakomplex och dess dyner. Sveriges geol. unders. Årsbok 20: 1—192.

Jauhiainen, E., 1970: Über den Boden fossiler Dünen in Finnland. Fennia 100: 3: 1—32.

Jennings, J. N., 1957: On the orientation of parabolic or U-dunes. Geogr. J. 123: 474—480.

Kádár, L., 1938: Die periglazialen Binnendünen des norddeutschen und polnischen Flachlandes. C. R. Congrès Intern. Géogr. Amsterdam. Actes du Congrés I: 167—183.

Klemsdal, T., 1969: Eolian forms in parts of Norway. Norsk Geogr. Tidsskrift 23: 49—66.

Kobendza, J. & R., 1958: Rozwiewane wydmy Puszczy Kampinoskiej. Résumé: Les dunes éparpillées de la Forêt de Kampinos. Wydmy śródladowe Polski I: 95—170.

Kobendzina, J. & Urbaniak, U., 1969: Bibliografia wydmowa Polski. Inst. Geogr. Polskiej Akad. Nauk, Prace Geogr. 75: 369—386.

Kujansuu, R., 1967: On the deglaciation of western Finnish Lapland. Bull. Comm. géol. Finlande 232: 1—98.

Liljequist, G. H., 1956: Meteorologiska synpunkter på istidsproblemet. Ymer 77: 59—73.

Lindroos, P., 1972: On the development of late-glacial and post-glacial dunes in North Karelia, eastern Finland. Geol. Survey Finland, Bull. 254: 1—85.

Louis, H., 1929: Die Form der norddeutschen Bogendünen. Z. Geomorph. 4: 7—18.

Lundqvist, G., 1943: Norrlands jordarter. Sveriges geol. unders. Ser C. 457: 1—166.

Lundqvist, G. and Nilsson, E., 1959: Highest shorelines of the sea and ice dammed lakes in late Quaternary age. Atlas över Sverige: 23—24.

Maarleveld, G. C., 1960: Wind directions and cover sands in the Netherlands. Biul. Perygl. 8: 49—58.

McKee, E. D., 1966: Structures of dunes at White Sands National Monument, New Mexico (and a comparison with structures of dunes from other selected areas). Sedimentology 7: 1—69.

Ohlson, B., 1957: Om flygsandfälten på Hietatievat i östra Enontekiö. Summary: On the drift-sand formations at the Hietatievat in eastern Enontekiö. Terra 69: 129—137.

— 1964: Frostaktivität, Verwitterung und Bodenbildung in den Fjeldgegenden von Enontekiö, Finnisch-Lappland. Fennia 89: 3: 1—180.

Pernarowski, L., 1966: Glacjalna i postglacjalna cyrkulacja atmosfery w świetle kierunku wiatrów wydmotwórczych. Summary: Glacial and postglacial atmospheric circulation in the light of directions of dune-forming winds. Czasopismo Geogr. 37: 3—24.

Poser, H., 1948: Äolische Ablagerungen und Klima des Spätglazials in Mittel- und Westeuropa. Naturwissenschaften 35: 269—276, 307—312.

— 1950: Zur Rekonstruktion der spätglazialen Luftdruckverhältnisse in Mittel- und Westeuropa auf Grund der vorzeitlichen Binnendünen. Erdkunde 4: 81—88.

Prusinkiewicz, Z., 1969: Gleby wydm śródlądowych w Polsce. Summary: The soils of inland dunes in Poland. Inst. Geogr. Polskiej Akad. Nauk, Prace Geogr. 75: 115—144.

Rudberg, S., 1968: Wind erosion — preparation of maps showing the direction of eroding winds. Biul. Perygl. 17: 181—193.

Seppälä, M., 1969: On the grain size and roundness of wind-blown sands in Finland as compared with some Central European samples. Bull. Geol. Soc. Finland 41: 165—181.

— 1971: Evolution of eolian relief of the Kaamasjoki—Kiellajoki river basin in Finnish Lapland. Fennia 104: 1—88.

Solger, F., 1910: Studien über Nordostdeutsche Inlanddünen. Forschungen zur deutschen Landes- und Volkskunde 19: 1: 1—89.

Stankowski, W., 1963: Rzeźba eoliczna Polski Północno—Zachodniej na podstawie wybranych obszarów. Summary: Eolian relief of North-West Poland on the ground of chosen regions. The Poznań Soc. Friends of Sciences. Publ. sect. geogr.-geol. 4: 1: 1—147.

Sundborg, Å., 1955: Meteorological and climatological conditions for the genesis of aeolian sediments. Geogr. Ann. 37: 94—111.

Tanner, V., 1915: Studier öfver kvartärsystemet i Fennoskandias nordliga delar III. Résumé: Etudes sur le système quarternaire dans les parties septentrionales de la Fennoscandie. Fennia 36: 1: 1—815.

Verstappen, H. Th., 1968: On the origin of longitudinal (seif) dunes. Z. Geomorph. 12: 200—220.

Wasylikowa, K., 1964: Roślinność i klimat późnego glacjału w środkowej Polsce na podstawie badań w Witowie koło Łeczycy. Summary: Vegetation and climate of the late-glacial in Central Poland based on investigations made at Witów near Łęczyca. Biul. Perygl. 13: 261—417.

Part V
FLUVIAL ACTION

Editor's Comments
on Papers 33 and 34

33   **LEFFINGWELL**
Excerpts from *The Canning River Region, Northern Alaska*

34   **KING**
Some Periglacial Problems

Rivers in periglacial regions have certain characteristics that depend on the nature of the periglacial environment. The rivers can be divided into the mountain rivers of the high-altitude periglacial zone, the local rivers of the high-latitude periglacial zone, which is likely to be also within the permafrost region, and the allogenous rivers that flow from nonperiglacial areas to their mouths in the periglacial zone. Mountain rivers tend to have a greater flow owing to the high precipitation characteristic of high-altitude areas; steep slopes also cause more rapid runoff. Such streams in a periglacial climate zone also have their flow increased by meltwater from glaciers and snow; the melt period tends to be lengthened in mountains as the snow line gradually moves up the slopes with the advancing season. These streams have one characteristic in common with those of other types of periglacial zone, which is the great difference in flow throughout the year; low flow is typical of the winter period and a period of very high flow occurs during the major melt season.

Arctic rivers, unlike mountain rivers, tend to have a low total discharge, owing to the low precipitation resulting from low temperatures. Precipitation in these areas is normally between 100 and 500 mm, but evaporation may be rather greater than is sometimes suggested, and desiccating winds can be strong. The local rivers of the periglacial zone only flow in the warm season. The Piassina River in the Taimyr Peninsula, for example, passes 83 percent of its total discharge in June, July, and August; flooding is not common because of low precipitation and the ability of the boggy ground to absorb much of the snow melt and summer precipitation, which is the most important in the year. Another example is the Mecham River near Resolute in the Canadian Arctic. This river remained frozen until June 20 in 1959; it then started to flow gradually and reached a peak discharge of 4.5

million m³/day by July 20, and remained at a volume of about 100,000 m³/day until freezing was complete by September 10. In this river 80 percent of the flow passed in a 10-day period.

The allogenous rivers behave differently because they bring a continuous flow into the periglacial zone from the temperate zone that they traverse. These rivers are exemplified by the Yukon and Mackenzie in North America, and the Ob, Lena, and Yenisei in Siberia, all of which are long, large rivers. Their flow is related to the presence of ice in their beds. The runoff increases steadily from source to mouth. It is this buildup of discharge that causes the breakup that is characteristic of these rivers in the spring. The blocks of ice released from upstream jam against the still fast ice farther north, causing the blockage, which builds up until the volume of water is sufficient to float the ice off and allow it to pass downstream to the Arctic coast.

Mackay (1963) records the process of this important aspect of fluvial geomorphology in a periglacial region as it affects the Mackenzie River, one of the most important periglacial rivers of North America. He deals with the processes associated with both the breakup in the spring and the freeze-up in the autumn. The Lena provides an example of a Siberian River of rather similar type, being a river characteristic of a mixture of forest and steppe country before it enters the periglacial zone. This river has a very large discharge; the mean annual value is 13,900 m³/sec at Kyusyur. The Yenisei has a value of 17,400 m³/sec. Marked flooding occurs in the spring breakup, or *débâcle,* the French term for breakup. This occurs in May in the middle reaches and June in the lower reaches; the level of the river may rise by up to 10 m or more, and large areas are flooded over which distinctive sheets of alluvium are laid down. During the breakup the riverbanks may be eroded by the ice floes forced against them by the vigorous flow; the surge that results when jammed ice floes give way also causes erosion. Under these conditions a load much greater in caliber than could normally be carried can be easily carried downstream, thus explaining the very large boulders that sometimes occur in these conditions.

This situation contrasts strongly with that characteristic of minor streams, which may be grossly overloaded as a result of all the efficient mass movement processes that operate on the periglacial hillsides, and which deliver a heavy load to the generally feebly flowing periglacial river, which only has a very short period during which there is sufficient water to enable it to evacuate the load supplied from the slopes. Thus local periglacial rivers tend to

399

accumulate more material than they can transport. They are, therefore, often aggrading. This is particularly true of those that have glacial sources and which are supplied by ice-meltwater streams with the heavy load characteristic of proglacial streams. Under these conditions the streams aggrade, building up sandur deposits or valley trains when they are confined within a valley floor. Thus each of the three types of periglacial river has its own characteristic regime that can be related to their geomorphic activity, but they have in common great variation of flow throughout the year.

One feature associated with periglacial rivers is the formation of icings, or "aufeisen" as they were called by E. Leffingwell. Paper 33 is his description of these periglacial features in the Canning River region of northern Alaska; it is one of the best and earliest descriptions of the features characteristic of this type of periglacial environment. The type of aufeisen described by Leffingwell is formed during the freeze-up process and subsequently during the winter season as water overflows. When the river starts to freeze up, ice first forms at the surface and the base of the river so that the flow is gradually confined. The flow is confined, so the stream overflows its banks often under hydrostatic pressure due to continued flow under the ice. The water freezes into candle ice that can cover quite large areas at times.

Normally, aufeisen, which are also called "naled," rarely cover more than a few hectares, but some have been reported as large as 2185 hectares and up to 9.1 m thick. In arctic Alaska they have been reported up to 4 m thick, 5 km wide, and 50 km long. Shumskii (1964) reported one aufeis 27 km long, 10 m thick, and containing 5×10^8 m³ of ice. The ice forms in layers as each outburst of water freezes into candles that are perpendicular to the surface of the layer. The layers of candle ice are sometimes interdigitated with layers of snow. The more massive aufeisen can survive melting during the summer season so that they may grow in size over a period of time.

Not all aufeisen are associated with rivers of the allogenous type. Some form in local rivers as the freeze-up process takes place, and some form where perennial springs produce water at intervals throughout the winter period. Examples of this type of aufeisen are seen in Baffin Island, where they are associated with deep kettle holes. The deep kettle holes in gravelly outwash deposits have no visible outlet, but below them in some instances aufeisen are found. The kettle holes are deep enough to prevent the water in them from freezing solid in the winter, and as the pressure rises water is forced out at the spring head below the kettle hole; this water freezes to form the aufeisen in the shallow gully below the kettle hole. Again the volume of ice was great enough for some to survive over the summer melting period.

There has been considerable discussion concerning the effectiveness of fluvial processes in a periglacial environment. Peltier (1950) considered that fluvial action was limited to the removal of material supplied by solifluction and frost weathering; in some instances it does not even seem capable of achieving this, and aggradation is characteristic of some valleys in periglacial regions. On the other hand, the detailed study of periglacial processes made by Rapp (1960) in the periglacial area of northern Scandinavia at Kärkevagge revealed that the most effective process in removing material from the hill slopes was running water carrying material in solution. The running water evacuated material rapidly the whole length of the slope, thus achieving more than solifluction and creep, which moved a far greater volume of material, but only for a very short distance compared to the movement by running water. This type of process is not, however, strictly fluvial.

The character of river channels in periglacial areas reveals some of the processes that go on in the river system. The channels normally show braiding, which is characteristic of stream with a widely varying load and discharge, especially when they have easily eroded banks. Where the load reaching the river channel contains much coarse debris, such as that derived by frost shattering, the banks are likely to be erodible; the coarse bed load also leads to braiding, as the coarser material is deposited after the summer period of high discharge. The material forms banks around which the river divides, and which grow by deposition. The sediment is transferred downstream in a series of jumps, with a great deal being held in storage at any one time.

The difference in behavior of rivers under alternating periglacial and temperate regimes is discussed in Paper 34, by W. B. R. King, which draws attention to the formation of terraces in river valleys under these conditions. During the periglacial phase the headstreams of such rivers as the Thames in southern England tend to aggrade their beds, owing to the change in regime, which affects both discharge characteristics and the supply of load. The load increase as a result of the deterioration of the vegetation and the more effective periglacial slope processes, which produce material such as head. Meanwhile, with ice advancing elsewhere; the sea level falls; and while the upper part of the river is aggrading, the lower part is adjusting to the low glacial sea level and so is tending to incise. With the advent of temperate conditions the situation is reversed. Aggradation tends to take place in the lower reaches of the river as sea level returns to the high interglacial level. In the upper reaches, however, as the slopes become better vegetated and the regime of the river returns to the more uniform temperate system, load is decreased as a result of better vegetal protection and less efficient slope processes; so the river tends to degrade its bed, forming terraces of the material aggraded during the preceding periglacial stage. There is a complex region where these two divergent sets of processes may converge and perhaps overlap.

33

Reprinted from *U.S. Geol. Survey Prof. Paper 109*, 158–159, 199–200 (1919)

THE CANNING RIVER REGION, NORTHERN ALASKA

E. de K. Leffingwell

[Editor's Note: In the original, material precedes this excerpt.]

AUFEIS (RECENT).

The heavy deposits of ice that are formed over the flood plains of Arctic rivers have been described in many of the recent reports upon Alaska. Middendorff, however, first described and explained this phenomenon in the middle of the last century, after several years of observation in northern Siberia.[4] The writer can add very little either to his description or his explanation.

Deposits of this kind of ice are called "glaciers" by miners and even by some geologists. "Flood ice" has also been used, but does not convey the proper impression. Middendorff introduced the term aufeis, and the writer has adopted it for this report.

The process of formation of aufeis is as follows: During the winter the flow of the rivers is locally impeded by the formation of anchor and frazil ice,[5] or the shoal places may freeze solidly to the bottom. The water coming from the upper stretches of the river, being thus impeded, will rise and flood the adjacent land. When the river is entirely frozen over, as is the rule in the Arctic, the hydraulic pressure is sufficient to bulge up and fracture the ice at weak places. The escaping water is soon coated with ice, and the flow is gradually restricted by freezing, until sufficient hydraulic pressure is set up to enable the water to burst through again. This flooding and freezing goes on all winter, or at least until the winter flow of water is so reduced that it can pass through the gravels beneath the ice. Thus a deposit of ice may be built up, much after the manner of an alluvial deposit.

If the winter flow is sufficient the aufeis may reach a considerable thickness, so that it may override the ordinary banks of the river and spread out over the whole flood plain. The greatest deposit seen by the writer was about a mile wide and 3 or 4 miles long. The thickness in the last part of June was about 12 feet.

In autumn the river is covered with thick ice before the flow is retarded sufficiently to set up hydraulic pressure. Acting under this pressure the water forces up the domes and ridges of ice, which are a conspicuous feature of aufeis deposits. These elevations are as a rule less than 10 feet high; about 15 feet is the maximum. As a rule their shape is oblong, though ridges over a hundred feet long have been noted. There is invariably a fracture along the crest of the mounds, whence water occasionally flows. The writer has never seen these mounds in the process of formation, but early in November the Canning was dotted with them. The natives say that they have seen them rising early in the autumn, accompanied by an outflow of water. The prospector Arey confirms this report.

After this first process of formation of the mounds the water escapes more quietly. As soon as the newly flooded area is frozen over hydraulic pressure is again set up, but it has only a few inches of ice to fracture. Consequently there is but slight disturbance of the surface.

---

[4] Middendorff, A. T. von, Sibirische Reise, Band 4, Theil 1, pp. 439–457, 1859.

[5] Barnes, H. T., The formation of ice, London and New York, 1906.

[Editor's Note: Certain plates have been omitted owing to limitations of space.]

With the advent of warm weather the flooding waters, no longer freezing, cover the whole deposit of aufeis. Soon the drainage is concentrated into troughs, which have been melted along the lines of greatest flow. As these troughs are cut downward those most favorably situated grow at the expense of the neighboring streams, until by the time the actual river bed is reached the water is concentrated into one or two streams flowing at the bottom of ice canyons. Abandoned channels upon the aufeis of the Canning are shown in Plate XX, *C*.

The ice is gradually undermined by the river, so that large blocks break off with loud reports and fall into the water. Navigation at this time would be very dangerous, for there is danger from falling ice and of being swept under the ice by the current. All of the ice within reach of the river is cut out before the summer is over, but that upon the high bars may remain until September or may possibly last over a second winter. By the 1st of July the aufeis of the Canning was removed from the stretch north of the mountains; a week earlier that near the forks was almost intact. The mounds often remain some weeks after the thinner deposits have melted away.

Okpilak River contained very little aufeis. There was a steep valley train of ice that had been built out a couple of hundred yards from the lower end of the West Fork Glacier. A similar deposit floored the bottom of a valley that stretches eastward from Mount Michelson. At the time of the writer's visit the Hulahula had two areas of aufeis outside of the mountains, and within the mountains as far as the forks most of its floor was covered with ice. Sadlerochit River had one area of aufeis north of the mountains, but above this there were only a few patches confined to the side streams. On the Canning aufeis occurs nearly everywhere from the forks to the coast, the greatest development being near the forks and below Shublik Springs.

[*Editor's Note:* Material has been omitted at this point.]

Ice formed in bodies of fresh water generally has a structure which is vertically prismatic, each prism being an optically distinct crystal. The axes of the crystals are all parallel to each other and perpendicular to the freezing surface. McConnel says:[5]

Some of the ice in the St. Moritz Lake is built of vertical columns from a centimeter downward in diameter and in length equal to the thickness of the clear ice; that is, a foot or more.

Tolmatschow says:[6]

Ice which forms upon the surface of water exhibits a parallel growth of long-stalked crystals. In thawing, this ice decomposes into a series of irregular prisms, which may be several inches long. * * * At first the freezing may be complicated, but later it becomes simpler and regular and gives an ice which as a whole is characterized by its prismatic structure, whereby it is easily separated from snow ice.

Drygalski[7] and Tarr and Rich[8] also described actual examples of prismatic ice, in which they are entirely in accord with the observations just quoted.

Fresh-water ice shortly before its disintegration in the spring is composed of vertical prisms which give way when stepped upon. There are many references to this phenomenon in the literature of northern countries.

Not only is the ice from standing fresh water prismatic but the ice of flowing water is also. Ordinary river ice does not differ notably from standing water ice. The heavy deposits of aufeis, described elsewhere (p. 158), show the same general structure as pond ice.

The writer has seen little reference in the literature to the freezing of water against inclined or vertical surfaces. Hess[9] states that in a vessel whose sides are good conductors of heat, ice forms chiefly with the axes of the crystals perpendicular to the cold surface. In icicles, though the central core may be an optically single crystal, the outer portions, if they are added to after a pause in the freezing, show a radial arrangement of crystals whose axes are perpendicular to the surface.

The well-known lines of air bubbles leading from all sides toward the middle of a block of artificial ice are indicative of the prismatic structure mentioned by Hess. The writer has examined specimens of ice from a mass that had formed from water flowing over a steep slope. They were composed of short irregular prisms with the greatest dimensions perpendicular to the surface. They were about 1 inch long and from one-fourth to one-half inch in diameter.

The generally accepted opinion is that the air content of fresh-water ice is small. The circumstances under which ice is formed, however, may play a great part in the quantity of air included in it. If the material on the bottom of the pond gives off much gas during the freezing, then the air content may be high. The air from such a source usually appears in the form of rather large bubbles. Some bubbles are flattened out horizontally, showing that they were included in the ice while pressing against its lower surface. Other bubbles are elongated vertically, as though they had been caught between the points of downward-growing prisms. A concentration into horizontal rows indicates pauses in the freezing.

[5] McConnel, J. C., and Kidd, D. A., On the plasticity of glacier and other ice: Roy. Soc. Proc., vol. 44, p. 334, 1888.
[6] Tolmatschow, I. P. von, Bodeneis vom Fluss Beresovka: Russ. k. mineral. Gesell. Verh., 2d ser., vol. 40, p. 418, 1902.
[7] Drygalski, Erich, op. cit., pp. 405–419, 485–487.
[8] Tarr, R. S., and Rich, J. L., op. cit., p. 226.
[9] Hess, H., Die Gletscher, pp. 11, 12, Braunschweig, 1904.

[*Editor's Note:* Material has been omitted at this point.]

34

Reprinted from *Proc. Yorkshire Geol. Soc.*, **28**, Pt. 1, 43–50 (Apr. 1950)

SOME PERIGLACIAL PROBLEMS

BY W. B. R. KING

(Presidential Address, December, 1949).

The topic of Glacial Drifts is not new to this Society, but I feel it is a subject in which many of you are keenly interested. Recently the details of the Drift succession of the North of England and particularly our own county have been closely studied, and various challenges concerning the orthodox interpretations have been made. That we should be shaken out of complacency is all to the good and that we should frequently re-examine from every angle the facts on which our interpretations are based, is the best way to arrive at the truth. Today, I intend to study some of the evidence in the periglacial areas and to see how this helps us to interpret the succession in the critical areas which were sometimes under the ice-sheet and at other times free from ice.

Whatever our views on the succession of events during the Ice Age may be, we must admit that in pre-Glacial times the climate was temperate, became cold and is now again temperate. Let us consider the effect a sequence such as this must have, in the area which, during the greatest advance of the ice, still remained periglacial.

The problem has been dealt with by several workers, but let us first consider the matter as it affects the various reaches of the valley system.

As a large ice-cap grows it is obvious that there must be a marked eustatic lowering of sea-level and probably at the same time an isostatic lowering of land surface under the ice-cap. In the ice-cap area therefore, movements of the strand-line will be highly complex and this is well seen in W. B. Wright's study of his isokinetic theory regarding the movements at the end of the ice age in Scandinavia. In the periglacial area however, the isostatic effects will not be felt or certainly much less so than the eustatic, therefore we can say with some certainty that the period of maximum extension of the ice-cap will coincide in time with a lowering of sea-level in non-glaciated areas. The result as far as the lower reaches of the periglacial rivers are concerned must be that lowered base-level results in active erosion of the alluvial tract, leaving terraces on either side of the river, with new alluvium being laid-down under control of the new base-level. With return of the water to the oceans on amelioration of the climate this over-deepened valley will fill with alluvial deposits up to the former level.

In actual fact it would appear that there has been in recent geological times a tendency for sea-level to have been *in general* falling, so that the glacially controlled fluctuation is superimposed on an overriding fall of sea-level. Thus, the newer alluvial infilling does not reach up to the pre-existing level but leaves that as a terrace rising above the summit level of the newer alluvial flat.

In these cases the higher topographical surface generally envisaged as the "Terrace" is more ancient than a lower "Terrace". This does not mean that the age of the gravel forming the terrace can be determined from a study of altitude alone, for clearly the upper part of the more recent 'fill' may be higher than some of the material deposited in the early stages following some previous marked period of 'cut'.

Before considering the conditions in the middle reaches, we will look at the probable sequence of events in the upper reaches. Since we are dealing with a periglacial area, the upper reaches will be influenced by the climatic changes and all the effects which accompany the increase of cold, but if the river system is of fair size, it is probable that the rejuvenating effects of the glacially controlled eustatic change in base-level will not reach to the upper reaches, although the effect of a general rise of land relative to sea-level must in time, work through the whole system. Under these conditions the sequence of events, will be similar to those envisaged by Paterson and Zeuner. These may be summarised as follows : increasing cold results in lowering of the tree-line leaving the ground unprotected and with the alternating frost and thaw of the periglacial regions, marked increase of solifluxion takes place. The river valleys therefore tend to become choked with coarse alluvial deposits and a period of 'fill' or aggradation ensues. As the climate ameliorates, the protective cover of trees re-establishes itself, solifluxion transportation on the slopes decreases, and even if the rainfall remains constant, the fact that débris does not reach the river in such quantities must result in the river becoming progressively less over-loaded, until it reaches the state when it can cut into the accumulations of alluvium, and so sink its bed leaving a terrace of ill-sorted débris above the new alluvial flat. Should there be a series of glacial and interglacial episodes, these would be represented in the periglacial area by a series of terraces, the higher being the older, regardless of any general lowering of ocean base-level, the height of the erosion scarps between the terrace features giving some measure of the duration of the mild interglacial episode. This, of course, is the well-known basis of the Penck and Bruckner scheme for the Deckenschotter and Terraces of Germany.

The point to be emphasised here, is that aggradation in the upper parts of the system takes place during the maximum of the cold climate, while under these conditions, erosion is the main

theme in the lower reaches of the same system. If this is true,
then there must be a critical reach of the river coinciding more
or less with the point where the rejuvenation resulting from the
lowering of sea-level under glacial control has extended up the
river system. If this point is within the zone where aggradation
is taking place, then clearly a highly complicated state of affairs
may ensue and the battle between erosion and aggradation may
move up and down the river over considerable distances. It is
the presence of this 'battle zone' which complicates the study of
river terraces so greatly, and is apparently so frequently ignored
in general schemes which attempt to correlate sea-levels, terraces
and glacial moraines.

What, therefore, is the best approach by which correlation
between glacial deposits, outwash gravels, terrace deposits, etc.,
can be correlated? As in most stratigraphical work, we can have
a petrological or palaeontological approach; for preference these
should be combined but the overriding principle of order of
superposition must be satisfied. In dealing with true glacial
deposits we must take due note of the caveats put forward by
Mr. R. G. Carruthers about the need to examine critically the
undersurfaces of boulder-clay sheets before we accept the order
of superposition as giving the order of age in every case. In the
same way it is essential to make sure that solifluxion has not
caused wholesale slipping of terrace gravels to lower levels during
a period of renewed river cutting, thereby causing fossils belonging
to a high terrace to occur at a lower level and thus invalidating
apparently good evidence regarding the position of river level at
that period. Yet again, in loose alluvial deposits it is frequently
hard to be certain whether a deposit is banked against a cliff of
earlier gravel or is all part of a single deposit.

The petrological approach has usually been made either
through a study of heavy minerals or from a study of erratic
boulders. Both these appear to have their limitations. For
relatively short-distance identification of glacial or river deposits,
a study of heavy minerals appears to be of considerable value,
but the work of Baak on the sands of the North Sea shows that
for long-distance correlation, sands of a single age may possess
marked local distribution of their mineral content which might
vitiate any long-range correlation by the means of heavy mineral
assemblages.

The investigation of boulder assemblages in glacial drift,
has been carried on for many years and all agree that from a
study of both boulders and matrix a general view of the direction
in which the ice-sheets moved can be obtained. Similarly, the
gravels resulting from the destruction of a sheet of boulder clay,
show their parentage by their contained pebble content; perhaps
the incoming of flint into some of the Upper Thames and Avon
gravels is one of the best examples of the value of this type of

evidence. The abundance of Scottish and Lake District rocks in the later terraces of the Severn and their virtual absence in the earlier terrace is of the highest significance in the working out of the relationship of terrace gravels and ice-sheets, as has been done by Wills.

In dealing with the palaeontological approach we immediately find that the odd half-million years since the first ice-sheets appeared in Britain is too short to be able to observe any real evolutionary changes in the fauna. On the other hand it is clear that there are numerous cases where the fauna of a given horizon is closely controlled by climate acting probably in the main through the flora. We therefore have a series of facies-faunas and, as many workers have shown, the really big break in fauna is late in the Pleistocene, although before this time the majority of workers recognise an alternation of "warm" and "cold" faunas in the sequence. Most of these workers consider that there is ample evidence on faunal succession alone for saying that there were marked oscillations in the climate of the periglacial areas during the Pleistocene.

In one group of 'fossils', however, rapid evolutionary changes can be seen—this is in flint implements. Man himself perhaps behaved more like the larger Pleistocene mammals, evolving as far as his skeleton was concerned only slightly, some races perhaps being almost facies indicators and dying out with change of facies, but the actual bones of these early men are so rare that they are of little use as fossils in stratigraphical studies. The flint implements they made, however, do show marked and progressive evolution, and with care, I feel can, and should, be used as fossils for zonal purposes, provided they are used with the same controls as other fossils. This means they must be collected carefully, their relative age being determined by superposition, and care exercised that they are not derived. When series have been built up on this firm basis, they may be used in the same way as ammonites for determining the age of a deposit. Perhaps, however, it would be a more accurate simile to liken their use to that made of the Coal Measure freshwater lamellibranchs, for it is seldom safe to decide a horizon from a single find. Variation within the forms of a single horizon may be great, but when a good series is examined it is found that the experts nearly always agree on the position in the general series where it should be placed. Certain features, like prepared striking platforms, may be accepted as an indication that the find cannot be earlier than a certain stage. It is to be expected that occasional men were more proficient than the average of the time and thus would make higher grade implements, but it is probable that they would use the technique of the period and not a new one. A whole tribe might learn to make particularly good implements and thus give the appearance of belonging to a later period, but, used with common sense, implements may be of the greatest value for relatively short distance correlation.

In dealing with glacial and periglacial deposits we have an approach denied to most stratigraphical problems. The disadvantage of the sediments being laid down on the land, and therefore patchy, is offset by the fact that, being subjected to subaerial weathering, we have a factor which can be of value for giving relative age to the deposits. In the U.S.A., this has been highly developed and the state of weathering of drift and loess sheets is used as a criterion of age. The 'granite weathering ratio' of granite boulders and the depth and extent of weathering in the various drifts does appear to be of the greatest value.

In this country these factors do not seem to have been much used and the division into Older Drift and Newer Drift favoured by many workers rests more on the state of the topographical features, rather than on the state of weathering. Within the newer drift, however, the depth below surface of decalcification of the limestones can certainly be used if done with care.

The importance of the state of 'loamy weathering' of the loess sheets has been emphasised by Zeuner, as giving sure indication of the duration and climate of the time following the deposition of a loess sheet. In N.W. Europe the extensive sheets of loess afford a possibility of correlation over wide areas provided the loess sheets can be safely identified. In this respect the identification by flint implements has indicated the possible pit-falls due to the contemporanity of cultures of different types in areas separated by considerable distances and by geographical barriers. Nevertheless, for comparatively short distance correlation the loess may be of the greatest value, since it is independent of any normal base-level.

Let us now see if we can apply some of these theoretical considerations to the drifts and river deposits of this country.

The criteria discussed above can clearly be best applied in a periglacial area where later advancing ice-sheets would not disturb the pre-existing deposits. Such an area is the Thames drainage basin, where the succession of events has been worked out in recent years by a whole host of investigators. In the Lower Thames, from the Goring Gap to the sea, and also in the Upper Thames around Oxford, there is evidnece of a series of Terraces, in general, the higher being the older. In the lower reaches, at any rate, their formation was, in general, controlled by a falling base-level, although evidence of marked aggradation at times is to be found. The contained fauna gives indications of climate and this is supplemented by evidence of the sediments. In general, a clear case can be made out for alternations of cold conditions, such as would suggest the presence of true ice-sheets not far away, with climates suitable for a fauna containing *Hippopotamus* and *Elephas antiquus*. The presence of numerous flint implements enables the succession to be equated with the sequence of human industries. It is here, however, that a difficulty arises for there is no firm agreement regarding the identification of the implements with those from the sequence found in the type area of the Somme, or even with those from other localities in Britain.

One of the most striking features of the Lower Thames is the well-defined 100 foot terrace. This yields an abundant fauna and numerous implements. The feature maintains a height of approximately 100 feet above present river level over most of the area, and is, on all counts, in keeping with the idea of aggradation in the lower reaches of the sytem due to a gradual rising of sea-level in a temperate to warm climate, following a glacial episode.

Other rivers of England tell a similar story. In the Lower Severn Valley, the Kidderminster Terrace gives evidence for a period of aggradation, following a glacial episode during a period of 'warm' climate, as indicated by the fauna. Here, however, the terrace is some 35 to 65 feet above the present river.

Again, in the Cambridge area, there is evidence for a marked period of aggradation following a glacial episode and the deposits contain a 'warm climate' fauna and rare (usually rolled) flint implements, but here the 'warm' gravels start about present river level and reach to some 60 feet above it. It may be, however, that recent finds will modify the accepted views relating to these lower level 'warm' gravels.

In the Great Ouse basin, near Bedford, a similar terrace gravel is recorded at about 40 feet above river level.

In France, in the Somme valley, the well-known 30 metre Terrace at St. Acheul has many of the features of the 100 foot Terrace, but it is claimed by some that the state of evolution of the flint implements at this type locality, is less advanced than that at Swanscombe in the Thames Valley.

When we study the higher reaches of the Thames around Oxford, the discrepancy in level, seen to some extent in the Cam and Ouse, becomes more alarming. Dr. Arkell has recently claimed that the low-level gravels of the Summertown Terrace, with a base about present river level, is where the equivalent of the Swanscombe representative of the 100 foot Terrace must be sought.

But we have seen that in a period of warm climate, we would expect to get cutting in the upper reaches and aggradation (controlled in part by raising sea-level) in the lower reaches, at any rate, until equilibrium became more or less adjusted to the new conditions. Should the critical area where the change from cut to fill, coincide with a marked restriction of the river valley, it would accentuate the break in the river regime. Thus applying this to the Thames, the Chalk Escarpment at Goring may mark the place where on one side there was very little deposition, or even marked erosion in the early stages of the Summertown

Terrace at Oxford, while considerable gravel deposits were being accumulated in the Lower Thames.

If this is anywhere near the truth, then the height of a terrace above present river level, can only be of diagnostic value in places at the same relative position in the river system, and apart from the actual estuarine tracts, altimetric methods may be of little use in correlation. Thus we are thrown back onto more normal stratigraphic methods of correlation—of these, fauna (including implements), the relationship between gravels and glacial or periglacial (solifluxion and loess) deposits, and state of weathering appear to be of most value.

To sum up, we may see where this leads us. In the South of England, the gravels containing the Middle Acheul type of implement with a 'warm' fauna appear to occur after the true glacial deposits which are represented, firstly, by the North Sea Drift, and the First Welsh Drift of the Midlands, and secondly, by the next set of Drifts, which include the Chalky Boulder Clay of North London, the Morton Drift and the Second Welsh Drift of the Severn Valley.

On the South Coast, gravels of this age appear to be con-contemporaneus with the Goodwood Raised Beach, which indicates a sea-level of over 100 feet above present O.D. This agrees with the view that a high sea-level was responsible for the formation of the 100 foot Terrace of the Lower Thames during a warm climate.

When we go further north, we find in the Cam and Ouse valleys the 'warm' climate gravels still with Middle Acheul implements, but they tend to be at a lower level, although in the same relationship with the Boulder Clays noted above.

A number of workers have claimed that these gravels are overlain by a later true glacial Boulder Clay or Boulder Clays, the latest of which is the Hunstanton Brown Boulder Clay. For this there appears to be good evidence.

Coming to Yorkshire unfortunately we have so far no Lower Palaeolithic implements to help us, but we do have the 'warm' fauna, here associated with a definite sea-level in the famous Sewerby section.

This 'warm' fauna is unfortunately known from at least two horizons, and can usually be dated accurately only when associated with flint implements. The lower is with Middle Acheulian and the upper with Levallois (formerly early Mousterian) cultures. There is some evidence that the Sewerby beach belongs to the later of the two, and the overlying Drifts are therefore later than most of the East Anglian and Midland Drifts.

There is much accumulative evidence along these lines that the Yorkshire Drifts are relatively recent in age, although it must be noted that these Holderness drifts have fairly degraded surface

features and so are older than the fresh morainic drifts of the
northern mountain areas.

Many matters are still outstanding, and it behoves us to
subject each exposure to critical examination again and again,
to test out new and old theories against observed fact.

Part VI
MARINE ACTION

Editor's Comments
on Papers 35 and 36

35 HUME and SCHALK
The Effects of Ice-Push on Arctic Beaches

36 NICHOLS
Excerpts from *Characteristics of Beaches Formed in Polar Climates*

There are long stretches of coastline in the periglacial regions, including all the north coast of Europe and Asia, and North America, where the many Canadian Arctic Islands add greatly to the length of coast. There are also the islands of the Spitzbergen group and others off the coast of the USSR. In the southern hemisphere the coastline of Antarctica is largely formed of the ice sheet and ice shelves, but there are limited areas of solid rock coasts. The coasts have, however, been relatively little studied and much remains to be learned about marine processes in the periglacial environment. Some general principles apply, however, and some distinctive features have been recognized and described. Papers 35 and 36 have been included to illustrate the type of feature found on polar coasts and to indicate the operation of the processes in this environment.

One of the most important ways in which periglacial marine processes differ from those in other areas is owning to the presence of sea ice for much of the year. For most of the year the sea ice is frozen fast to the shore, and even in summer for a long period the ice remains offshore as floes, which have the effect of damping wave action. Thus periglacial beaches are essentially low-energy beaches on which wave action is negligible all winter and only effective for short periods of the summer. Paper 35, by J. D. Hume and M. Schalk, illustrates, from direct observation on the north coast of Alaska near Barrow, the relative importance of the normal wave action processes and those directly attributable to the periglacial environment. The authors show that only when storms occur during the short time of open water will their effect be maximized, but under these conditions major changes may take place over a short time. In this particular storm the amount of movement was equivalent to 20 normal years.

Paper 36, by R. L. Nichols, is a study of the beaches of the Antarctic. The author lists 13 features that are associated with polar beaches. Some are the result of the low-energy-wave environment, such as poorly rounded beach stones; the poorly sorted nature of beach material is also characteristic. On some Baffin Island beaches the sand and finer-sized sediment showed a positive ϕ skewness, indicating a tail of finer grains, which is the reverse of the normal beach negative skewness that results from a tail of coarser grains. The difference is a reflection of the low-energy environment in the sheltered polar waters compared to the very vigorous foreshore environment on normal beaches. As well as being poorly rounded and sorted, periglacial beach deposits contain much more fine sediment and are much less consolidated than beach deposits in other environments. They also have recognizable minor features, which include mounds, pitting, and ridges. The mounds appear to form around small icebergs that ground on the foreshore as the tide falls; the pitting, which is very common and widespread, is the result of the melting out of lumps of ice that become incorporated in the accumulating beach deposits. Such pitting was seen on the rounded boulder field of the Åland Islands, a feature that has already been mentioned as probably of marine origin. The pits are roughly circular, usually less than 5 m in diameter, and up to about 80 cm deep.

Perhaps the most conspicuous feature of periglacial coasts is the presence of ice-pushed ridges. These ridges are formed as the sea ice melts during the spring. When it is driven onshore by waves and wind, it pushes up the beach sediments to form a conspicuous ridge, the height of which is increased by the included blocks of ice. At Marble Point a ridge rising 2.74 m was built 30.5 m offshore; it had an ice core and only a thin veneer of gravel 30 to 60 cm thick and included layers of sand. Leffingwell has described the ice-rafted ridge along the north shore of Alaska in the Canning River region. The coast in this area is usually overridden about 5 to 6 m by ice, but exceptionally the ice may be pushed 60 m inland. The ridges formed in this way rarely exceed 2 m when the ice within them has melted, and they are frequently destroyed by summer waves. The grounding of heavy ice can produce shoals in shallow water offshore and strand large boulders in the shallow offshore zone. It can also striate the stones over which it drags other material. Ice rafting is an important process along some coasts. Tricart (1970) suggests the term ice-floe ramparts or *rides de banquise* to describe these ridges.

Tricart is of the opinion that sand and finer sediments are

particularly abundant on periglacial beaches, partly because wave action is not adequate to disperse the fine material and partly because periglacial rivers carry a large load of material in this size category. In his opinion, this fact accounts for the prevalence of lagoons and marshes on periglacial shores and also for the large number of deltas that are being formed in this environment. On the other hand, he considers coastal erosion under the action of frost to be very effective. He cites the rapid destruction of cliffs, on which the yearly freeze–thaw cycle is very effective because of the large moisture content. The foot of the cliff is especially effectively attacked, and thus steep or vertical cliff profiles are common in these environments. The perpendicular cliffs of granite in Graham Land exemplify this process. The debris derived from the cliffs is broken up by freeze–thaw, as the weak wave action cannot remove it effectively. Thus periglacial coasts differ in many important respects in the processes and their relative effectiveness in comparison with coasts in other environments. Shore platform formation in solid rock under the action of shore ice is probably particularly effective, and Tricart suggests the term "cryogenetic shore platform" to describe this feature; the French equivalent is *plate-forme littorale de gélivation.*

35

Reprinted from *Amer. Jour. Sci.*, **262**, 267–273 (Feb. 1964)

THE EFFECTS OF ICE-PUSH ON ARCTIC BEACHES

JAMES D. HUME* and MARSHALL SCHALK**

ABSTRACT. Pack ice along the northern Alaskan coast is commonly pushed on to the beaches where it striates, planes, gouges, and, after melting, leaves deposits in the form of mounds or ridges. These ice-pushed ridges are commonly 2 feet high but may reach 15 feet. The deposits probably do not form more than 10 percent of the beach material above sea level; 1 to 2 percent is typical.

During the summers of 1954 to 1959 and the winter of 1956-57, Schalk observed the effects of ice-push during studies of the Alaskan coast between Point Lay and Tangent Point. Hume joined him in 1959. Since then, Hume has been studying sedimentation near Barrow, Alaska, and has continued ice observations.

The work has been supported by the Arctic Institute of North America under contractual arrangements with the Office of Naval Research. Logistic support was provided by the Arctic Research Laboratory directed by Max C. Brewer. The authors have been assisted in the field by George H. Denton, Geoffrey W. Smith, and Patricia W. Hume. Arthur H. Stukey, Jr. has done the drafting.

In the past, several geologists and explorers have reported on ice-push effects. The most often mentioned include McClintock (1856), Armstrong (1857), Leffingwell (1919), Stefansson (1921), and Johansen who was quoted by Kindle (1924). More recently, Washburn (1947), Rex and Taylor (1953), and Nichols (1953, 1961), have mentioned the phenomenon. All of these reports have tended to be descriptive with few, if any, actual measurements. The authors felt such measurements should be made. In particular, the amount of sediment added to the beaches by the ice should be determined.

The coast which was studied lies between 250 and 350 miles north of the Arctic Circle (fig. 1). Approximately nine months of the year, the Arctic Ocean is frozen. During the other three months, the ice near the coast melts, breaks into floes, and drifts away from shore to varying degrees. Occasionally, the wind or currents may force the ice, either loose pack ice or solid winter ice, up on the beach. This may happen at any time of the year. The ice may move over the beaches as more or less unbroken, smooth sheets, or it may buckle and break up into blocks. Regardless of the method of motion, the ice usually gouges some beach material, pushes it up on the beach, and deposits it,

In 1961, Hume observed the effects of an almost ideal case of smooth, gliding ice movement near the old Eskimo village of Nuwuk on Point Barrow (fig. 2; pl. 1-A). The ice advanced up to 140 feet over a low, sandy beach with a width that approached 400 feet. In its advance, the ice planed off and striated the beach. At the terminus, an almost continuous line of low mounds was left by the ice-push. At the flanks of the sheet, the ice buckled, possibly because the beach slopes were steeper.

Buckling of the ice seems more common than smooth flow. When this happens, the buckling usually causes gouging, transportation, and deposition of the

* Department of Geology, Tufts University, Medford, Massachusetts.

** Department of Geology, Smith College, Northampton, Massachusetts.

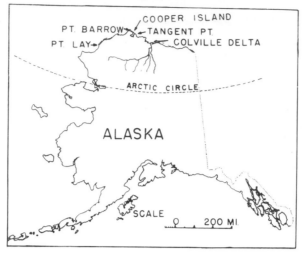

Fig. 1. Location map.

beach sediments in irregular mounds or ridges which have been called ice-push, ice-pushed, or ice-shoved mounds or ridges. Buckled ice can reach considerable heights. Between Tangent Point and Point Lay, 20 or 30 feet would be considered a high ice ridge. The gravel ridges themselves are not nearly as high. The lateral extent of the ice-pushed ridges also varies considerably.

During the spring and early summer of 1960, three individual ice-pushed ridges were left along the beach at Barrow. The youngest, nearest the shoreline, was the highest and most extensive (pl. 1-B). It extended along approximately two-thirds of the coast between Point Barrow and Barrow Village, 8 miles to the southwest. Normally, the ice melts and the gravel is left, first as a veneer over the melting ice, finally as a low, irregular, discontinuous ridge of sand and gravel a few feet high which is usually destroyed by waves. In this case, the melting proceeded slowly under the insulating gravel, and the remnants of some of the mounds, still containing some ice, were recognizable one year later.

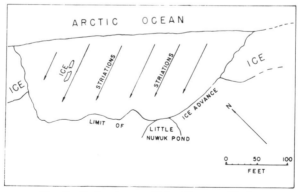

Fig. 2. Striated area, Point Barrow, Alaska.

PLATE 1

A. Striated area, Point Barrow, Alaska, July 5, 1961. 12 foot rod. Point Barrow radar reflector in background.

B. Ice-push near the Arctic Research Laboratory, Barrow, Alaska, July 10, 1960. 7 foot rod.

PLATE 2

A. Ice-pushed mounds near the Arctic Research Laboratory, Barrow, Alaska, June 30, 1961.

B. Ice-pushed mounds, Cooper Island, Alaska. August 21, 1959.

In order to obtain exact measurements of the size and quantity of the deposits left by the ice, two examples were chosen, one on the beach in front of the Arctic Research Laboratory, 3 miles toward the Point from Barrow Village, the other on Cooper Island about 25 miles east of Barrow (pl. 2, A and B). The first case was chosen because it was thought to be typical, the second because it was the largest known to the authors along that part of the Alaskan coast.

Five profiles (fig. 3) from the 1961 ridge system at Barrow were made at 20 foot intervals after some melting had occurred. The profiles were surveyed with transit and stadia rod, depth to the ice core being determined by a steel probe. The ridges were typical in the sense of size and position. The one selected for measurement extended inland approximately 25 feet, had a height of about 6 feet above sealevel, and was about 1000 feet long. It was one of a series of ridges (new) again extending along two-thirds of the coast between Barrow Village and Point Barrow. Two samples of the ice core were taken and found to be about 20 percent sediment by volume. After complete melting the ice-pushed debris would have formed an irregular ridge about two feet high.

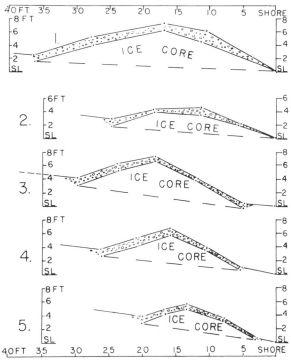

Fig. 3. Profiles of ice-pushed mounds near the Arctic Research Laboratory, Barrow, Alaska, June 29, 1961.

As these deposits were near the shoreline, they probably would have been destroyed by waves later in the summer. In this particular case they were excavated for fill before melting was complete.

Three profiles were made using the same method at Cooper Island (fig. 4). Inclement weather prohibited more field work. The profiles were made at intervals of about 275 feet across areas of extensive ice-pushed ridges. *No ice core was present.* The mounds were about 5 feet high, 30 feet wide, and individual ridges extended for several hundred feet parallel to the beach. The complex of ridges extended for approximately one mile parallel to the ocean shore, the side from which the ice came. These older mounds were found from 110 to 340 feet inland. Younger, smaller mounds were present along the shore. The amount of sediment added to the island by ice, in the area of extreme ice-push, was about 10 percent of the mass above sea level. This compared with about 3 percent for similar beach areas near Barrow.

Unfortunately it could not be determined definitely whether the older mounds had been pushed inland or left inland by prograding of the beach. The presence of scattered tufts of grass on the ice-push ridges suggested that the ridges were several years old, probably 10 or more. If the ridges had been formed prior to prograding, it was thought that they must have been a great deal older than 10 years to allow time for beach advance. In this case, the

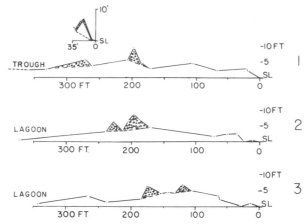

Fig. 4. Profiles of ice-pushed mounds, Cooper Island, Alaska, September 5, 1961. The insert is profile 3 of figure 3.

grass cover would have been more complete and would have resembled that between the ridges and the present high tide mark. It was therefore felt that the ridges had probably formed after most, if not all, of the beach growth and that the ice in this area had been pushed inland over the beach for several hundred feet. The presence of ice wedges in the beach and their absence in the ridges supported this conclusion. Striations, if they had ever existed, had been destroyed.

Ice-push seems to be a common phenomenon along the northern Alaskan coast. Point Barrow, lying athwart a major direction of ice movement, seems to be subjected to great ice action. The old Eskimo village of Nuwuk on the tip of Point Barrow was reported to have been abandoned because ice overrode it. Cooper Island also seems to be an area of extreme ice-push because of its proximity to a major pass in the barrier islands. Normally, the ice-pushed ridges and mounds will not be preserved. However, if the ice action terminates well inland or at sufficient elevation, the mounds may be preserved. Prograding of the beach may also protect the deposits. Ice-pushed mounds and ridges, when found, should be examined carefully for an ice core. A core might conceivably last for several years, depending on summer temperatures.

The Cooper Island ridges, without an ice core, were the largest found by the authors. David M. Hopkins (personal communication, 1962), however, has seen ice-pushed ridges 12 to 15 feet high on the north-central shore of Selawik Lake, Seward Peninsula, Alaska. As these ridges had grass growing on them, they were several years old and could not have had an ice core. These ridges were part of an extensive system of ice-pushed features occurring around approximately one-half of Selawik Lake.

Most of the Alaskan coast north of the Arctic Circle is probably subject to ice action at one time or another. During any one year, it is doubtful if more than 5 percent of the coast is attacked, however. Certain areas, such as Barrow, Cooper Island, and Selawik Lake, are for various reasons subject to much more intensive ice action. Barrow and Cooper Island, for instance, receive ice-push deposits over the majority of the beach almost every year. In most cases, the deposits are temporary and are destroyed by the first storm waves, especially if the storm results in a higher than normal tide. (Normal tide range at Barrow is less than one foot; a storm in October, 1954, probably the worst in over a hundred years, pushed up a storm tide of at least 9 feet at Barrow.) In some areas, namely Barrow and Cooper Island, the deposits may persist because wave erosion is retarded by the presence during most of the year of pack ice. In some cases, also, the ice may push sediment even beyond the reach of major storm waves. However, even in areas where ice-push is pronounced, it is doubtful if the deposits ever amount to more than 10 percent of the sediment above sea level. Probably a more typical figure for ice-push deposits along this section of the Alaska coast would be 1 or 2 percent.

REFERENCES

Armstrong, A., 1857, A personal narrative of the discovery of the North-West Passage: London, Hurst and Blackett, 616 p.

Kindle, E. M., 1924, Observations on ice-borne sediments by the Canadian and other Arctic expeditions: Am. Jour. Sci., 5th ser., v. 7, p. 251-286.

Leffingwell, E. deK., 1919, The Canning River region, northern Alaska: U. S. Geol. Survey Prof. Paper 109, 251 p.

McClintock, F. L., and Haughton, S., 1856, Reminiscences of Arctic ice-travel in search of Sir John Franklin and his companions: Royal Dublin Soc. Jour., v. 1, p. 183-250.

Nichols, R. L., 1953, Marine and lacustrine ice-pushed ridges: Jour. Glaciology, v. 2, p. 172-175.

————— 1961, Characteristics of beaches formed in polar climates: Am. Jour. Sci., v. 259, p. 694-708.

Rex, R. W., and Taylor, E. J., 1953, Littoral sedimentation and the annual beach cycle of the Barrow, Alaska, area: Stanford Univ., Office of Naval Research, Final Rept., sec. 1, Contract Nonr 225 (09), 67 p.

Schalk, M. S., and Hume, J. D., 1961, Sea-ice movement of beach material in the vicinity of Point Barrow, Alaska [abs.]: Jour. Geophys. Research, v. 66, p. 2558-2559.

Stefansson, V., 1921, The friendly Arctic: New York, Macmillan Co., 784 p.

Washburn, A. L., 1947, Reconnaissance geology of portions of Victoria Island and adjacent regions, Arctic Canada: Geol. Soc. America Mem. 22, 142 p.

36

Reprinted from *Amer. Jour. Sci.*, **259**, 694–704, 708 (Nov. 1961)

CHARACTERISTICS OF BEACHES FORMED IN POLAR CLIMATES

ROBERT L. NICHOLS

Department of Geology, Tufts University, Medford, Massachusetts

ABSTRACT. Beaches formed in polar climates are characterized by features that generally are not found in beaches formed in nonpolar climates. Arctic and Antarctic beaches commonly show one or more of the following features: (1) They rest on ice. (2) They are pitted. (3) They have ridges and mounds formed because of ice push and/or deposition from stranded ice. Those formed by ice push are commonly associated with beach scars. (4) They have beach ridges that terminate abruptly because ice was present when they were formed. (5) Ice-rafted fragments are found on them. (6) They have poorly rounded beach stones. (7) Frost cracks and mounds, stone circles and polygons, and solifluction deposits are found on them. (8) They are associated with striations formed by sea ice and icebergs. (9) The beach ridges may have short erosional gaps that were formed by meltwater streams. (10) The beaches are associated with ice-contact features (proglacial deltas, eskerlike features) and glaciomarine deposits. (11) Ventifacts are present. (12) They contain cold-water fossils. (13) They may contain the soft parts of marine organisms.

Ancient beaches that have one or more of these characteristics may have formed in a polar climate.

INTRODUCTION

Ridges, swales, cusps, wave deltas, and other features are found on beaches formed in both nonpolar and polar climates. However, beaches formed in polar climates have certain characteristic features which formed because of the presence of permafrost and land and sea ice. "Polar climates" is not used in a strict climatological sense in this paper. It includes the tundra, icecap, subarctic, and even highland climates of Finch and Trewartha (1936, p. 245-270), as well as the climates of those regions where there is or has been abundant sea ice, lake ice, water-terminating glaciers, or deeply frozen ground.

ACKNOWLEDGMENTS

The field work on which this paper is based was carried out during the 1958-1959 and 1959-1960 field seasons and it was supported by the Arctic Institute of North America and the Antarctic Program of the National Science Foundation. The Commanding Officer, U. S. Navy Construction Battalion Reconnaissance Unit, provided some of the data. The writer greatly appreciates the stimulation and assistance received from Donald G. Ball, Metcalf and Eddy, Engineers, Boston, Massachusetts. Clifford A. Kaye, U. S. Geological Survey, and A. L. Washburn, Yale University, critically read the manuscript and made many valuable suggestions. George Denton of Tufts University ably assisted the writer in the field.

BEACHES RESTING ON ICE

Most and perhaps all of the active part as well as the inactive and slightly elevated part of a beach approximately 2000 feet in length a few miles north of Marble Point between the Wright Glacier and Gneiss Point, McMurdo Sound, Antarctica is resting on ice (fig. 1; pls. 1-A, 2-A). The gravel blanketing the ice is in places several feet thick. Two to three feet of gravel is interbedded with the ice (fig. 2). In places the gravel blanket is thinner than the

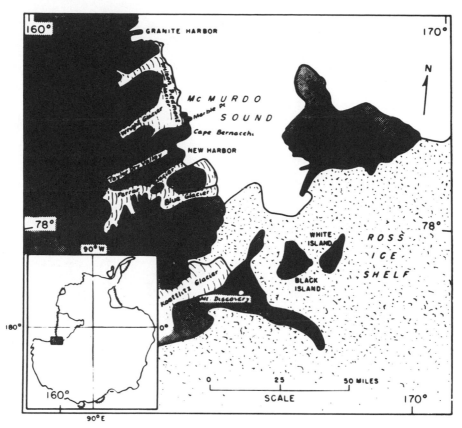

Fig. 1.   Index map showing the location of McMurdo Sound, Antarctica.

active layer,[1] for fresh cracks resulting from the collapse of the gravel indicate that the ice below the gravel has recently melted. In other places the beach gravel is probably thicker than the active layer and, if so, the gravel blanket here protects the ice and prevents it from melting. The writer believes that the ice is not a group of growlers and/or floes but is, instead, a continuous slab. The essential horizontality of this ice, its position with respect to the shoreline, the fact that elevated beaches are found more than 40 feet above it, and the interbedded gravel suggest that it may be an ice foot[2] that took more than one year to form, and not a stagnant piece of glacial ice (Wright and Priestley, 1922, p. 310, 313; pls. 213, 220).

A beach approximately 1500 feet long, 300 feet wide, and 50 feet above sealevel rests on clean ice at Marble Point, McMurdo Sound (fig. 1, pl. 2-B). Bulldozing, blasting, and digging prove that the ice extends beneath an area 600 feet long and 70 feet wide and probably extends beneath most of the beach. A hole dug in the beach showed that clean ice was present at a depth of

[1] Layer of ground above the permafrost which thaws in the summer and freezes in the winter.

[2] A low narrow fringe of ice found in polar regions which is attached to the coast. It is unmoved by the tides, and its flat upper surface is only slightly above sealevel. In winter it joins the sea ice and the land; in summer it is between the land and the open ocean. It is composed of sea ice, frozen snow, and frozen spray.

PLATE 1-A

A beach which rests on ice. Beach is between the Wright Glacier and Gneiss Point, McMurdo Sound, Antarctica. Snowdrift ice slabs are on the left, a beach deposited on an ice foot on the right.

PLATE 1-B

A pitted elevated beach resting on ice between 45 and 55 feet above sealevel. Most of the pits are 10 to 50 feet in diameter, and some are as much as 6 feet deep. Marble Point, McMurdo Sound, Antarctica. This is the same beach as that outlined by white ink on plate 2-B.

approximately 20 inches; 43 inches of ice were penetrated, and the bottom of it was not reached. A hole blasted through the beach and into the ice by the U. S. Navy showed that the ice is essentially clean. The hole penetrated 7 to 8 feet of ice but did not reach the bottom of it. Observations made during the digging suggested that the ice was not a group of isolated, unconnected bergy bits, growlers, or floes, but was a continuous and essentially flat-topped ice slab. Lenses of ice are found immediately above the zone of permanently frozen ground in the Marble Point area. They are not universally present and they are not continuous, for a lens may be seen on one side of a pit but not on the other. Few are more than one foot thick; most are only a few inches thick, and they are commonly dirty. The extent, continuity, thickness, and purity of the buried ice prove that it is not a lens of ground ice. It is either a stagnant, detached, and buried piece of glacial ice or a buried ice foot. The altitude of the beach proves that both the beach and the buried ice are old. The ice has been preserved because the gravels that bury it are nearly as thick as the active layer.

PITTED BEACHES

Pitted beaches do not generally form in temperate climates. They are common in parts of Antarctica. Many of the highest beaches in McMurdo Sound are pitted. The highest beach immediately south of Gneiss Point has scores of pits. The pits are subcircular, less than 15 feet in diameter, and 1 to 2 feet deep. The highest beach at the south end of the mainland near Spike Cape, about 10 miles north of Marble Point is also pitted.

The pits on the highest beaches may have been formed in the following way: During the maximum stage of glaciation the glaciers in McMurdo Sound terminated in deep water. With the retreat of the ice and the elevation of the Earth's crust consequent on deglaciation, a time finally came when the glaciers terminated in water so shallow that beaches could be built against them. The highest beaches were formed at this time from the glacial and marine deposits immediately in front of the glaciers. If they were built against terminal ice cliffs, blocks of ice could fall from the glaciers onto the beaches, and growlers, brash, and floes could be washed up onto the beaches. If some of this ice was buried by beach gravels that were thinner than the active layer in this kind of material in this area, the ice could melt at a later date and the pits could be formed. The lower younger beaches of the area, which could not have been formed against the glaciers, are in general not pitted; this suggests that blocks that fell from the glaciers were more important in the formation of the pits than was ice that was floated onto the beaches.

Not all of the highest beaches in McMurdo Sound are pitted. Pits may be absent because: (1) No ice was buried by beach gravels. (2) The ice was buried by beach gravels, which have always been thicker than the active layer. (3) The glaciers retreated landward more rapidly than the ocean retreated seaward while the beaches were being formed, so that pits formed by the melting of the ice that fell from the glaciers onto the beaches were filled by gravel washed into them by wave action.

Pits, hillocks, and trenches characterize the active as well as the slightly elevated part of the beach that rests on ice between the Wright Glacier and

PLATE 2-A

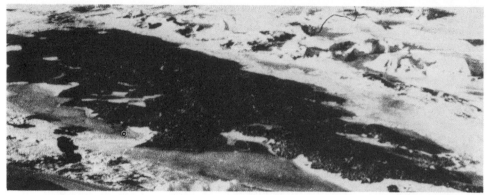

A part of the pitted beach between the Wright Glacier and Gneiss Point, McMurdo Sound, Antarctica.

PLATE 2-B

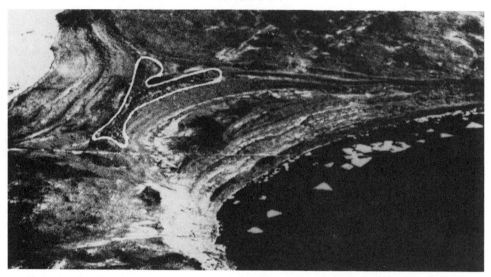

An aerial photograph showing the elevated pitted beach, outlined in white ink, and the lower elevated beaches with ridge and swale topography at Marble Point, McMurdo Sound, Antarctica.

Gneiss Point (pl. 2-A). The pits are as much as 5 feet deep and 20 feet across. They formed as follows: The surface of an ice foot is irregular. The surface of the pitted beach was initially more or less smooth. The gravels, therefore, were thickest where the surface of the ice foot was lowest, and thinnest where it was highest. Where the gravel was thinner than the active layer, the ice below the gravel melted. Where the gravel was thicker than the active layer, the ice did not melt. Pits formed where the gravel-buried ice melted most rapidly. Hillocks formed between closely spaced pits. Fresh cracks in the gravel indicate that ice is melting and that pits are currently forming. The beach will continue to be pitted even after the ice has completely melted because it is unlikely that the deposition of a layer of gravel of varying thickness on either a regular or irregular pre-beach surface would produce a smooth surface. The trenches are about at right angles to the elongation of the beach. One was 50 feet long, 18 feet wide, and 3 to 6 feet deep. Outlet streams from the small ponds behind the beach frequently run across the beach, as indicated by the thin, fine-grained fluviatile deposits in some of the trenches. The streams are not strong enough, however, to form trenches in the beach by removing beach gravel. However, the pond water at times is undoubtedly above 32°F, and it seems likely that the outlet streams have melted the ice immediately beneath the gravels and in this way formed the trenches.

Pits, hollows, hillocks, and mounds are found on the beach at Marble Point, which rests on several feet of clean ice. The area containing the pits is about 1500 feet long and 200 to 400 feet wide. It is in most places between 45 and 55 feet above sealevel, and it contains hundreds of pits (pl. 2-B). The pits are round, oblong, or irregular (pl. 1-B). Most are single, but two or more may coalesce. The single pits range in diameter from 13 to 40 feet. One oblong compound pit is 115 feet long. Most of the pits are less than 4 feet deep, although some are as much as 6 feet deep. Most are dry, but during the summer a few contain water, which comes from the melting of the ice. These pits, like those on the pitted beach between the Wright Glacier and Gneiss Point, were formed by the differential melting of the ice beneath the beach gravels.

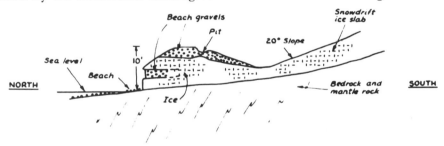

Fig. 2. A diagrammatic sketch showing the relationship of the beach to the ice foot.

If an ancient beach that rests on ice is not pitted, the active layer since the formation of the beach has never been as thick as the beach gravels. If the gravels of a pitted beach that rests on ice are thicker than the present active layer, the beach has probably been in existence during a warmer climate than that at present.

The U. S. Navy, in carrying out engineering studies, removed with a bulldozer between 1 and 2 feet of sand and gravel from a part of the elevated pitted beach. A few weeks later, collapse depressions developed in the sands and gravels because ice beneath them melted, and well-developed thermokarst topography (Péwé, 1954, p. 329-338) was formed. The removal of the gravel by bulldozing brought the ice well within the active zone. One depression was 13 feet long, 9 feet wide, and 15 inches deep (pl. 3-B). Some of the depressions contained water from the melting ice; others had damp sand at the bottom. The ice was in most places only a few inches below the bottom of the depressions. Ridges of sand and gravel with medial tension cracks as well as mounds with radial tension cracks also formed because of differential collapse.

PLATE 3-A

An ice-cored gravel ridge approximately 120 feet long and more than 9 feet high surrounded by sea ice and located about 100 feet offshore near Marble Point, McMurdo Sound, Antarctica. The core of ice, the veneer of gravel, and two layers of interbedded sand are shown.

PLATE 3-B

Collapse depressions in beach gravel at Marble Point, McMurdo Sound, Antarctica, resulting from the differential melting of ice beneath the gravel following the removal of a few feet of gravel by bulldozing. Same area as that outlined by white ink on plate 2-B.

This pitted beach resembles pitted outwash, but the material in the pitted beach is much better sorted than the outwash deposits in the Marble Point area. The pitted area is immediately below higher elevated beaches, whereas pitted outwash is usually found above the marine limit. Also, the bifurcation of the mainland side of the pitted area (pl. 2-B) would be expectable if the area were an elevated pitted beach, but would not be expectable if the pitted area were outwash.

Pitted beaches are formed not only by the burial of an ice foot and growlers, floes, and small blocks of ice by beach deposits, but also by the burial of the snout termini of glaciers. The Wilson Piedmont Glacier nearly reaches the strandline for a distance of approximately 400 yards immediately south of the deglaciated mainland near Spike Cape about 10 miles north of Marble Point, McMurdo Sound. A gravel ridge about 400 yards long, up to 7 feet high, and in most places less than 30 feet wide is found between the ocean and the terminus of the glacier. Although the ridge is more or less continuous, it has gaps that are scores of feet long. In other places small isolated gravel hillocks as much as 4 feet high take the place of the ridge. Pits and mounds are found on the ridge. The gravel is unweathered and somewhat sorted, and the fragments are somewhat rounded. The similarity of this gravel to that found in the nearby beaches and the absence of morainal material on the nearby glacier prove that the gravel is a beach deposit and not morainal material. It is generally only a few feet thick and rests on ice. The nearness of the glacier indicates that the ice is a stagnant piece of glacial ice. The thickness of the upturned floes immediately offshore indicates a steep offshore profile. The steep offshore profile and the snout terminus of the glacier enabled the waves to throw beach gravel derived from glacial deposits onto the foot of the glacier. Following the deposition of the gravels a small amount of glacial retreat took place. The gravels protected the ice beneath them, and an ice-cored gravel ridge with pits and hillocks was formed.

Pitted beaches indicate that land and/or sea ice or perhaps permafrost was present when the beaches were formed and are therefore diagnostic of polar climates. Some of the ancient beaches in New England, Scandinavia, the Great Lakes region, and elsewhere are probably pitted.

ICE-PUSHED AND ICE-DEPOSITED RIDGES

Marine ice-pushed ridges and mounds formed by sea ice, growlers, and bergy bits are found on beaches formed in polar climates. The material in the ridges and mounds is derived from the beach scars that are associated with them. Ice-pushed ridges and mounds are common on both the modern and elevated Arctic beaches (Washburn, 1947, p. 78-80; Nichols, 1953b), and the writer has seen them in the Antarctic.

Somewhat similar ridges and mounds are formed by the deposition of material from ice stranded on polar beaches. A gravel ridge approximately 120 feet long and projecting as much as 9 feet above sealevel was seen about 100 feet offshore near Marble Point. The sides in places slope at approximately 30°, but in most places they are gentler. The ridge is surrounded by water during a part of the summer and by sea ice during the remainder of the year

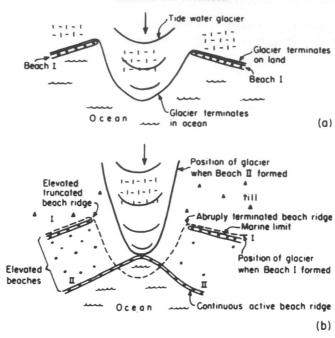

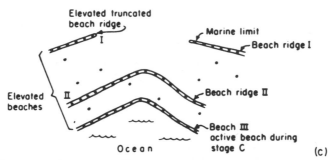

Fig. 3.   Diagrams showing the formation of elevated beach ridges which terminate abruptly because of the former presence of glacial ice.

(pl. 3-A). A bulldozer cut made at the end of the ridge showed that the ridge had a core of ice and was only veneered with gravel. The gravel, on the average, was between 1 and 2 feet thick; the degree of rounding of the fragments and the gray color indicate that it is modern beach gravel. Interbedded in the ice were two thin layers of sand (pl. 3-A). The bottom of the ice was not seen, and other layers may be present. This ice mass is probably a detached, rafted, and stranded ice foot. The interbedded sand suggests that it may have been more than one year in forming. The active layer in this area, in gravels similar to those on this ridge, is 2 or more feet deep. The ice core is, therefore, slowly melting. When it has melted away, only a small ridge of gravel will remain, which will probably not be associated with a beach scar. Gravel ridges and mounds formed in this way are also found on nearby elevated beaches. Many others were probably destroyed by wave action.

Ridges and mounds composed of talus, landslide, fluviatile, eolian, and solifluction material can also be formed in polar climates, as these types of deposits may collect at times on an ice foot (Brown, 1887, p. 358). It may not always be easy to differentiate a ridge or mound formed by ice push from one formed by deposition from an ice foot or from floating ice of other origin. However, those ridges and mounds that are not associated with beach scars or are not composed of beach deposits are probably not due to ice push. As the small masses of ice responsible for these features cannot be drifted very far into warm waters, the presence of either type of ridge on an ancient beach in general indicates that the beach was formed in a polar climate as defined in this paper.

Depressions and hollows that do not have ridges or mounds associated with them are found on polar beaches (Nichols, 1953c, p. 61). The absence of ridges and mounds and the presence of normal beach faces both above and below these depressions and hollows suggest that gravel may have been removed from the beaches. Joyce (1950, p. 646) has noted that an ice foot becomes on disintegration a powerful transportation agent as beach material frozen to

PLATE 4

A beach which terminates abruptly in the distance against a thick and high snowdrift ice slab.

the underside is rafted away. Many scars were probably formed in this way. Scars formed in this way are found only on polar beaches, as the ice foot is not developed in nonpolar climates.

TRUNCATED BEACHES

Some beaches terminate abruptly against cliffs, knobs, and upland areas; others end in deep water. Once-continuous elevated beaches may stop abruptly because they have been in part destroyed by fluviatile or glacial erosion or because of burial by talus, landslide, solifluction, or eolian deposits. Beaches that are continuous for long distances and that terminate abruptly against glaciers or snowdrift ice slabs are common in the Antarctic (pl. 4). If, following the formation of such a beach, glacial retreat and uplift consequent on deglaciation take place, an elevated beach that terminates abruptly will be formed (fig. 3). These are also found in Antarctica. In most cases it should be easy to distinguish between abrupt truncation resulting from the former presence of ice and abrupt truncation resulting from other causes. If the highest beach ridge terminates abruptly and is pitted, it seems likely that not only the pits but also the abrupt termination of the beach are due to the presence of ice when the beach was formed.

PRESENCE OF ICE-RAFTED FRAGMENTS

Ice-rafted fragments are found on modern and elevated beaches in the Arctic, the Antarctic, and elsewhere (Washburn, 1947, p. 80-82; Coleman, 1922, p. 11-12). The ice-rafted origin of a fragment is fairly certain if all of the following conditions are met: (1) the fragment is much larger than the largest roundstones on the beach that have been transported and deposited by waves and shore currents; (2) it rests on the beach deposits and does not project through them from till or other deposits below; (3) the beach has not been over-run by glacial ice; (4) the fragment is angular or subangular and striated; (5) similar fragments are found up to the marine limit but not above it; (6) it is lithologically unlike the local till or bedrock. Ice-rafted fragments can be deposited on beaches formed in nonpolar climates. However, if the ice-rafted fragments are both large and numerous, the beach was probably formed in a polar climate. In some cases it might be difficult to differentiate fragments transported in the roots of trees and in other special ways from ice-rafted fragments.

[*Editor's Note:* Material has been omitted at this point.]

REFERENCES

Blackwelder, Eliot, 1940, The hardness of ice: Am. Jour. Sci., v. 238, p. 61-62.

Brown, Robert, 1887, Our Earth and its story: London, Cassell & Co., Ltd., 376 p.

Coleman, A. P., 1922, Physiography and glacial geology of Gaspé Peninsula, Quebec: Canada Geol. Survey Bull. 34, 52 p.

Crosby, I. B., and Lougee, R. J., 1934, Glacial marginal shores and the marine limit in Massachusetts: Geol. Soc. America Bull., v. 45, p. 441-462.

Flint, R. F., 1947, Glacial geology and the Pleistocene epoch: New York, John Wiley & Sons, Inc., 589 p.

Finch, V. C., and Trewartha, G. T., 1936, Elements of geography: New York, McGraw-Hill Book Co., Inc., 782 p.

Gould, L. M., 1928, Report on physical geography, *in* Putnam, G. P., The Putnam Baffin Island expedition: Geog. Review, v. 18, p. 27-40.

Joyce, J. R. F., 1950, Notes on ice-foot development, Neny Fjord, Graham Land, Antarctica: Jour. Geology, v. 58, p. 646-649.

Leffingwell, E. de K., 1919, The Canning River region, northern Alaska: U. S. Geol. Survey Prof. Paper 109, 251 p.

Mathiassen, Therkel, 1933, Contributions to the geography of Baffin Land and Melville Peninsula: Report of the Fifth Thule Expedition, 1921-1924, v. 1, no. 3, 102 p. Copenhagen, Gyldendalske Boghandel, Nordisk Forlag.

Nichols, R. L., 1953a, Geomorphologic observations at Thule, Greenland and Resolute Bay, Cornwallis Island, N.W.T.: Am. Jour. Sci., v. 251, p. 268-275.

————, 1953b, Marine and lacustrine ice-pushed ridges: Jour. Glaciology, v. 2, p. 172-175.

———— 1953c, Geomorphology of Marguerite Bay, Palmer Peninsula, Antarctica: Ronne Antarctic Research Expedition, Technical Report No. 12, Office of Naval Research, Dept. of the Navy, Washington, D. C., p. 1-151.

O'Neill, J. J., 1924, The geology of the Arctic coast of Canada west of the Kent Peninsula: Canadian Arctic Expedition, 1913-1918, Rept., v. 11, pt. A, 107 p.

Péwé, T. L., 1954, Effect of permafrost on cultivated fields, Fairbanks area, Alaska: U. S. Geol. Survey Bull. 989-F, p. 315-351.

———— 1959, Sand-wedge polygons (tesselations) in the McMurdo Sound region, Antarctica—a progress report: Am. Jour. Sci., v. 257, p. 545-552.

Simpson, G. C., 1923, Meteorology tables, British Antarctic Expedition, 1910-1913: London, Harrison & Sons, Ltd., 835 p.

Suess, H. E., 1954, U. S. Geological Survey radiocarbon dates, I: Science, v. 120, p. 467-473.

Teichert, Curt, 1939, Corrasion by wind-blown snow in polar regions: Am. Jour. Sci., v. 237, p. 146-148.

Tolmachoff, I. P., 1929, The carcasses of the mammoth and rhinoceros found in the frozen ground of Siberia: Am. Philos. Soc. Trans., n.s., v. 23, pt. 1, p. 1-74.

Trefethen, J. M., and Harris, J. N., 1940, A fossiliferous esker-like deposit: Am. Jour. Sci., v. 238, p. 408-412.

Washburn, A. L., 1947, Reconnaissance geology of portions of Victoria Island and adjacent regions, Arctic Canada: Geol. Soc. America Mem. 22, 142 p.

Wright, C. S., and Priestley, R. E., 1922, Glaciology: London, Harrison & Sons, Ltd., British Antarctic (Terra Nova) Expedition, 1910-1913, 581 p.

CONCLUSION

The contributions selected for inclusion in this volume have been chosen partly to illustrate the very wide range of processes that operate only within or most efficiently within the periglacial environment. These processes are responsible for many intriguing distinctive features that give this environment its own fascination from the geomorphic point of view, and which pose real problems to the development of these extensive and important areas, where may resources remain to be exploited. The importance of the periglacial environment and the processes that operate within it can be assessed by the increasing attention that geomorphologists have paid to it in the recent past. An increasing literature is being devoted to its problems and processes. This includes a number of books, such as Washburn's *Periglacial Processes and Environments* (1973), J. Tricart's *Geomorphology of Cold Environments* (1970), and the *Illustrated Glossary of Periglacial Phenomena* by Hamelin and Cook (1967). Several journals are also now devoted entirely or partially to the periglacial regions and their processes, including *Biuletyn Peryglacjalny, Boreas, Arctic and Alpine Research,* and *Quaternary Research.*

BIBLIOGRAPHY

Ahlmann, H. W. 1919. Geomorphological studies in Norway. *Geog. Ann. 1*, 1–148, 193–252.

Andersson, J. G. 1906. Solifluction, a component of sub-aerial denudation. *Jour. Geol. 14*, 91–112.

Ball, D. F., and Goodier, R. 1970. Morphology and distribution of features resulting from frost action in Snowdonia. *Field Studies 3*, 193–217.

Berg, T. E. 1969. Fossil sand wedges at Edmonton, Alberta, Canada. *Biul. Peryglac. 19*, 325–333.

Black, R. F. 1952. Polygonal patterns and ground conditions from aerial photographs. *Photogramm. Eng. 18*, 123–134.

———. 1954. Permafrost—a review. *Bull. Geol. Soc. America 65*, 839–855.

———. 1969. Climatically significant fossil periglacial phenomena in northcentral United States. *Biul. Peryglac. 20*, 225–238.

Bostrom, R. C. 1967. Water expulsion and pingo formation in a region affected by subsidence. *Jour. Glaciol. 6*, 568–572.

Brown, R. J. E. 1960. The distribution of permafrost and its relation to air temperature in Canada and the U.S.S.R. *Arctic 13*, 163–177.

———. 1970. *Permafrost in Canada*. Toronto, Toronto University Press.

Bryan, Kirk. 1934. Geomorphic processes at high altitudes. *Geol. Rev. 24*, 655–656.

———. 1946. Cryopedology—the study of frozen ground and intensive frost-action with suggestions on nomenclature. *Amer. Jour. Sci. 244*, 622–642.

Büdel, J. 1953. Die "periglazial"—morphologischen Wirkungen des Eiszeitklimas auf der ganzen Erde. *Erdkunde 7*, 249–266.

Capps, S. R. 1910. Rock glaciers in Alaska. *Jour. Geol. 18*, 359–375.

Clark, G. M. 1968. Sorted patterned ground: New Appalachian localities south of the glacial border. *Science 161*, 355–356.

Cook, F. A., and Raiche, V. G. 1962. Freeze–thaw cycles at Resolute, N.W.T. *Geog. Bull. 18*, 64–78.

Corte, A. E. 1962. Vertical migration of particles in front of a moving freezing plane. *Jour. Geophys. Res. 67*(3), 1085–1090.

———. 1966. Particle sorting by repeated freezing and thawing. *Biul. Peryglac. 15*, 175–240.

Costin, A. B. 1955. A note on gilgais and frost soils. *Jour. Soil Sci. 6,* 32–34.

Czudek, T. 1964. Periglacial slope development in the area of the Bohemian Massif in Northern Moravia. *Buil. Peryglac. 14,* 169–194.

Davies, J. L. 1969. *Landforms of cold climates.* MIT Press, Cambridge, Mass.

Dionne, J.-C. 1970. Fentes en coin fossiles dans la région de Québec. *Rev. Geog. Montreal 24,* 313–318.

Dylik, J. 1960. Rhythmically stratified slope waste deposits. *Biul. Peryglac. 8,* 31–41.

———. 1967. Solifluxion, congelifluxion and related slope processes. *Geog. Ann. 49A,* 167–177.

Eakin, H. M. 1916. The Yukon–Koyukuk region, Alaska. *U.S. Geol. Survey Bull. 631,* 88 pp.

Ekblaw, W. E. 1918. The importance of nivation as an erosive factor, and of soil flow as a transporting agency, in northern Greenland. *Proc. Natl. Acad. Sci. 4,* 288–293.

Embleton, C., and King, C. A. M. 1975. *Periglacial geomorphology.* Volume 2 of *Glacial and Periglacial Geomorphology,* 2nd ed., Edward Arnold, London.

Fraser, J. K. 1959. Freeze–thaw frequencies and mechanical weathering in Canada. *Arctic 12,* 40–53.

French, H. M. 1971a. Ice cored mounds and patterned ground, southern Banks Island, western Canadian Arctic. *Geog. Ann. 53A,* 32–38.

———. 1971b. Slope asymmetry of the Beaufort Plain, Northwest Banks Island, N.W.T., Canada. *Canada Jour. Earth Sci. 8,* 717–731.

Galloway, R. W. 1961. Periglacial phenomena in Scotland. *Geog. Ann. 43,* 348–353.

Geikie, J. 1894. *The great ice age and its relation to the antiquity of man,* 3rd ed. Edward Stanford, London.

Groom, G. E. 1959. Niche glaciers in Bünsow-land, Vestspitsbergen. *Jour. Glaciol. 3,* 369–376.

Hamelin, L.-E., and Cook, F. A. 1967. *Le périglaciaire par l'image; illustrated glossary of periglacial phenomena.* Les Presses de l'Université, Quebec.

Hay, T. 1936. Stone stripes. *Geog. Jour. 87,* 47–50.

Högbom, B. 1914. Über die geologische Bedeutung des Frostes. *Uppsala Univ. Geol. Inst. Bull. 12,* 257–389.

———. 1923. Ancient inland dunes of northern and middle Europe. *Geog. Ann. 5,* 113–242.

Hopkins, D. M. 1949. Thaw lakes and thaw sinks in the Imuruk Lake area, Seward Peninsula, Alaska. *Jour. Geol. 57,* 119–131.

Hussey, K. M. 1962. Ground patterns as keys to photointerpretation of arctic terrain. *Iowa Acad. Sci. 69,* 332–341.

Inglis, D. R. 1965. Particle sorting and stone migration by freezing and thawing. *Science 148,* 1616–1617.

Ives, J. D. 1966. Block fields, associated weathering forms on mountain tops and the Nunatak hypothesis. *Geogr. Ann. 48A,* 220–223.

Ives, R. L. 1940. Rock glaciers in the Colorado Front Range. *Bull. Geol. Soc. America 51,* 1271–1294.

James, P. A. 1971. The measurement of soil frost-heave in the field. *Brit. Geomorph. Res. Group Tech. Bull. 8,* 43 pp.

Kesseli, J. R. 1941. Rock streams in the Sierra Nevada, California. *Geog. Rev. 31,* 203–227.

King, C. A. M., and Hirst, R. A. 1964. The boulder fields of the Åland Islands. *Fennia 89*(2), 41 pp.

King, R. B. 1971. Boulder polygons and stripes in the Cairngorm Mountains, Scotland. *Jour. Glaciol. 10,* 375–386.

Leffingwell, E. de K. 1915. Ground-ice wedges; the dominant form of ground-ice on the north coast of Alaska. *Jour. Geol. 23,* 635–654.

Łoziński, W. 1909. Über die mechanische Verwitterung der Sandsteine im gemässigten Klima. *Acad. Sci. Cracovie Bull. Internat., Cl. Sci. Math. et Naturelles 1,* 1–25.

Mackay, J. R. 1963. Progress of break-up and freeze-up along the Mackenzie River. *Geog. Bull. 19,* 103–116.

———. 1971. The origin of massive ice beds in permafrost, western arctic coast, Canada. *Canadian Jour. Earth Sci. 8,* 397–422.

———. 1972. The world of underground ice. *Assoc. Amer. Geog. Anns. 62,* 1–22.

Mangerud, J., and Skreden, S. A. 1972. Fossil ice wedges and ground wedges in sediments below the till at Voss, western Norway. *Norsk. Geol. Tidsskr. 52,* 73–96.

Matthes, F. E. 1900. Glacial sculpture of the Bighorn Mountains, Wyoming. *U.S. Geol. Survey 21st Ann. Rept.* (2), 173–190.

Middendorff, A. T. von. 1859. *Sibirische Reise.*

Miller, R., Common, R., and Galloway, R. W. 1954. Stone stripes and other surface features of Tinto Hill. *Geog. Jour. 120,* 216–219.

Muller, S. W. 1947. *Permafrost or permanently frozen ground and related engineering problems.* J. W. Edwards, Ann Arbor, Mich.

Peltier, L. C. 1950. The geographic cycle in periglacial regions as it is related to climatic geomorphology. *Assoc. Amer. Geog. Anns. 40,* 214–236.

Penner, E. 1968. Particle size as a basis for predicting frost action in soils. *Soils and Foundations 8*(4), 21–29.

Péwé, T. L. 1966. Paleoclimatic significance of fossil ice wedges. *Biul. Peryglac. 15,* 65–73.

———. (ed.). 1969. *The periglacial environment.* McGill–Queen's University Press, Montreal.

Philberth, K. 1964. Recherches sur les sols polygonaux et striés. *Biul. Peryglac. 13,* 99–198.

Pissart, A. 1963. Les traces de "pingos" du Pays de Galles (Grand-Bretagne) et du plateau des Hautes Fagnes (Belgique). *Z. Geomorph.* NF *2,* 147–165.

———. 1964. Vitesse des mouvements du sol au Chambeyron (Basses Alpes). *Biul. Peryglac. 14,* 303–310.

Porsild, A. E. 1938. Earth mounds in unglaciated arctic northwestern America. *Geog. Rev. 28,* 46–58.

Poser, H. 1954. Die Periglazial-Erscheinungen in der Umgebund der Gletscher des Zemmgrundes (Zillertalen Alpen). *Göttinger Geog. Abh. 15,* 125–180.

Prentice, J. E., and Morris, P. G. 1959. Cemented screes in the Manifold Valley, North Statfordshire. *East Mid. Geog.*, No. 11, 16–19.

Rapp, A. 1960. Recent development of mountain slopes in Kärkevagge and surroundings, northern Scandinavia. *Geog. Ann.* *42*(2–3), 65–200.

————., and Rudberg, S. 1964. Studies on periglacial phenomena in Scandinavia. *Biul. Peryglac.* *14*, 75–90.

Rudberg, S. 1958. Some observations concerning mass movement on slopes in Sweden. *Geol. Fören. Stockholm, Förh.* *80*, 114–125.

Seppälä, M. 1972. The term "palsa." *Z. Geomorph.* NF *16*(4), p. 463.

Sharp, R. P. 1942. Periglacial involutions in northeastern Illinois. *Jour. Geol.* *50*, 113–133.

Shotton, F. W. 1960. Large-scale patterned ground in the valley of the Worcestershire Avon. *Geol. Mag.* *97*, 404–408.

Shumskii, P. A. 1964. Ground (subsurface) ice (Podzemnye l'dy). *Canada Nat. Res. Coun. Tech. Translation 1130,* 118 pp.

Smith, H. T. U. 1953. The Hickory Run Boulder-Field, Carbon County, Pennsylvania. *Amer. Jour. Sci.* *251*, 625–642.

————. 1965. Dune morphology and chronology in central and western Nebraska. *Jour. Geol.* *73*, 557–578.

Soons, J. M., and Greenland, D. E. 1970. Observations on the growth of needle ice. *Water Resources Res.* *6*, 579–593.

Sugden, D. E. 1971. The significance of periglacial activity on some Scottish mountains. *Geog. Jour.* *137*, 388–392.

Svensson, H. 1969. A type of circular lake in southernmost Norway. *Geog. Ann.* *51A*, 1–12.

Swinzow, G. K. 1969. Certain aspects of engineering geology in permafrost. *Eng. Geol.* *3*, 177–215.

Taber, S. 1943. Perennially frozen ground in Alaska: its origin and history. *Geol. Soc. America Bull.* *54*, 1433–1548.

Tedrow, J. C. F. 1969. Thaw lakes, thaw sinks and soils in northern Alaska. *Biul. Peryglac.* *20*, 337–344.

Te Punga, M. T. 1956. Altiplanation terraces in Southern England. *Biul. Peryglac.* *4*, 331–338.

Thorarinsson, S. 1951. Notes on patterned ground in Iceland, with particular reference to the Icelandic "Flás". *Geog. Ann.* *33*, 144–156.

Tricart, J. 1970. *Geomorphology of cold environments*, translated by E. Watson. Macmillan, London. (1963, (*Géomorphologie des régions froides,* Presses Univ. de France, Paris, 1963).

Troll, C. 1944. Strukturböden, Solifluktion und Frostklimate der Erde. *Geol. Rundschau 34*, 545–694.

Tufnell, L. 1969. The range of periglacial phenomena in northern England. *Biul. Peryglac.* *19*, 291–323.

U.S. Forest Service. 1968. *Snow avalanches*. U.S. Department Agriculture Handbook 194, 84 pp. rev.

Washburn, A. L. 1969. Weathering, frost action, and patterned ground in the Mesters Vig district, Northeast Greenland. *Medd. om Gronland 176*(4), 303 pp.

————. 1973. Periglacial processes and environments. Arnold, London.

————., Smith, D. D., and Goddard, R. H. 1963. Frost cracking in a middle-latitude climate. *Biul. Peryglac. 12,* 175–189.

Watson, E. 1971. Remains of pingos in Wales and the Isle of Man. *Geol. Jour. 7,* 381–392.

White, S. E. 1971. Rock glacier studies in the Colorado Front Range, 1967 to 1968. *Arctic Alpine Res. 3,* 43–64.

Williams, P. J. 1961. Climatic factors controlling the distribution of certain frozen ground phenomena. *Geog. Ann. 43,* 339–347.

————. 1963. Quantitative investigations of soil movement in frozen ground phenomena. *Biul. Peryglac. 11,* 353–360.

Williams, R. B. G. 1964. Fossil patterned ground in eastern England. *Biul. Peryglac. 14,* 337–349.

————. 1968. Some estimates of periglacial erosion in southern and eastern England. *Biul. Peryglac. 17,* 311–335.

Yardley, D. H. 1951. Frost-thrusting in the Northwest Territories. *Jour. Geol. 59,* 65–69.

AUTHOR CITATION INDEX

SUBJECT AND PLACE INDEX

Places outside the United States are mainly indexed under countries; within the United States, they are indexed under the state.

About the Editor

CUCHLAINE A. M. KING is Professor of Physical Geography and at present Head of Department of the Geography Department, Nottingham University, England. She teaches courses in geomorphology, physical geography, and oceanography.

Professor King received the degrees of B.A. (1943), M.A. (1946), Ph.D. (1949), and Sc.D. (1973) from the University of Cambridge, England, in geography and geomorphology. In 1972–1973 she held a Senior Visiting Fellowship of the National Science Foundation at the State University of New York, Binghamton. She is author and co-author of 12 books, including second editions, on various aspects of geography and geomorphology, with special emphasis on coastal and glacial geomorphology.